NATIONAL ENERGY PROFILES

NATIONAL ENERGY PROFILES

Edited by
Kenneth R. Stunkel

PRAEGER

PRAEGER SPECIAL STUDIES • PRAEGER SCIENTIFIC

Library of Congress Cataloging in Publication Data

Main entry under title:

National energy profiles.

    Includes indexes.
    1. Power resources. 2. Energy policy.
I. Stunkel, Kenneth R.
TJ163.2.N372   333.79   80-25046
ISBN 0-03-050646-8

Published in 1981 by Praeger Publishers
CBS Educational and Professional Publishing
A Division of CBS, Inc.
521 Fifth Avenue, New York, New York 10175 U.S.A.

© 1981 by Praeger Publishers

123456789 145 987654321

94790

Printed in the United States of America

Energy is the only life and is
    from the Body
and Reason is the bound and outward
circumference of Energy.

William Blake, "The Marriage of
                    Heaven and Hell"

PREFACE

The subject of energy is huge, tangled, fluctuating, and multi-dimensional, intersecting as it does with disparate fields like engineering, political science, physics, economics, climatology, and ecology. The focus can be as narrow as the technology and pricing of photovoltaic cells in solar devices, or as broad as the historical and evolutionary framework within which all energy conversion has taken place on earth. This volume falls somewhere in between these polarities, being concerned with the role, fitness, and prospect of nations as managers and policy makers in the domain of world energy resources.

No doubt one might quibble readily enough about the choice of countries that are the subjects of these national energy profiles. As is the case with many projects of this sort, the choice was settled, more or less, by who was willing to participate, work within the conceptual guidelines, and come up with a solid manuscript. After the dust of the editor's bush-beating offensive had dispersed, eight countries had been paired off with appropriate authors, falling roughly into two groups: developed (the United States, the Soviet Union, and Japan) and developing (the People's Republic of China, Brazil, Mexico, Nigeria, and Taiwan). While the geographic distribution of these nations is less than ideal, certain other criteria have been satisfied well enough.

In the first instance, the three developed countries have large populations—all of which rank among the world's seven most populous—and the world's most energy-intensive economies. Together they account for a substantial share of global economic activity and potential, with a corresponding impact on energy conversion. In the second instance, three of the developing countries (the People's Republic of China, Brazil, and Mexico) have a combined population more than twice that of the three developed states and a considerable short-term potential for industrialization, all of which adds up to a strong impact on energy and the environment, both now and in the future. As formerly very poor nations coming late to the industrialization process, their behavior with respect to energy policy, in light of the waste and blunders of the more developed countries, can be instructive. In the third instance, Nigeria, the People's Republic of China, the Soviet Union, and Mexico are all exporters of energy. Nigeria also has the distinction of being at once Africa's most populous nation and second only to South Africa in gross national product. As an oil-rich African state ruled by blacks and struggling for development, it seemed a logical choice to represent that part of the world.

Instead of Taiwan, it might have been better to have Indonesia or South Korea to illustrate the energy profile of a smaller developing Asian country, but such alternative choices were not in the cards. Fortunately enough, Taiwan has some interesting features that make it worthy of study, quite apart from the contrast its policies pose with those of the People's Republic of China. Among the things one might look for are a rapidly growing economy, a nuclear power program, energy dependence, and an unsettled position in regional and global politics. It would have been helpful to include countries from Western and Eastern Europe, but once again this option was not feasible.

Since the production and use of energy are fundamental to the welfare of all nations, and since solutions to energy supply and conversion problems must be tackled chiefly by individual countries in their own historical, political, and social settings, it is reasonable to hope that more volumes of national energy profiles will be organized by other editors. An excellent precursor of this volume is Leon Lindberg's The Energy Syndrome (1977), containing among other things a fine treatment of India that persuaded this editor to bypass South Asia. The eight studies in this volume can be viewed as a modest contribution to the "profile" style of exposition and analysis, as it bears specifically on mankind's presently unhappy relationship to dwindling and uncertain energy resources.

Several disciplines are represented among the authors of these profiles: political science, economics, history, and area studies. Differences in style, tone, approach, and choice of specific materials are therefore inevitable. The reader should not expect to find everyone marching to the same beat. Actually, the variations are all to the good, for they throw light on the problems of energy supply and use from somewhat different angles. Nevertheless, a conceptual framework was devised by the editor to promote some unity and coordination of purpose amidst the centrifugal tendencies of eight authors, all of whom have been asked to stay within a common format of topics, which appears as a topical index at the end of this book to facilitate cross-reference. As the chosen framework implies, energy policy and planning are best understood in a diversified context in which linear thinking is eschewed. Effective "solutions" or strategies bearing on energy are likely to have political, social, economic, and environmental dimensions in addition to the usual appeals to technology. The best kind of energy "expert" is inclined to systems analysis and recognizes that everything is connected to everything else. One must deal with complexly interacting variables associated with many realms of discourse, even if some variables are more important than others with respect to a given question.

The spirit of this book is frankly normative and policy oriented. The empirical analysis of the usual social science studies is mingled with a deliberate consideration of useful and meaningful goals, with self-conscious acceptance of the premise that certain preferred values ought to govern any rational, sensible energy policy. Thus it is better that energy resources be conserved rather than wasted; better that economic systems aim at thermodynamic efficiency rather than muddle through with careless, ignorant inefficiency; better that energy systems harmonize with nature rather than pollute and despoil it; better that energy resources such as fossil fuels be viewed as a trust for future generations rather than as a convenient bonanza to be squandered with due speed. The authors of these profiles have been encouraged to leaven good scholarship with thoughtful interpretation and bold judgment.

For the reader's convenience there are glossaries of technical terms and acronyms at the end of the book. There is also a conversion table to render more intelligible various ways of stating quantities of energy. There is no summary conclusion. The editor decided to abstain from writing one for two reasons. Only a long conclusion could do justice to the material and observations in the chapter, which might have distracted unnecessarily from the studies themselves and certainly would have violated the sound principle that books on heavy subjects ought to be kept as short as possible. On the other hand, a short conclusion would have contributed nothing of substance.

The authors of these profiles have tried to keep their discussions abreast of the latest developments. Needless to say, the world and its energy problems are not static, and there is always the risk that some delectable generalization will be scotched by the flow of events. Most of these studies reflect a data base through late 1979. Some reflect efforts at revision through the early months of 1980. All the authors are reconciled to the elusiveness of the latest facts.

CONTENTS

LIST OF TABLES

xiii

NATIONAL
ENERGY
PROFILES

INTRODUCTION:
ENERGY AND THE FUTURE OF NATIONS
Kenneth R. Stunkel

A quarter-century ago hardly anyone imagined that energy would become virtually overnight a major nemesis in world affairs. Nearly all dimensions of human life are touched, more or less, by the availability, type, use, abuse, and price of energy. The prosperity and stability of industrial states, the economic and social aspirations of developing countries, food production, population and economic growth, environmental balance, the amity and tensions of nations, the promise and limits of technology, the growing militarization of the world economy, the realistic options of unborn generations—all are related, directly or indirectly, to troublesome problems of securing and using huge amounts of energy in a very short span of time. The discovery and implementation of workable solutions to these problems are likely to tax the resources of institutions, the adaptability of societies, and the ability of policy makers to confront harsh realities. The alternatives to a conscious search for solutions are quite simply drift, periodic "crisis storms," and chaos. As E. F. Schumacher (1973) has put the case for energy, "it is impossible to overemphasize its centrality. It might be said that energy is for the mechanical world what consciousness is for the human world. If energy fails, everything fails."

ENERGY IN HISTORICAL AND GLOBAL PERSPECTIVE

Certain identifiable historical forces are associated causally with the sudden and dramatic pressure of mankind on energy supplies, and all of them are rooted in the career of Western civilization. Among these forces one must include nationalism and its creature,

1

the nation-state, both of which have spread beyond the confines of
Europe to embrace most of the world's people in the past century.
The energy crisis has arisen in and among nations, some 154 of them,
most of which belong to the United Nations. For better or worse,
the accommodation of humanity to energy imperatives will have to
take place in an international political framework of multiple sovereign-
ties. Other historical forces are the development of modern science
and technology, which joined forces decisively in the late nineteenth
century, and the industrial mode of production based on the combustion
of fossil fuels. Without the latter it is inconceivable that world population could have attained the present level of some 4 billion.

Along with the political, economic, and technological framework within which the contemporary crisis of energy has been fostered,
nineteenth-century European civilization also provided the conceptual
framework within which the peculiar nature of the crisis can be isolated and understood. The concept of exponential growth, first drama-
tized by Thomas Malthus, brought into focus the dangerous phenomenon
of successive doublings of some given quantity in ever shorter periods
of time. Pressure on world energy supplies is related, at least in
part, to the exponential growth of world population, schematized as
follows from United Nations data:

Numbers of People	Years of Growth	Year Reached
First billion	2,000,000	1830
Second billion	100	1930
Third billion	30	1960
Fourth billion	15	1975
Fifth billion	11	1986
Sixth billion	9	1995

Merely sustaining this widening flood of humanity requires
immense quantities of food and other forms of energy. In addition to
population there are other serious instances of exponential growth,
most notably energy use, industrialization, urbanization, and global
pollution. In 1776 the gross world product (GWP) is estimated to
have been some $150 billion. Two hundred years later it was in the
vicinity of $5 trillion, most of which had been accumulated since
1950, when GWP was $700 billion. From 1950 to 1973, global con-
sumption of energy rose at some 4 percent a year, a rate that, if
sustained, would lead to a 50-fold increase in a century. Obviously
an exponential curve in nature cannot be maintained indefinitely.
The significance of this growth pattern is that large numbers materi-
alize with frightening suddenness. On one day all seems to be well;

on the next day the roof falls in, leaving no time for thoughtful preparation or protective measures. Humanity had its entire history on
earth to prepare for the first billion people. The present generation
has 20 years to get ready for another 2 billion, roughly a billion each
decade, surely a challenge unprecedented in human experience.

The second concept is that of thermodynamic entropy, formulated by Sadi-Carnot, which spells out an inescapable limit on the
extraction of work from energy. Only energy in specific forms can
yield work, and the quantity of work available is always well below
the quantity of energy involved in the conversion process. Not only
is the efficiency of energy conversion limited by nature, the other
side of the coin is an irrevocable loss of energy in the form of diffused, unusable waste heat. The upshot of the Second Law of Thermodynamics is that all activity or process is moving in one direction
and produces irreversible effects in nature. No matter how sophisticated the technology of an energy system, it must always operate at
a deficit and must always generate pollution in the form of waste heat.
The clearest implication of this law is that man cannot do just as he
pleases. The cost of biological or economic activity is always greater
than the product of that activity (Georgescu-Roegen 1971). There are
insuperable constraints on achievable efficiencies. Among other
sobering consequences, here is the final answer to "money fetishism"
in economic theory, or the notion that price rises can compensate for
resource shortages. All economic systems are "solidly anchored to
a material base" and checked by a process that has a "unidirectional,
irrevocable evolution" (Georgescu-Roegen 1972).

The third concept is biological evolution, formulated in convincing terms by Darwin, which provides the means for grasping the
entire surface of the earth—from the deepest trench in the oceans to
the top of the atmosphere—as a complex biological "organism" developed over billions of years and distinguished by a multiplicity of
interdependent ecological niches. The whole biospheric film, only a
few miles deep, hovering between the sterility of space and the life-
lessness of the earth's crust, can be viewed as an intricate flow of
energy with which human beings are associated organically (Singer
1970). The energy flows and cycles of the biosphere are the foundation of all life on earth. What has become alarmingly clear in the
past few decades is that human life, unlike other forms, has the
power to interfere significantly with energy flow patterns. Climate
modification is only one of the most obvious interferences. The extinction of a plant or animal species through pollution or the destruction of habitat removes a unique mode of energy conversion from the
whole system.

The fourth concept is that of the earth as an isolated life-sup-
port system in space. The Apollo vision of earthrise from the moon

stretched our consciousness to the realization that the earth is a lonely "energy system" swinging through black reaches of inhospitable space. From the Apollo perspective, it is one world, finite and indivisible, the only setting available for the continued survival of its living inhabitants, all of whom are quite literally in one boat.

For all but a fraction of his brief life on earth, man has been unaware of the pattern traced by his relationship to energy stocks and flows. Human energy conversion in the past 100,000 years has been riding a discernible exponential curve, whose steepest ascent appears in the past 100 years, chiefly as a consequence of widespread industrialization. The significance of the curve's sudden, upward flight does not seem to be appreciated in the councils of nations. There is no public or official recognition that it is an abnormal, remorselessly temporary phenomenon. The time must come, probably sooner than most people expect, when the upward climb will cease and fall back precipitously to a lower plane or level off on a plateau. As Kenneth Boulding has observed, anyone who supposes that exponential growth can go on forever is either a madman or an economist.

In hunting and gathering communities of the paleolithic era, energy conversion was tied to food consumption, which amounted to some 2,000 kilocalories (kcal) a person per day, roughly the amount of energy discharged by a 100 watt bulb. The mastery of fire raised the conversion level to 4,000 kcal as humans began the exploitation of forests to provide fuel. The advent of agricultural societies 8,000 to 10,000 years ago pushed the level to about 12,000 or 15,000 kcal, which remains to this day the amount of energy used by 70 percent of the human race. On that modest energy base all the great civilizations of the past have been created and sustained for very long periods of time. The Chinese, for example, gaze back on more than 3,000 years of continuous and distinguished civilized life nourished by a self-sustaining agricultural economy.

The industrial revolution of the nineteenth century allowed a few countries—England, Belgium, France, Germany, and the United States—to achieve the unprecedented conversion level of 70,000 kcal a person per day, a low-technology phase of industrialization that lasted around 30 years. The development of electrical power and motor vehicles marked the onset of the high-technology phase. At that stage the level of energy conversion rose to astonishing peaks, the United States outdistancing all other nations with a current level of 230,000 kcal a person per day. For the remainder of the world's people to use energy on such a scale, global energy production would have to increase by at least a factor of seven (Hubbert 1971).

Global use of energy has been growing faster than world population, the former averaging 4 percent a year for the past 100 years compared to 2 percent for the latter. The demands of some countries

on energy supplies verge on the insatiable. The United States is
doubling its energy use every 14 years (for electrical power it is
every ten years). Should this momentum remain uninterrupted, total
American energy use by the year 2000 would equal total world use at
the present time. Prior to 1973, when the Japanese economy was
rocked by the oil price revolution, Japan's economic growth rate was
averaging about 10 percent a year. At that exceptional rate of growth,
it has been suggested that Japan, with only 3 percent of the world's
people, would have required nearly all of the world's available energy
and resources by 2000 A.D. (Hedberg 1972).

The shares apportioned from available energy supplies are
radically out of balance. The United States with a bit less than 6 percent of the world's people uses more than 30 percent of the world's
energy each year. Western Europe has twice the population of the
United States, yet its use of energy amounts to 1.25 billion tons of
oil equivalent a year compared to 1.85 billion tons for the United
States (see Darmstadter et al. 1977). The three largest national
economies—the United States, the Soviet Union, and Japan—have
about 15 percent of the world's people among them but manage to
secure 60 percent of the energy. Altogether there are about 25 industrial states with 30 percent of the world's people that consume
nearly 85 percent of the annual energy pie. The remaining 125
nations or so, embracing two-thirds of humanity, must scrape along
on what is left. It is a curiosity of our time that so many people in
affluent countries see nothing especially unsettling, anomalous, or
dangerous about this skewed distribution of energy use. There are
even some who argue boldly that it is a benign milestone of progress
(Singer and Bracken 1976).

In the longer perspective of history mankind is clearly at the
threshold of a precarious transition, for until recently, human societies have been relatively self-sufficient and self-sustaining in their
relationship to energy sources. As hunters, gatherers, fishermen,
and farmers, human beings drew from the inexhaustible well of primary energy supplied by sunlight, stored mainly by the photosynthesis
of green plants. Societies functioned on energy income and produced
wastes that could be absorbed safely and recycled by natural biological
systems. These low-entropy societies still exist and most of the
world's people live in them, from Chinese agricultural communities
to Kurdish herdsmen and Malaysian fishing villages.

Over the past 100 years the industrial type of society has established itself throughout the world. Most peoples and national governments associate the promise of a better life with industrialization,
technoscientific innovation, and speedy economic growth. Where the
industrial pattern of life is established one expects to find material
abundance, a technologically sophisticated infrastructure, advanced

productive capacity, stress on the economic foundations of existence, and deliberate, large-scale modification of the environment. It is widely assumed, uncritically for the most part, that everyone will be better off to the extent that key features of industrial civilization have been adopted and assimilated. Another way of articulating this trend is to say that nearly everyone wants to use more energy.

The 800 million people living in advanced industrial societies constitute a "world middle class," the criterion for membership in this exclusive group being a per capita income of $2,000 (Keyfitz 1976). This relatively affluent segment of humanity is sustained by economic and social orders almost wholly dependent on fossil fuels. High-entropy industrial societies are a parasitic genre of human culture feeding off nonrenewable, fixed stocks of energy. In dramatic contrast to virtually all great civilizations of the past, the industrial type—communist, capitalist, or "mixed"—thrives on optimistic assumptions that endless growth is both possible and good, that affluent material standards of life, defined by high levels of productivity and consumption, can prevail everywhere, that the secret of universal wealth is an alliance of unimpeded economic growth with technological innovation, and that nature can absorb the wastes of energy conversion on scales of indefinite magnitude. The world middle class is growing at some 5 percent a year, which means that by the year 2000 the circle of affluence could embrace 1.8 billion people out of a world total of 6 billion. Such an index of relative world affluence would represent a quantum leap in the acceleration of energy conversion in the space of a single generation. Amory Lovins believes that "the present energy growth rate cannot be maintained for long," yet every government on earth is proceeding on the assumption not only that can it be maintained but that it can be prodded to a level of conversion six times greater than now by the first decade of the twenty-first century. In effect, "governments have apparently acceded to quick depletion of cheap energy reserves on the tacit assumption that a new source will turn up in time to maintain ever faster growth" (Lovins 1974). In an orgy of extravagance, limited supplies of energy are being devoured with the fulsome expectation that unlimited supplies will materialize just as the last morsels are gulped down.

PROSPECTS FOR WORLD INDUSTRIALIZATION

Industrial civilization has been around only for 100 years or so, an insignificant droplet of time in the longer perspective of history and prehistory, yet it is already uncertain that the existing scope of industrial activity can be sustained much longer, quite apart from the naive expectation that it can be expanded painlessly to accommodate

another 2 billion people in the next two decades. The unequivocal
short-term threat to industrial societies is the depletion of key resources before capital investment and technological advances can
provide substitutes. The development of energy resources is not
keeping pace with spiralling consumption: "in 1950 the production
and consumption of energy was in virtual balance for the developed
countries as a whole. . . . By 1973 production in the developed countries had nearly doubled but consumption had far outrun it and the
deficit had swollen to a third of consumption" (Keyfitz 1976). This
tendency for demand to outrun available resources has become the
unacknowledged Achilles heel of industrial economies.

At the threshold of the 1980s it is clear enough that neither
resources nor institutions have given an adequate livelihood to nearly
two-thirds of humanity, while at the same time the materially privileged one-third has moved into an era of insecurity. The signs of
strain are unmistakable and pervasive: inflation, unemployment,
social unrest and malaise, shortages of capital, and a perceptible
decline of environmental quality. While it appears that modern man
has already bitten off more than he can chew, 2 billion more people
will be crowding onto the stage in the next 20 years in need of food,
clothing, shelter, jobs, medical care, and energy. How that much
more can be provided on top of what is still needed does not seem to
be a priority question with politicians and decision makers. As
Kingsley Davis wryly observes, "the human species is now in the
preposterous situation of using an extremely advanced technology to
maintain nearly four billion people at a low average level of living
while stripping the world of its resources, contaminating its water,
soil, and air, and driving most other species into extinction, para-
sitism, or domestication" (Olson and Landsberg 1973).

There are, of course, dissenters from this admittedly pessimistic viewpoint. For those who are attracted to modest assurances
that the present is not so bad and the future bright, there are Irving
Kristol and the Wall Street Journal or Norman Podhoretz and Commentary. For those who prefer solace in the grand manner, there
is Herman Kahn and the Hudson Institute, in whom the cornucopian
vision can be found in all its uninhibited splendor (Kahn et al. 1976).
Kahn believes that by 2176 A.D. there will be a world population of
15 billion, a GWP of $300 trillion (compared to the present $5 trillion),
and an average per capita income of $20,000. Moreover, pollution
will be so negligible that people will drink water straight from the
rivers. He sees the future as "incredibly bright." Expanding production is the key to riches and environmental quality for nearly all.
These heady figures are scaled down somewhat in a more recent book
(Kahn 1979), but not enough to make a serious difference in the magnitude of growth Kahn thinks is feasible. There are no obstacles

invulnerable to appropriate technology. Nothing stands in the way except the contingencies of bad luck and incompetent management and the defeatism of zero and slow growth advocates. In short, a utopian material condition for quadruple the present world population depends on an unremitting commitment to growth and technological progress. Anything less is likely to bring disaster. The gap between rich and poor is an indispensable mechanism driving the world engine of economic development. The poor will get richer inevitably as the rich get richer. The present course of industrial civilization toward ubiquity and growth at all possible speed is the only rational alternative.

This brief introduction is not the suitable place to examine in detail the nature and implications of such elephantine technocratic thinking, but it is appropriate to take a stand of sorts. Confronted with Neo-Malthusian and cornucopian "scenarios," in their assorted guises, it is understandable that many people are inclined to accept an optimistic viewpoint congenial to their aspirations and expectations. Kahn is telling a disturbed world what it wants to hear, that civilization is headed toward a nearly utopian state, that more growth and not less is the medicine for our social and economic ills, and that all is likely to be well if we brace up and think positively. This counsel of business as usual from an "authority" cannot fail to please a broad spectrum of citizens and policy makers.

Unfortunately a choice between viewpoints on the basis of one's fondest hopes is not the best way to steer either thought or behavior into line with compelling realities. The responsible approach is not to take sides with authorities but to examine as much of the evidence as one can and reach an independent judgment. After following that difficult and time-consuming procedure, the present writer, for what it is worth, has concluded that a better fit to reality can be found in the analyses of Lester Brown (1978a; 1978b) and William Ophuls (1977) than in those of Kahn and his associates. If the reader wants to know what is happening on this planet, he would do well to give the publications of the Worldwatch Institute priority over those of the Hudson Institute.*

In the short term one can dismiss cornucopian visions of the future, for the insistent question at this moment is how to provide basic necessities for human beings already alive and for those expected to be born in the next decade or so. The Ehrlichs (1970)

---

*The reader might try also two novels of Doris Lessing: Memoirs of a Survivor (1974) and Briefing for a Descent into Hell (1971).

suggest that lifting the 3.6 billion people of 1970 to a U.S. standard of living would take, among a great many other things, 75 times the annual extraction of iron, 100 times the copper, 200 times the lead, 75 times the zinc, and 250 times the tin. If the population anticipated for the year 2000 were to be supported at a U.S. standard, all of those figures would have to be doubled. The extraction and processing of minerals on such a scale would drain off prodigious quantities of energy—utterly beyond present means—and probably overwhelm the waste-absorbing capacity of the biosphere. Far from most human beings living in mass consumer societies any time soon, it is doubtful that basic needs can be provided through the next generation, and uncertain just how much sustainable industrialization can be tolerated by the earth's biological systems. Intimately associated with these questions is the underlying puzzle of energy: how much and what types can be made available at reasonable cost in a given period of time (U.S. Senate Committee 1978).

As long as world expectations for economic growth remain undisciplined and open-ended, prospects for energy conversion in the next 30 years are not good. Nearly all short-term industrialization must rely on fossil fuels. While other sources of energy are available in practice and principle, it is still an inescapable reality, for a variety of complex reasons, that none of them can make a substantial difference until the advent of the twenty-first century, and even then one must assume that technological developments proceed without a hitch and that catastrophes—economic, military, or ecological— do not intervene. With proper attention to capitalization and safety, nuclear fission and solar energy, especially the latter, could serve as transitional bridges out of the petroleum era for some countries. Other potential sources of energy, such as nuclear fusion, magnetohydrodynamics, or solar collecting space satellites hooked to the earth by laser beams, are irrelevant as solutions to the depletion[*] of fossil fuels in the remainder of this century (Hayes 1978a; Browne 1979).

A rational approach to energy must concern itself with sources readily available with the help of existing technology and capital. A rational approach will be governed by prudence and a proper sense of rate and magnitude, because large-scale energy projects cannot

---

[*]The concept of depletion has more to do with cost than with the quantities of fuel in the earth's crust. When the cost of extracting fossil fuels exceeds the value of the energy they yield, depletion can be said to have taken place.

be implemented overnight, even should technology, raw materials, and capital be readily on tap, nor can energy that might be available tomorrow be used for the tasks of today. In short, the only energy sources that matter are the ones capable of making a difference in the near and medium terms. It is no easy matter cutting through a jungle of energy studies to realistic issues and options. The trouble with most such studies is that they are linear in method, dwelling on narrow schemes for more energy through technological manipulation or innovation. An endless proliferation of speculative scenarios has the effect of paralyzing rather than stimulating thought. Energy "experts" carry on too often as though capital, environmental resilience, and social adaptiveness are infinitely flexible variables. The more specialized among them are frequently naive about political and economic dimensions of the energy crisis. Lovins (1974) may have a point in suggesting that energy shortages can be relieved by burning energy studies. Short of that extreme measure it is wise to use them with caution and skepticism.*

Understanding the gravity of the fossil fuel "fix" demands close attention to several unrelenting truths. First, the earth's stock of fossil fuels is essentially nonrenewable. Whatever the quantity tucked away, it is ineluctably finite, for the rate at which it is being withdrawn far exceeds the rate at which it is being stored by natural processes. Between 1959 and 1969 the world exploited a quantity of petroleum equal to the amount produced in the whole of human history prior to 1959. Geological processes, on the other hand, are slow and deliberate, much as though drops of water are being deposited in a well while entire buckets are being hauled out. In the next million years the pool of available fossil fuels will increase by less than 1 percent; in the 100 years since oil was first discovered, man has used 20 percent of the deposits laid down in the past 500 million years. Second, it is relatively unimportant how much fossil fuel actually exists if it cannot be recovered at acceptable cost by technology on hand (Commoner 1976). Thus most of the cheap, easy wells on U.S. soil have been drilled. Technical difficulties and expense mount as one goes deeper. When the cost of extraction tops the value of the energy obtained, as measured by its ability to do useful work, the exploitation of fossil fuels will cease. In other words, the unavoidable limitation of net work profit will put fossil fuels beyond the reach of industrial civilization. That limitation is

---

*An energy study one can use with confidence is Wilson Clark's comprehensive Energy for Survival (1974).

edging into view with alarming rapidity. Third, estimates of "reserves" usually omit all discussion of the environmental consequences of extracting and using fossil fuels, yet these consequences are part of the price paid for energy.

What all of this amounts to is the operation of a law of diminishing returns in man's relationship with energy supplies drawn from fixed stocks. Planners who exult in the hundreds of years of energy conversion supposedly implied by untapped reserves of coal need to consider that a descent from high grades to lower grades of fuel binds us more tightly in the grip of diminishing returns. It is one thing to compute optimistically the earth's reserves of oil, coal, and natural gas. It is quite another to get it out of the ground and make use of it.

The upper limit of recoverable crude oil reserves appears to be in the neighborhood of 2 trillion barrels, while proven reserves—those recoverable with present technology and costs—stand at some 600 billion barrels. That sounds like a mighty pool of oil until one ponders the effect of consumption rates. Recoverable reserves amount to 500 barrels for each person in the world. Lester Brown illustrates how far that much oil might go:

> An American with a large automobile that averages 10 miles per gallon and that is driven ten thousand miles per year requires just over forty barrels of oil per year. At this rate, an individual's share of remaining oil would be exhausted in just twelve years. This assumes, of course, that all ultimately recoverable reserves will materialize, that the world's remaining reserves are shared equitably, that there is no further increase in population, and that no oil is saved for future generations (Brown 1978b).

At the 7 percent rate of economic growth that prevailed in the world through the 1960s, world reserves would be gone in a mere 20 years (McKelvey 1977). In one of the more thoughtful and convincing energy studies, an international conclave headed by a professor from the Massachusetts Institute of Technology distinguished between a slowing of petroleum production and a decline of production, both of which can create a supply-demand imbalance with unpleasant economic consequences. The former type of imbalance may occur by 1981 (some argue that it is already here); the latter is expected by the mid-1990s. In either case, the world "must drastically curtail the growth of energy use and move massively out of oil into other fuels with wartime urgency. Otherwise we face foreseeable catastrophe" (Wilson 1977). The world ratio of reserves to production has been dropping and shows no signs of recovery. Should oil consumption go on doubling every ten years, a doubling of world oil reserves—an unlikely contingency in any event—would stave off physical depletion for no more than another decade. A technological escape hatch, the extraction of oil from shale and tar sands, has turned out to be illusory. Many private companies working shale deposits in Wyoming and Colorado have given up because of the expense, and much the same disappointment has plagued companies trying to exploit Athabascan tar sands. While world coal reserves are 17 times the petroleum reserves, there is small chance of more than a fraction of it being used before prohibitive costs and serious environmental impacts close in. An example of a coal barrier is the fact that much of America's supply lies beneath prime farmland, which would have to be mined by the stripping technique, and that it would be of little help in the transportation sector, which is dominated by petroleum and uses 20 percent of the country's energy budget (Commoner 1976). In the case of natural gas, closely associated with oil deposits, the expected lifetime of reserves is a bit ahead of petroleum.

Despite these known limitations to fossil fuel consumption, industrialized societies, and increasingly the developing ones, are dealing with energy as though it were a free good that can be burned up with endless impunity and, what is worse, be substituted for all other forms of capital. Virtually every scheme for dispelling scarcity implies vast expenditures of energy in the form of fossil fuels, which constitute 97 percent of world primary energy use (wood, nuclear fission, hydropower, solar, and geothermal account for most of the other 3 percent). Are we contaminating fresh water supplies? Build nuclear desalting plants to process fresh water from the sea. Are we squandering handy lodes of mineral ore? Go after nodules at the bottom of the sea, crush minute deposits from the earth's primal rock, or shoot men into space to mine the asteroids. Are urbanization, irrational farming practices, and pollution shrinking the globe's indispensable stock of arable land? No matter. Raise the productivity of what is left with petroleum-based fertilizers, pesticides, and mechanization. Are ecological systems being destroyed by strip mining? Repair the damage with landscaping projects. It is in just this manner that the high grade and natural are replaced by the low grade and unnatural by waving the magic wand of free energy.

Well before the critical frontier of physical depletion is reached, large amounts of fossil fuel, a subsidy whose days are numbered, will have to be invested in the development of renewable energy capable of sustaining the promethean economic activity of the industrial world and supplying the basic needs of billions. One has to imagine an energy infrastructure based on solar, geothermal, nuclear, tidal, wind, wave, hydro, and other like sources doing all the work currently shouldered by fossil fuels. How long would it

take to perfect and install such an infrastructure? How much would
it cost? What social and economic adjustments would have to be
made from one country to the next? How much fossil fuel in a given
economy would be necessary to engineer a transition from capital to
income sources, without serious disruption of living standards?
These questions ought to be at the top of most national agendas, for
it is certain that a transition from capital to income sources of energy
cannot be made without generous drafts on the fossil fuel subsidy.

It has been argued that while problems may be growing exponentially, so is knowledge. Know-how and better technology will
subdue the crises of energy, food, resources, and the environment.
For the technocratic temperament "the assumption is that in the aggregate resources are infinite, that when one flow drys up there will
always be another, and that technology will always find cheap ways
to exploit the next resource" (Daly 1977). This illusion that machines
can be redesigned to deplete one resource after another has reached
something of an apotheosis in the view of economist Robert Solow
(1974) that with "backstop technology" like breeder reactors, and "at
some finite cost, production can be freed of dependence on exhaustible
resources altogether." Perhaps small populations living in agricultural societies can get on without nonrenewable resources, but large
populations living in industrial societies must have such resources
even to produce the backstop technology. Since there are no material
factors in the economic process other than natural resources, Solow's
notion that the world can get along without them "is to ignore the difference between the actual world and the Garden of Eden" (Georgescu-
Roegen 1975).

Technology cannot rescue people from the consequences of
stubborn folly and nonadaptive behavior any more than a physician
can save a recalcitrant chain-smoking alcoholic from disease or
death. Technology is no substitute for wise political action, judicious
economic reform or courageous social change. The technological
"fix" cannot replace the will to act, a conscious choice of goals and
priorities, the virtues of discipline and restraint, or a mature acceptance of limits. No doubt technology must be part of any large-
scale solution to the problem of energy, but any responsible policy
maker will take into account its limitations (Stunkel 1973).

First, most scientific knowledge is irrelevant to problems of
energy and growth. Moreover, it is a moot point that such knowledge
is growing exponentially. Just as certain areas of physics—optics
for example—have exhausted their theoretical potential and have become tools of the engineering trade, so one can expect the same ripe-
ness to overtake fields of inquiry bent on mitigating the limits to
growth (Stent 1969).

Second, technology applied to the energy crisis is constrained by the laws of nature. Not infrequently one hears the technocrat carrying on as though the laws of physics can be used to circumvent the laws of biology, or as though physical laws are temporary inconveniences, soon to be breached by human ingenuity. For the technocratic imagination nothing is impossible, an outlook that belongs in the realm of magic, sorcery, and witchcraft. Solow's idea that industrial societies can function without exhaustible resources is worthy of Merlin the Magician. All technology must come to terms with principles of impotence, the most dramatic of which is the Second Law of Thermodynamics. The creation and deployment of new technologies to meet the demands of growth made possible by other technologies will mean further inputs of energy, more entropy in the form of waste heat, and further depletion of high-grade energy and mineral resources.

Third, technological proposals must be paid for. Innovators must explain where the capital is going to come from, or someone must do the explaining (Lovins 1977). In 1972 it was widely believed that nuclear reactors would supply cheap, abundant electrical power. The sudden rise of petroleum costs in 1973 added glitter to the nuclear option. The Atomic Energy Commission expected in 1974 that the total energy supply of the United States in 1972 would be provided by a thousand 1,000-megawatt nuclear facilities by the 1990s. The cost of a 1,000-megawatt generator in 1972 was $225 million, which converts to an oil equivalent of $25 a barrel, a level of expense reckoned on to drop as numerous plants were built. In fact the costs escalated at a breathless 26 percent a year, triple the U.S. rate of inflation, and reached $900 million for a reactor of the same size in 1977, four times the price tag of 1972. The reason was the need for safety devices and design changes as the early plants manifested flaws while going through their paces. This unanticipated cost revolution has blunted the market for nuclear reactors and reduced the competitiveness of atomic power against fossil fuels. Indeed, Commoner (1979) has argued that all the nuclear power plants in the country could be shut down without adding more than a half dollar to the average utility bill, providing that idle fossil fuel plants were brought into full and efficient operation.

A further sobering example of technology running aground on the shoals of finance is the notion, still commonly held, that synthetic oil and gas can be produced with economic success from coal and shale and substantially relieve dependence on foreign petroleum. A study of this question has suggested that the total coal production of the United States in 1975 would have yielded synthetic fuels amounting to only one-tenth of the nation's present total energy supply (Wilson 1977). The processing of that much coal for such a minuscule return would require the construction of 95 facilities—gasification, liquefaction, and shale oil plants—at a cost of over $70 billion. Moreover, the resulting fuel would be more expensive by several times than fuel from OPEC.

Fourth, indefinite technological progress in the conversion of energy is not possible, for a time will come when efficiencies can no longer be improved. Thus the limit of efficiency for the conversion of fossil fuels to electricity seems to be 60 percent. There is no way around this thermodynamic constraint.

Fifth, technological solutions can be environmentally disruptive. An uninhibited technological offensive by countries with industrial capacity might well generate enough energy to transform the earth into a body resembling the moon, given enough time. It is not commonly understood that technology usually substitutes one kind of pollution for another, or that no technology is capable of functioning on earth with zero environmental impact. All activity in nature produces irreversible effects. An instance of this misunderstanding turned up on a PBS television show called "Nova," in which hydrogen was discussed as a "pollution-free" source of energy.

Sixth, technology must be socially and politically acceptable. Whatever the authentic merits of nuclear power, the fact remains that public sentiment against it is very strong in some industrial countries, notably Japan and the United States. Public misgivings have been intensified by the Three Mile Island incident, a force likely to influence the future of nuclear power as a "technological solution" to the energy crises even should its defenders be right that "taken as a whole, the accident at Three Mile Island generally confirms what we have been told about nuclear power," namely that it is safe (McCracken 1979). The record of nuclear power in Japan has produced an especially apprehensive climate of opinion, illustrating as it has a disconcerting tendency to obey Murphy's Law: "If something can possibly go wrong it will."

The secular faith in the omnipotence of technology trails in the wake of industrialization, assuming a conscious guise in the Baconian tradition of the West that all things possible can be effected, or it glides silently just below the surface of modernization in most developing countries (McDermott 1969). The misfortune is that public policy cannot make the best use of technology if it is not recognized as being part of the problems associated with growth. Without changes in social behavior, economic expectations, and industrial structures, technology can do no more than buy some time before unpleasantness sets in. Nor will technology be of maximum assistance without a proper understanding of energy constraints and potential. Even modest industrialization for all of the world's nations does not seem likely if the energy base for development is to be fossil fuels.

What satisfaction can be taken in an industrial establishment that cannot be sustained because of scarce, expensive, nonrenewable energy?

There is a deep irony in all of this. The more industrialization poor countries can achieve, while the affluent ones continue to grow, even at slower rates, the shorter the life span of usable fossil fuel reserves, and the more rapidly all industrialized countries will be faced with the painful question of sustainability. When that inevitable moment comes, for different nations at different times, the shift to renewable energy sources in advanced industrial states may need up to 50 years of planning and implementation. In every country with an industrial sector it will be necessary to drum up a subsidy of fossil fuel and capital adequate to develop energy conversion systems based on renewable sources that are comparable in magnitude to the ones already in existence.[*] As Rufus Miles (1976) has argued, one should not take for granted the ability of political and social management systems to cope with a crisis of such novelty and sweep. Increased interdependence is a concomitant of increased energy use, and interdependence means a high level of complexity reliant for its stability on economic and political systems designed and controlled by fallible men: "the more complex and interdependent the systems and subsystems, the more vulnerable they become to design failures." Where the complexity of interdependence systems outstrips human capacity to manage them, breakdowns in the systems will outrun the resourcefulness of problem solvers.

An example of management breakdown in a highly interdependent social and economic environment is the apparent inability of government on all levels in the United States to confront and deal with the basic arithmetic of oil production and consumption and to accept the necessary restraints of a genuine national energy policy. The U.S. scene presents an especially discouraging spectacle in the failure of perception. There is deep public anger over the suggestion that an era of cheap, abundant energy is over for good and that the future can never be like the past. A nation that uses nearly one-third of the world's energy with 6 percent of the world's people seems unable to admit gracefully that a 30-year energy binge has run its course. In the meantime, "energy policy" flounders over relatively trivial issues

---

[*]A recent and important study by the Harvard Business School (Stobaugh and Yergin 1979) points firmly to conservation and solar energy as the best means of transition.

like how to maintain a cheap, plentiful supply of gasoline from week to week.

THE REALM OF THE POOR

Apologists for economic inequity delight in observing that the poor have always been with us. More up-to-date apologists, like Herman Kahn, justify inequities by citing them as a necessary condition for global economic progress; without poverty the poor can never improve their lot, or something to that effect. Most of these apologists are insensitive to the unprecedented and inauspicious features of poverty at this stage of human history.

The sheer number of the poor has never been so great, nor have they ever multiplied with such prodigious abandon. Of the world's 4 billion people, one-third are desperately poor, which means that basic human needs (the biophysical requirements for survival) are not being met (McHale 1978). Below the level of the Third World countries (for example, Morocco, Malaysia, South Korea, the People's Republic of China, Saudi Arabia, Brazil, Jamaica), whose prospects have improved with exportable resources and strong leadership, are the Fourth World countries, about 40 of them with an aggregate population of more than 900 million, mostly in Asia and Africa. These people are truly, in the phrase of Frantz Fanon, the "wretched of the earth." On a larger scale, 2.5 billion people in the poorest countries share an annual product falling below $250 per capita (on the average) and share, with 60 percent of the world's population, about 9 percent of the GWP.

In this 1979 International Year of the Child it is worth remembering that one-third (1.4 billion) of the earth's people are children under the age of 15, an appalling number of whom are afflicted with energy starvation in a variety of forms. Nearly half the world's 50 million annual deaths are children under the age of five, due to malnutrition, disease, and the neglect of parents struggling for survival. Millions live in Third and Fourth World slums and shantytowns with no access to educational, medical, or sanitary facilities. The obvious response to this pathetic situation in our age of "progress" is sufficient energy to meet their "ground floor" needs. The poignant distresses of children in the modern world—superb material for the anger of a new Charles Dickens—will become even more visible in the next 20 years as their numbers grow by 0.5 billion. This calamity of the helpless young draws into perspective the relative importance of American motorists not being able to "top off" their tanks with cheap gasoline (McHale 1979).

Projections of world population to the beginning of the twenty-first century vary from 5.8 to 6.5 billion, depending on one's assumptions, the most strategic being the effectiveness of family planning in Third and Fourth World countries. Most of the 2-billion-person increase will be in the poorer countries where life is already hard, insecure, and unrewarding. Many of these countries have population growth rates of 2 or 3 percent a year, which result in rapid doublings that no country can long sustain. At Mexico's present rate of growth (3 percent annually), the combined populations of the USSR and China will be exceeded in the next century. It is quite wondrous that some optimists are comforted by a slight reduction of the world demographic growth rate: from 1.98 percent in the period 1967-77 to 1.88 percent in 1975-77. A wit forced these data into perspective by remarking that jubilation over a decline that small is a bit like reacting to a 200-foot tidal wave shrinking to 190 feet (Intercom 1978). However slightly diminished, a tidal wave of humanity looms on the horizon. The point of this is the huge and certain future demand for energy to provide basic necessities and jobs, a demand whose size and sudden arrival have no analogue in human history.

In the second place, the gap between rich and poor has never been so spectacular. There is no precedent for the consumption levels that divide Mali and Bangladesh from Japan and the United States. The average energy use of an American is 40 times that of a Nepalese. Energy use in the United States is twice the total for Asia (excluding Japan), Africa, South America, and the rest of North America, regions that account for half the world's population. In 1974 per capita kilograms of coal equivalent, an American consumed 11,484 to 1,269 for Mexico, 646 for Brazil, 632 for the People's Republic of China, and 94 for Nigeria. Even should all the developing countries raise by 50 percent their use of world resources in the next 20 years, their share would still be a mere 15 percent of world resource use, surely a gloomy prognosis for the near-term future of the poor (Leontief et al. 1977).*

In the third place, never have the poor been so resentfully aware of their poverty and so aggressive about righting the balance. Affluent countries perceive the energy crisis as a threat to present and future advantages of material prosperity and security. On the other side of the tracks are the poor, whose ranks are swelling by some 65 million a year, faced with the immediate imperative of

---

*Unless the poor do much better, 24 out of 25 people will live in poverty by 2025 A.D.

supplying basic needs. Thanks to films, picture magazines, and tourism, the comparative wealth of the privileged nations is in full view. The phenomenon of rising expectations has spilled out of the veteran industrial states. At this historical moment there is no reason to believe that billions in the realm of the poor will act with courteous resignation and "go gently into that good night." Indeed, some traditionally poor countries blessed with valuable natural resources have already struck the first substantial blows in favor of a New International Economic Order. Since the 1973 oil embargo, one of the greatest redistributions of wealth in world history has taken place. The 13 countries comprising OPEC raised the price of oil from $3.29 a barrel in 1973 to $12.38 in 1976 (in 1979 the price was hiked to a range of $18.00 to $21.00 a barrel) with an increase of revenues from $23 to $90 billion, and it looks as though price adjustment upward will be an annual event. Other poor countries with needed resources—from phosphate rock and bauxite to bananas and timber—have begun to play the politics of raw materials. While the rich will get richer in the short term, evidence is that they cannot expect to get rich as fast as in the past. When the industrialized states in due time are up against the problem of sustainability, it appears likely that many poorer countries will be forced to scale down or abandon altogether the goal of large-scale industrialization and rest content with the satisfaction of basic needs.

A small clutch of once indigent countries has achieved a favorable margin of power and leverage in the world economy, but most Third and Fourth World peoples have in fact been victimized by the escalation of energy prices. There is no price differential for energy that gives the impoverished a handicap in the global economic game of growth and survival. One wonders, for example, how food production is to keep pace with population growth when nitrate fertilizer production must increase by 100 percent in a fairly short time. To provide 800 billion kilograms of nitrogen fertilizer by 2000, around 20 percent of present total world energy use may be needed. Food for the poor, it seems, is due to become even more of a luxury than it is now.

Since most of the population growth in the balance of the century will occur in less developed countries, decisions about energy are no less urgent for them than for the developed ones. Fortunately the poor nations are in a position to work out a healthier relationship with energy resources than has come to pass among those far ahead in the modernization process. As Denis Hayes (1978a) puts the case: "The Third World will make the transition [that is, to solar energy] because they have little choice. With little capital and inadequate reserves of conventional fuels, more and more poor nations are realizing they cannot afford a major commitment to a petroleum based economy when world oil production is expected to peak around 1990." The finite energy base of industrial states is slipping away, endangering the viability of economic and social systems heavily committed to interdependence, and illustrating the danger of reliance on fossil fuels. Even if the affluent lack the good sense to restrain energy demand and conserve, the poor can still make headway toward self-sufficient, renewable, locally controlled, relatively nonpolluting energy sources capable of providing a foundation for the provision of basic needs. They are, in effect, well placed to learn from the discomfiture of the rich.

The rising cost of centralized, capital-intensive energy systems suggests that poor nations are well advised to avoid the "hard technology" approach to energy supply. A "soft," intermediate technology seems the wiser alternative (Hayes 1978b). A capital-intensive technology like nuclear power, for example, would be a mistake in most cases. If started today, a modest reactor in the range of 900 megawatts would cost roughly $1 billion. According to World Bank figures, capital investment required for nuclear plants is twice that for oil-, gas-, or coal-fired plants ($480 compared to $240 per kilowatt hour). The smallest reactor (made in Canada) is 600 megawatts, while those made in France, West Germany, and the United States are no smaller than 900 megawatts. It is estimated that not more than 20 developing countries out of more than 100 could be ready for the smallest reactor by 2000, and that only five will be in that class by 1980.

It has been pointed out that developing nations, for reasons of "dignity" and national prestige, may not want soft technologies. An objective solution to energy needs is less important in this context than political and ideological resistance to what may be perceived as a second-rate, poor man's technology. Moreover, they may choose energy- and capital-intensive routes to economic development because such routes are associated historically with the wealth and power of postindustrial states. Such attitudes might prevail temporarily, but the harsh insistence of various constraints—capital shortages, fossil fuel depletion, environmental ravages—probably is going to deflect and moderate them (Benoit 1976).

The great issue here is the kind of development paradigm best suited to developing countries at this stage of world history. The dominant pattern since 1945 has been economic growth through industrialization. The central assumption of this paradigm is that income will "trickle down" to the poor as industrial capacity grows. As Kamenetzky (1979) says, "it was believed that the faster a country industrialized and modernized, the sooner the standard of living would rise." This is essentially Herman Kahn's program for the world's poor in all of the Hudson Institute studies. It is also the view of most economists in the West and of not a few everywhere else. The result of this paradigm, however, has been extraordinary surges of consumption among the well-off in both developed and developing countries, a "phenomenon, combined with the population explosion among the poor," which "is exhausting the earth's resources, fueling inflation, spreading environmental illness and possibly even causing climate changes" (Kamenetzky 1979). The answer to these palpable failures of the growth and industrialization paradigm is a basic needs paradigm, a shift from traditional notions of modernization to an outlook concerned with fundamental human needs, equity, and sustainability. The emphasis in such a paradigm would be on simple, appropriate technologies, small scales of production, intensive use of local talent and initially unskilled people, exploitation of local institutional resources, and the nourishment of self-reliance and participation. The leaders of developing countries ought to reconsider seriously the course of linking their own economic aspirations to those of developed countries.

The example and influence of rich industrial states have been dubious in still another direction. It is an ominous and melancholy sign that military expenditures in the poorest nations expanded between 1961 and 1975 at twice the rate of living standards, as indicated by average per capita GNP (Sivard 1977). Squandering tight capital on socially useless armaments is a self-destructive and perverse means of achieving "national security" in countries whose obvious perils are not foreign armies but hunger, disease, unemployment, and social unrest. The spread of the "military mystique" owes much to the bloated military establishments of several big industrial states and their eagerness to profit in a feverish arms trade, but the greater culpability lies with misguided leaders in Third and Fourth World countries who cannot get their priorities straight.

THE ENVIRONMENTAL DIMENSION

It is fairly commonplace to find intelligent, well-educated people resisting fiercely the idea that physical limits stand in the way of the growth ethic (Leiss 1975). Nature is viewed as an object to be shifted about and manipulated at will, not as an organic, finite context within which all human activity is circumscribed. This attitude toward nature is similar to the faith in technology in that the world is perceived as a place where anything can happen, projecting us once again into an ethos of magic.

Probably the most disconcerting observation one can make is that an unlimited, cheap source of energy—the dream of technocrats and growth advocates—would merely hasten the irreversible breakdown of the earth's biological systems, already under considerable pressure from oceanic pollution, widespread species extinction, soil deterioration, and atmospheric contamination. Pollution is not just a "side effect," a marginal phenomenon to be subordinated quietly to really important matters like economic growth. It is an inevitable consequence of all economic activity, an outcome of physical and biological laws, and can never be eliminated completely. All energy conversion is polluting, especially the kinds associated with industrial production. A pollution-control technology controls pollution by creating other varieties of the same problem; in order to avoid producing its own side effects, it would have to be 100 percent efficient, which is thermodynamically impossible. Some of the larger questions facing humanity are these: How much and what kinds of energy conversion are possible before global strains on the biosphere become intolerable? With respect to wastes generated by energy conversion, what is the carrying capacity (maximum sustainable yield, a product of size and regenerative powers) of the earth? The concept of carrying capacity does not represent a static value. If we grant that human impact on the global environment is currently 5 percent of carrying capacity, then a 5 percent rate of growth each year would exceed it by 100 percent by the year 2036. An adjustment would have to be made long before that extreme was reached, because considerable elasticity in biological systems is necessary for their sustainability, and it is gratuitous to assume that 95 percent of the earth's carrying capacity is intact (MIT 1971). Evidence is good that carrying capacity in many biological systems has already been stretched or overshot (Commoner 1971).

Deforestation is a good example of environmental pressure resulting directly from energy use. Wood is a basic source of energy in poor countries where it provides warmth and power as well as building material; indeed, 80 percent of the wood used is burned as fuel (Eckholm 1979). Exponential population growth in the past half-century has increased the need for food and fuel. Thus agricultural expansion and the collection of firewood have eaten away at forests. Global demand for construction materials and various paper products have added to the pressure. Eckholm reports that heavily forested regions, constituting some 20 percent of the earth's land area, are vanishing at the rate of 11 million hectares a year (1.2 percent of the total), an area about the size of Cuba. Forests are a major component of the earth's biological energy flow mechanism, protecting land from erosion, holding water in the soil, and providing a habitat for myriad plant and animal species. As the "public service" functions of the biological system are degraded, it becomes necessary to substitute industrial energy that is nonrenewable and biologically corrupting.

The most fundamental limit to energy conversion, and therefore to industrialization, is the extrusion of waste heat into the atmosphere. Should world energy use go on doubling every 15 years
and come from fixed sources (for example, fossil fuels) rather than
energy income (for example, solar), the heat emitted would surpass
the quantity of energy driving the atmosphere in less than 100 years.
On a more local scale, electric power generation in the United States
releases heat at the rate of 0.05 watts per square foot of total land
area, on the average. Should electric power capacity continue to
double every ten years, within 90 years the released heat would
amount to 25 watts in contrast to the 19 watts delivered by the sun.

Still other environmental constraints on energy conversion
are the availability of land and water. Thus a coal-fired electric
power plant with a capacity of 3,000 megawatts needs a land site of
900 to 1,200 acres and a source of fresh water. Nuclear power
plants also sit on large tracts of land and are even harder on water
supplies for the operation of cooling systems. As urban populations,
agriculture, industry, and power-generating facilities compete for
fresh water, it is seldom remembered that world supplies of that
precious substance are fixed. Not a drop can be added to the amount
flowing through the earth's hydrological system, yet it can be degraded beyond the point of usefulness for industry or people. Should
the power, liquefaction, and gasification plants currently projected
for the United States be realized, water demand would exceed the
amount now used in the entire country by a factor of three to four.
All solutions to problems of energy supply call not only for heavy
inputs of energy but also for large amounts of water; as with the
energy subsidy, the water subsidy is taken for granted (Brown 1978b).
The development of pollution-control systems, in which technocrats
and cornucopians place so much confidence, is only a partial expedient, for "pollution control is . . . only a temporary tactic that will
allow growth of production to continue for just a little longer; it is
not a genuine solution to the problem of pollution, even under the
most optimistic assumptions about our technological capacities and
energy supply" (Ophuls 1977).

Apart from much-discussed effects of energy conversion on
human health, such as Japan's Yokkaichi asthma, there are effects
of a potentially more devastating nature. Two serious threats directly
attributable to the combustion of fossil fuels are inadvertent climate
modification and acid rains, both of which endanger agricultural productivity at a moment when stocks of arable land are shrinking and
world population is surging toward the 6 billion mark in a generation.

In the past century man has interfered with atmospheric composition and other variables that control heat balance. Carbon dioxide
content, for instance, is expected to double by the close of the century

as a direct result of fossil fuel combustion. Students of climate dispute the precise outcome of such compositional changes, but they incline to agree that almost any effect of heating or cooling is likely to be malign for agricultural production in the temperate zone (Woodwell 1978; National Academy of Sciences 1977). Acid rains are produced by the interaction of sulfur dioxide with atmospheric moisture, which forms sulfuric acid. The acidity of rainfall has increased 100 times in the eastern United States and is a grim problem in Sweden, Norway, Germany's Ruhr Valley, Britain, and Japan. It decimates marine life in fresh water lakes and undermines the organic basis of soil fertility (Brink et al. 1977). Both of these phenomena are aggravations of industrialization in an age of fossil fuels. They are certain to grow worse as the industrial mode of production spreads.

Environmental issues have been politicized by Third and Fourth World spokesmen eager to justify unrestrained industrialization and energy use. It is held that poverty is the most objectionable form of "pollution" and must take precedence over so-called environmental balance. Johan Galtung, an influential student of world order, has sided with the "poverty is pollution" advocates without being either a technocrat or a Marxist himself. He argues that The Limits to Growth—a now-famous computer simulation study that projected global disaster within a half century should demographic, economic, and environmental trends continue—is "a discredit to the Club of Rome" and "a dangerous study," because it puts all of us in the same boat when inequities and "structural violence" demonstrate that everyone decidedly is not in the same boat. He goes on to maintain that alarm over resource shortfalls and environmental decay has appeared only as inconveniences have begun to plague rich countries. The affluent have discovered, with painful suddenness, the shortages and pollution that have been routine for most of the world's people. Galtung approves of a Third World manifesto proclaimed at the Stockholm Conference in 1972, which asserts in part "that holding economic growth per se responsible for environmental ills amounts to a diversion of attention from the real causes of the problem which lie in the profit-motivation of the systems of production in the capitalist world" (Galtung 1973).

Thus with an ideological flourish the ecological crisis is laid at the feet of greedy capitalists. Industrial systems operated and controlled by the "people," it is implied, cannot despoil either resources or the environment. Needless to say, the Chinese, the Soviets, and other parties anxious to get on with ambitious development programs smile upon this line of argument, for the underlying premise is that pollution and resource depletion are a consequence less of natural law than of unsound economic and political systems (Kim 1978). Politics rather than nature is responsible for the energy crisis

as well. Warnings about energy shortfalls and environmental stresses
are dismissed as "ecologism" and "scientific imperialism" fabricated
by capitalist regimes worried about the survival of their affluence in
the face of Third and Fourth World demands for equity.

However forceful the moral animus behind these arguments,
the logic is bizarre and mischievous. The principles of thermodynamics and ecology are not rescinded merely because a large segment
of the world's poor has awakened to the fact that a handful of industrial
countries uses most of the world's resources. Evidence is that the
industrial mode of production has much the same impact on resources
and the environment everywhere—with minor differences, such as the
difference in relative efficiency of energy conversion in Western
Europe and the United States—regardless of political, social, or economic systems (Kelley, Stunkel, and Wescott 1976). It is the industrial
process itself, whatever the ideological bunting, that produces irreversible effects in nature on a grand scale. It is surely better that
economic life serve the many rather than the few, but an economy of,
by, and for the people will cause pollution and reduce the resource
base as surely as one in the hands of oligarchs or stereotyped capitalists. "Poverty pollution" is not identical with the kind of pollution
that causes biophysical imbalance. Those who obscure this distinction, for whatever reasons, are doing the poor a disservice. If poor
countries accept laws of nature that make nuclear fission possible,
as do India and China, then it seems a bit arbitrary for them to set
aside laws that limit energy conversion and physical growth. The
conviction that ideology can bypass the laws of nature, or that the
laws of nature are a function of ideology, is a form of word magic.
Poor countries relying on such incantation will find the ground cut
from under them as nature quietly ignores the illusions of political
man.

Kenneth Boulding has made useful distinctions among three
conceptual approaches to world order: structural, dialectical, and
evolutionary. A structural approach will emphasize fixed patterns
and static forms, as in the mechanistic model of economic activity
developed by nineteenth-century classical economists and Galtung's
notion of structural violence. The dialectical approach sees the
world in terms of struggle and conflict between classes and nations,
as with current variants of Marxism. The evolutionary approach
looks on the world as "a disequilibrium system consisting of the
ecological interaction of innumerable species, interacting under conditions of constant change" (Boulding 1977). The structural and dialectical perspectives, whatever their value for analyzing social
injustice, economic inequity, and political oppression, are virtually
useless for thinking about the subtle relationship between energy conversion and biophysical imbalance, which cannot be reduced to social

contexts. The evolutionary approach helps us "to learn from the observed working principles of a world that for 3 billion years has been patiently designing stable energy-consuming systems in accordance with physical law" (Lovins 1974). Transcending the value systems of human societies are certain predispositions to coherence in the evolutionary process such as order, complexity, diversity, and interdependence—all characteristic of higher evolutionary development.

Modern industrial societies are promoting biophysical disorder, simplicity, uniformity, and diminished interdependence—all indicative of an earlier, more primitive stage of development. A reversion to these pristine features bodes ill for both survival and creativity, the two ultimate standards of evolutionary success. Creation means forming higher levels of order from matter and energy. The opposite process—destruction—points in the direction of biological poverty and physical entropy. As presently organized, with their parasitic relationship to nonrenewable energy and unwillingness to bring productive systems into line with nature's waste-absorbing and recycling potentials, industrial societies are alienated from evolutionary standards of survival. Most politicians, economists, and energy experts, capitalist or otherwise, apparently do not realize "that there will be a profound difference in attitude—indeed, a profoundly different value system—between those who understand the history of evolution and the interacting processes of the biosphere, and those who do not" (McHarg 1970).

A social order concerned with survival and the creation of new, more complex cultural forms would subordinate economic values and productive systems to biological values and ecological systems, realizing that the former depend on the latter (Commoner 1976). The nonadaptive behavior of industrial societies in choosing to cling to the opposite hierarchy of priorities is a form of economic vandalism that militates against sustainability (Paradise 1969). In a sense the most urgent need for public policy on energy is to act in harmony with the predispositions of the evolutionary process. In that frame of reference there are no tradeoffs, alternative lifestyles, or capricious desires that can be tolerated if they reverse conditions of order, complexity, diversity, and symbiosis.

TOWARD A WORLD ENERGY POLICY

At present it seems that no country has a long-term, comprehensive energy strategy. Most decisions on energy questions are being made piecemeal on faulty assumptions. National energy studies reflect little more than a preoccupation with short-term fuel tactics—how to meet specific demands for fuel at the lowest cost—only peripherally related to coherent energy policy. On the whole there is no public awareness that what is needed at this stage of human history is a "Copernican revolution" away from energy capital toward energy income. The management of energy resources has become one of the major conditions of world order, yet attitudes and behavior are appropriate largely to a bygone age. A humane world order founded on peace, social justice, economic well-being, and environmental quality is not attainable as long as perceptions of the energy crisis remain so archaic and ineffectual (Michael 1970).

It may seem pointless to propose a framework for a world energy policy when individual nations have made such a poor showing. However, in fact there already exists a kind of world model for energy policy that can be abstracted from various national "policies," which add up to a collective desire to sustain the highest possible growth rates at the lowest cost for fossil fuel, without shifting substantially to other forms of energy. There is another model, or body of principles, that might be endorsed by international bodies like the World Bank or the United Nations, in the hope of persuading governments to act with more foresight and responsibility. These principles might be taken as a provisional world energy policy, a model or ideal to be encouraged, approximated, and discussed among all nations. Formulating such an "ideal type" has more to recommend it, given the terrible speed with which events seem to be unfolding, than waiting patiently to see what individual countries under stress might come up with. If a consensus could be achieved on basic principles, then international discourse, as a bare minimum, might be able to proceed with more focus and precision. What, then, might be the organizing principles of a world energy policy?

All nations, developed and developing, would cooperate to attain the following ten objectives, motivated by self-interest and hopefully by a desire for world order.

The basic criteria for all energy systems would be thrift, renewability, simplicity, and safety (on the latter, see Lawrence 1976).

Energy sources and systems for their conversion would be tailored to specific needs and tasks. Petroleum would no longer be wasted for space heating, since there are more efficient ways of doing the same job. In heavily populated, rural countries, the emphasis would be on decentralized "soft" technologies that are manageable and affordable by local communities; and so forth.

Thermodynamic efficiency, conservation, and recycling would displace the current patterns of inefficiency, waste, and depletion of nonrenewable resources. As Denis Hayes (1976) has explained so well, the cheapest, easiest, and safest way to produce energy is to conserve it. Americans waste half the energy they use, a profligacy that is inexcusable and avoidable. Prevention of half that waste would amount to an energy bonanza at half the cost of producing an equivalent amount of virgin energy (Widmer and Gyftopoulos 1977; Ford et al. 1975).

The world's fossil fuels would be recognized as an irreplace-able "gift" of nature to be used with the utmost restraint and wisdom from at least three crucial perspectives. First, from the perspective of providing a limited but indispensable subsidy for the development of self-sustaining, renewable, relatively nonpolluting energy sources, moving ultimately in the direction of a predominantly "solar civilization" (Hayes 1978c; Williams 1978). Second, from the perspective of future generations for whom fossil fuels can have a multitude of valuable uses (over 300 from medicine to fertilizer) other than combustion to produce heat or power. Third, from the perspective of minimizing undesirable environmental effects, particularly climate modification.

Systems of production would be modified in the direction of harmonizing with natural systems. Economic and social systems would become resource and environment oriented rather than consumption and market oriented. A thermodynamic theory of economics would supplement or even replace the older economic theories, Marxist and capitalist, based on assumptions of indefinite growth (Georgescu-Roegen 1974). Economic thinking would move deliberately toward a body of knowledge and theory aimed at clarifying an economics of sustainability (Daly 1977).

The first global priority in the production and distribution of energy would be the satisfaction of basic human needs. A sobering example of how grotesquely distorted existing priorities are is the sad outcome of the United Nations' Decade of Disarmament in the 1970s. World military spending doubled from $200 billion in 1970 to $400 billion in 1978.* Exports of major weapons to Third and Fourth World countries tripled. These massive expenditures for arms and armies divert resources, especially energy, away from the option of meeting basic human needs; "for the estimated cost of a new mobile intercontinental missile (the MX), 50 million malnourished children in developing countries could be adequately fed, 65,000 health care centers and 340,000 primary schools built" (Sivard 1978).

---

*The United States and the USSR, with 42 percent of world GNP, account for 64 percent of military spending.

Nations would embrace a more realistic conception of "national security" and allocate sparse resources and capital accordingly. Energy depletion threatens national well-being as surely as a hostile military force (Brown 1977). Despite accumulating stresses associated with energy, worldwide funding for research and development in the energy field is only one-sixth as much as it is in weapons research (Sivard 1977). The U.S. Defense Department, exclusive of the defense industry, uses the equivalent of 237 million barrels of oil a year, enough to supply a small nation for all its needs. It is doubtful that the expenditure of $1 trillion on defense since 1945 has made Americans more "secure" in an ecological age.

Means would be developed for the assessment and monitoring of new technology in order to anticipate and control undesirable effects (National Academy of Sciences 1969). At least two key standards would be applied as a matter of course to all technological schemes: the ability of human beings to manage and control their effects, and the reversibility of those effects, apart from the unavoidable consequence of entropy, which at least can be held to a minimum.

In the formulation of energy policy, whatever its ramifications, a number of distinctions would be made routinely:

- Between demand and need. What people want, or think they want, must not be confused with actual needs.
- Between energy prices and energy costs. The money cost of fossil fuels does not take into account damage to health, property, and the environment resulting from its combustion or conversion into chemical substances foreign to nature.
- Between energy conversion and social welfare. Americans do not live three times better than Frenchmen because they use three times more energy.
- Between the reversible and irreversible effects of energy conversion.
- Between net and negative energy returns (the net-gross distinction). One must receive more energy from a development scheme than one puts into it.
- Between energy policy and a fuels policy.

Finally, there would be a conscious sense of responsibility to future generations, whose prospects for high civilization and a humanely tolerable existence are poor without sufficient access to usable energy and a viable biosphere. The present generation of energy users are custodians of the future on a scale never duplicated in the past. In line with this immense power to shape the lives of the unborn, there should be hard thinking in all responsible circles about how many people the earth can support for how long and at what level of living. Undoubtedly energy would be the master concept in settling the question of "optimums" for those three imponderables (Lazlo et al. 1977).

A last word about the relationship of energy to issues of freedom. As the energy crisis deepens, the fragility of human liberties will appear in a new light as yet only dimly perceived. The source of that fragility in the context of energy is the reluctance of free men and women to control their appetites. Edmund Burke speaks to us clearly on this point, almost prophetically: "Society cannot exist unless a controlling power upon will and appetite be placed somewhere, and the less of it there is within, the more there must be without. It is ordained in the eternal constitution of things, that men of intemperate minds cannot be free" (quoted in Ophuls 1977). The treasured liberties of the West have been achieved during a historic period stretching from the Renaissance to the third quarter of the twentieth century. A preoccupation with freedom of all sorts has been associated closely with a 400-year expansion of wealth and physical resources, beginning with the exploration and settlement of the New World. A long phase of expansion seems to be ending and contraction has begun. In the midst of economic hardship, paralysis, or collapse, traditional notions of freedom in the liberal mold, as well as solicitude for the rights of minority groups, are unlikely to fare well. The Draconian methods of the authoritarian state may become necessary as millions in industrial societies use their freedom to exceed the carrying capacity of the world's energy commons. Civil liberty and struggles for economic survival are unpromising companions. Unless freedom can search out goals other than consumption, waste, and economic growth, the upshot could be a sloughing away of all freedoms that matter. Perhaps the surest way to bring on the demise of industrial civilization is for consumer-oriented nations to go on behaving as though all their problems can be resolved by riding an upward curve of energy conversion. The final shock of recognition might come in a sudden, jarring reversion to simpler conditions of life, not through voluntary choice and discipline but rather through exigencies of nature and blind circumstance (Davis 1979).

REFERENCES

Benoit, Emile. 1976. "The Coming Age of Shortages." Bulletin of the Atomic Scientists, January.

Boulding, Kenneth. 1977. "Twelve Friendly Quarrels with Johan Galtung." Journal of Peace Research 14, p. 1.

Brink, R. A., J. W. Densmore, and G. A. Hall. 1977. "Soil Deterioration and the Growing World Demand for Food." Science, August 12.

Brown, Lester. 1977. Redefining National Security. Washington, D.C.: Worldwatch Institute Paper No. 14.

———. 1978a. The Global Economic Prospect. Washington, D.C.: Worldwatch Institute Paper No. 20.

———. 1978b. The Twenty Ninth Day. New York: W. W. Norton.

Browne, Malcolm. 1979. "Fusion Power: Is There Still an Eldorado for Energy?" New York Times, April 15.

Budnitz, Robert and John Holdren. 1976. "Social and Economic Costs of Energy Systems." Annual Review of Energy. Vol. 1. Palo Alto, Calif.: Annual Reviews.

Clark, Wilson. 1974. Energy for Survival: The Alternatives to Extinction. New York: Doubleday.

Commoner, Barry. 1971. The Closing Circle. New York: Alfred A. Knopf.

———. 1976. The Poverty of Power: Energy and the Economic Crisis. New York: Alfred A. Knopf.

———. 1979. The Politics of Energy. New York: Alfred A. Knopf.

Cook, Earl. 1971. "The Flow of Energy in an Industrial Society." In Energy and Power. San Francisco: W. H. Freeman.

Daly, Herman. 1977. Steady State Economics. San Francisco: W. H. Freeman.

Darmstadter, Joel, Joy Dunkerley, and Jack Alterman. 1977. How Industrial Societies Use Energy. Baltimore: Johns Hopkins University Press.

Davis, W. Jackson. 1979. The Seventh Year: Industrial Civilization in Transition. New York: W. W. Norton.

Eckholm, Erik. 1979. Planting for the Future. Washington, D.C.: Worldwatch Institute Paper No. 26.

Ehrlich, Paul and Anne Ehrlich. 1970. Population, Resources, Environment. San Francisco: W. H. Freeman.

Ford, K. W. et al. 1975. Efficient Use of Energy, Report No. 25. New York: American Institute of Physics.

Galtung, Johan. 1973. "The Limits to Growth and Class Politics." Journal of Peace Research 1, p. 2.

Georgescu-Roegen, Nicholas. 1971. The Entropy Law and the Economic Process. Cambridge, Mass.: Harvard University Press.

——. 1972. "Economics and Entropy." The Ecologist, July 7.

——. 1974. "Mechanistic Dogma and Economics." Methodology and Science 7, p. 3.

——. 1975. "Energy and Economic Myths." Southern Economic Journal, January.

Hayes, Denis. 1976. Energy: The Case for Conservation. Washington, D.C.: Worldwatch Institute Paper No. 4.

——. 1978a. "The Coming Energy Transition." The World Tomorrow. Washington, D.C.: World Future Society.

——. 1978b. Energy for Development: Third World Options. Washington, D.C.: Worldwatch Institute Paper No. 15.

——. 1978c. Rays of Hope: The Transition to a Post-Petroleum World. New York: W. W. Norton.

Hedberg, Hakan. 1972. Japan's Revenge. London: Plenum.

Hubbert, M. 1971. "The Energy Resources of the Earth." In Energy and Power. San Francisco: W. H. Freeman.

Intercom. 1978. Washington, D.C.: Population Reference Bureau, December.

Kahn, Herman. 1979. World Economic Development: 1979 and Beyond. New York: Morrow Quill.

INTRODUCTION / 33

Kahn, Herman et al. 1976. The Next 200 Years: A Scenario for America and the World. New York: William Morrow.

Kamenetzky, Mario. 1979. "Development for the People." Bulletin of the Atomic Scientists, June.

Kelley, Donald, Kenneth Stunkel, and Richard Wescott. 1976. The Economic Super Powers and the Environment: The United States, the Soviet Union, and Japan. San Francisco: W. H. Freeman.

Keyfitz, Nathan. 1976. "World Resources and the World Middle Class." Scientific American, July.

Kim, Samuel. 1978. "China and World Order." Alternatives, May.

Komanoff, Charles. 1979. "Doing Without Nuclear Power." New York Review of Books, May 17.

Lawrence, W. W. 1976. Of Acceptable Risk: Science and the Determination of Safety. Los Altos, Calif.: Kaufman.

Lazlo, Ervin et al. 1977. Goals for Mankind. New York: E. P. Dutton.

Leiss, William. 1975. The Domination of Nature. New York: George Braziller.

Leontief, Wassily et al. 1977. The Future of the World Economy. Oxford: Oxford University Press.

Lessing, Doris. 1971. Briefing for a Descent into Hell. New York: Alfred A. Knopf.

———. 1974. Memoirs of a Survivor. New York: Octagon.

Lovins, Amory. 1974. "World Energy Strategies." Bulletin of the Atomic Scientists, May.

———. 1977. "Cost-Risk Benefit Assessments in Energy Policy." George Washington Law Review, August.

———. 1978. Soft Energy Paths. New York: Colophon Books.

McCracken, Samuel. 1979. "The Harrisburg Syndrome." Commentary, June.

McDermott, John. 1969. "Technology: The Opiate of the Intellectuals." New York Review of Books, July 31.

McHale, John and Magda McHale. 1978. Basic Human Needs: A Framework for Action. New Brunswick, N.J.: Transaction Books.

————. 1979. World of Children. Washington, D.C.: Population Reference Bureau.

McHarg, Ian. 1970. "Values, Process, and Form." In R. Disch, ed., The Ecological Conscience. Englewood Cliffs, N.J.: Prentice-Hall.

McKelvey, V. E. 1977. Energy Sources: An Overview of Supplies. Washington, D.C.: U.S. Geologic Survey.

Michael, Donald. 1970. The Unprepared Society: Planning for a Precarious Future. New York: Harper and Row.

Miles, Rufus. 1976. Awakening from the American Dream: The Social and Political Limits to Growth. New York: Universe.

MIT. 1971. Man's Impact on the Global Environment. Cambridge, Mass.: MIT Press.

National Academy of Sciences. 1969. Technology: Process of Assessment and Choice. Washington, D.C.: Panel of Technology Assessment.

————. 1977. Energy and Climate. Washington, D.C., July.

Olson, Mancur and Hans Landsberg, eds. 1973. The No-Growth Society. New York: W. W. Norton.

Ophuls, William. 1977. Politics and the Ecology of Scarcity. San Francisco: W. H. Freeman.

Orr, David and Marvin Soroos, eds. 1978. The Global Predicament: Ecological Perspectives on World Order. Chapel Hill: University of North Carolina Press.

Paradise, Scott. 1969. "The Vandal Ideology." The Nation, December 29.

Pirages, Dennis. 1978. The New Context for International Relations: Global Ecopolitics. North Scituate, Mass.: Duxbury Press.

Schumacher, E. F. 1973. Small Is Beautiful. New York: Harper and Row.

Singer, Max and Paul Bracken. 1976. "Don't Blame the U.S." New York Times Magazine, November 7.

Singer, S. F. 1970. "Human Energy Production as a Process in the Biosphere." The Biosphere. San Francisco: W. H. Freeman.

Sivard, Ruth. 1977. World Military and Social Expenditures. New York: Institute for World Order.

———. 1978. World Military and Social Expenditures. New York: Institute for World Order.

Solow, Robert. 1974. "The Economics of Resources and the Resources of Economics." American Economic Review, May.

Starr, Chauncy. 1971. "Energy and Power." In Energy and Power. San Francisco: W. H. Freeman.

Stent, Gunther S. 1969. The Coming of the Golden Age: A View of the End of Progress. New York: Natural History Press.

Stobaugh, Robert and Daniel Yergin, eds. 1979. Energy Future. New York: Random House.

Stunkel, Kenneth. 1973. "The Technological Solution." Bulletin of the Atomic Scientists, September.

U.S. Senate, Committee on Energy and Natural Resources. 1978. An Analysis of U.S. and World Energy Projections Through 1990, December.

Widmer, Thomas and Elias Gyftopoulos. 1977. "Energy Conservation and a Healthy Economy." Technology Review, June.

Williams, Robert, ed. 1978. Toward a Solar Civilization. Cambridge, Mass.: MIT Press.

Willrich, Mason. 1975. Energy and World Politics. New York: The Free Press.

Wilson, Carrol, ed. 1977. Energy: Global Prospects 1985-2000. New York: McGraw-Hill.

Woodwell, George. 1978. "The Carbon Dioxide Question." Scientific American, January.

THE NATIONAL ENERGY PROFILES

1

THE UNITED STATES
David J. Rosen

Energy has been the silent (and nearly invisible) partner in the development of the United States. The enormous growth and development of U.S. industry, with its concomitant high-consumption lifestyle and world power, rest, in large measure, on the secure and continuous flow of low-cost energy. While the availability of inexpensive energy has allowed (and even encouraged) the development of the characteristic American way of life, it has also led to a series of social, economic, and political choices that "lock" the United States into the heaviest per capita energy dependence of all the nations of the world. Although the United States contains only 6 percent of the world's population, it accounts for about one-third of the world's annual energy consumption. This high rate of consumption has placed an enormous stress on the U.S. environment, which has been called upon to supply the needed resources and to absorb the wastes of the high-energy society. It also poses serious current and future problems for the American society, economy, and polity.

THE PATTERN OF ENERGY USE

While Americans have historically felt secure in their supply of energy, patterns of energy supply and consumption have changed with alterations in America's circumstances. The rich forests of North America supplied the nation's principal energy source from colonial times until after the Civil War. The reliance on wood as the primary energy source during the nation's first century resulted in the deforestation of much of the Eastern seaboard. The relatively primitive forestry measures employed during this period caused substantial erosion and a high susceptibility to fires, but also cleared large expanses of land for the nation's rapidly growing population and industrial enterprises. The reliance on wood was supplemented by the use of available natural phenomena: winds, rushing water, and the like.

With the expansion of railroads and steel production in the 1880s, coal (with its higher energy content) came to supplant wood as the primary energy source in the United States. The world's first oil well was drilled in Titusville, Pennsylvania, in 1859,* but the use of petroleum did not begin to assume major importance until the introduction of large-scale use of the automobile some 60 years later. With increasing reliance on the automobile and the widespread substitution of petroleum products for coal, petroleum replaced coal as the primary energy source in the United States shortly after World War II. The growth in the use of natural gas, which is often geologically related to petroleum, paralleled the growth in petroleum consumption and became the second most important source of energy in the 1960s.[1]

While the mix of energy sources changed with developments in technology and shifts in end use, total energy consumption was growing exponentially. Although there were minor fluctuations in the annual increases, the mean annual increase for the first 70 years of this century was 3.1 percent. While this rate may appear modest, when it is sustained it produces a doubling in annual consumption every 22-23 years. From the founding of the Republic until 1940, annual U.S. energy consumption had grown to 23.9 quadrillion (1×10^{15}) Btus or quads. By 1960 annual consumption reached 44.6 quads and 12 years later it topped 72 quads. Projections made in the early 1970s foresaw annual energy consumption in excess of 150 quads by the end of the century. During this same period, 1940-73, per capita energy use nearly doubled from 180 million Btus to 359 million Btus[2] (see Table 1.1).

Important changes in the consumption of energy are not restricted to changes in the mix of sources or the rate of use. Increasingly in recent years the United States has shifted from decentralized energy generation in which fuel was consumed near the point of end use to a more highly centralized energy system in which electricity is produced by the burning of fossil fuels (or to a lesser extent by hydro or nuclear power) and transmitted, often over great distances,

----

*While Titusville was the first oil well, petroleum had been used for centuries throughout the world where oil seepage allowed the surface collection of the resource.

TABLE 1.1

Annual Energy Supplies for the United States, 1940–2000
(values in quads of energy)

	1940	1950	1960	1970	1977	1985	2000
Petroleum liquids							
Domestic	7.5	13.5	16.8	22.8	20.0	17.0	15.0
Imported	—	—	3.3	6.7	17.4	22.0	17.0
Gas							
Domestic	2.7	6.2	12.5	21.2	19.1	15.0	9.0
Imported	—	—	0.2	0.8	1.0	2.0	1.0
Coal	12.5	12.9	10.1	12.7	13.0	19.0	32.0
Hydro	0.9	1.4	1.7	2.7	2.1	3.0	3.5
Nuclear	—	—	—	0.2	2.7	6.0	11.0
Other	1.4	1.2	1.0	1.0	1.8	2.7	6.5
Gross total	25.0	35.2	45.6	68.1	77.7	86.7	95.0
Annual growth in quads	—	1.02	1.04	2.25	1.60	1.13	0.55

Source: Ernest Hayes, "Energy Resources Available to the United States, 1985–2000," Science 203, no. 4377 (January 1979).

to the point of eventual end use. This shift is indicated by the growth in electrical consumption over the past two decades at an average rate of more than 7 percent—a growth rate twice as rapid as that for all energy consumption. At a 7 percent annual growth rate, electricity consumption has had a ten-year doubling time. The shift to electricity resulted from a trend of declining consumer costs for electricity (owing to increased efficiency in the generation and transmission of electricity) and the greater convenience of electricity to the end user, who is not required to maintain facilities for the storage and combustion of fuel.

This increasing centralization is evident not only in the growing reliance on electricity generated off-site, but in structural changes within the electrical-generation industry. While there were 6,500 separate electrical generating and distributing systems in 1917, the number had been reduced to 233 by 1974. In practice the concentration is greater than these figures would indicate, as the creation of power pools and electric reliance councils permitted the interconnection and cooperative planning among the electric utilities in a region. Paralleling the growth in the size of electrical systems has been the increasing size of generating units. The earliest commercial generating units could produce no more than 7.5 kilowatts. By 1930 new units were producing 200,000 kilowatts. In the 1970s, units in the 1,000-1,200 MW range had become common, with projections of 3,000 MW units by 1990.[3]

An alternative perspective on changes in the American consumption of energy is provided by an examination of the shifts in the uses to which energy has been put. In the nineteenth century, approximately three-quarters of the energy consumed was used in the home, for cooking, space heating, and the like. While the residential use of energy has increased substantially in absolute terms during the twentieth century, it now accounts for less than one-fifth of the total U.S. energy budget. The most drastic increases in energy consumption have occurred in the production and marketing of goods and the transport of goods and people. Over 50 percent of the energy currently consumed in the United States is devoted to the industrial sector, while another 25 percent is used as fuel for transportation. Of the remaining energy consumed in residential and commercial locations, more than one-half is used to heat and cool buildings[4] (see Table 1.2).

The remarkable energy intensity of American society becomes clear when it is viewed from a comparative perspective. While the stark fact that the United States with about 6 percent of the world's population accounts for one-third of global energy consumption is impressive, it reflects the enormous economic disparities between rich and poor nations as much as energy consumption differences.

TABLE 1.2

Energy End Use in the United States

End Use	Percentage of Total
Transportation	24.9
Space heating	17.9
Process steam (for industry)	16.7
Direct heat (for industry)	11.5
Electric drive (for industry)	7.9
Feedstocks, raw materials	5.5
Water heating	4.0
Air conditioning	2.5
Refrigeration	2.2
Lighting	1.5
Cooking	1.3
Electrolytic processes (for industry)	1.2
Other	2.9
Total	100.0

Source: "Patterns of Energy Consumption in the United States," Report to the Office of Science and Technology (Washington, D.C.: U.S. Government Printing Office, 1972).

A more relevant comparison can be drawn with other industrialized countries. A detailed comparison between the United States and Sweden shows that the latter provides a standard of living comparable to the United States on most indexes while consuming only two-thirds as much energy per capita.[5] In 1972 the United States consumed 8.35 tons of oil equivalent per capita, compared to 4.12 tons in West Germany, 3.81 tons in the United Kingdom, and 2.90 tons in Japan.[6] While some of the difference in consumption reflects the higher gross domestic product (GDP) in the United States, each of these countries uses less energy (between one-half and three-quarters) to produce each dollar of GDP than does the United States.[7]

Historically, increases in U.S. economic productivity have been linked to increases in energy consumption. As long as energy was inexpensive and abundant, industry could use large quantities of energy to increase the industrial output per worker (and per dollar). Compounding this advantage, increases in the thermodynamic and mechanical efficiency of energy use were ensuring increased production per unit of energy consumed.[8] This trend of increasing energy efficiency reversed in 1966 (as a result of the attainment of thermodynamic efficiency limits in some processes and the increasing utilization of relatively inefficient automobiles and electrification), largely eliminating the advantage of substituting energy for other components of the production process.* In addition, some recent analysis suggests that the historical correlation between energy use and levels of employment is not a causal relationship as is generally assumed, but rather stems from the fact that a growing economy consumes more energy <u>and</u> creates more jobs (with the mix between the two sensitive to cost and availability of the two factors).[9]

A society's primary concern should not be its rate of energy consumption or economic output but the quality of life it provides. A 1973 national survey found that only 35 percent of the American public believed that the quality of life had improved during the preceding ten-year period (despite significant growth in both GNP and energy consumption). Fully 45 percent felt that the quality of life had declined during that period, while 15 percent detected no change. A portion of this dissatisfaction may reflect the social and environmental costs of an energy-intensive society.[10]

---

*In the mid-1970s this trend was again reversed and appeared to resemble its pre-1966 slope. Future trends are difficult to forecast at this point, but will probably be materially shaped by energy policies adopted by government.

DOMESTIC ENERGY POLICY

Prior to 1974 comprehensive energy policy was an alien concept in the United States. So pervasive was the assumption of perpetual energy abundance, U.S. decision makers perceived no need to plan for long-term management of energy resources. Rather, energy policies were developed ad hoc to serve a range of goals (high profits for energy producers, increased economic productivity, the avoidance of costly labor disputes, and the like) rather than addressing overall energy conditions. Reflecting the interrelationships between energy and many other policy areas, some of the governmental decisions that were to have the greatest impact on the U.S. energy situation were not even perceived as energy policies. Thus decisions to encourage the use of the private automobile and to spur the suburban exodus failed to take account of the far-reaching energy implications of such fundamental changes in American society. While these energy implications may not have been recognized, they fostered patterns of energy supply and consumption that affect current behaviors, circumstances, and options, and resulted in an infrastructure that severely limits future energy choices.

Common to nearly all pre-1974 energy actions was a pervasive optimism about energy supplies and a faith that their rapid exploitation would be beneficial. Gerald Garvey suggests that these attitudes are rooted in the American frontier experience.

> The seemingly limitless bounty of the continent fostered reckless, wasteful habits; first manifested in the frontiersman's appropriation of land and forests, then carried over into the eras of coal and petroleum development. Such a tradition of wastefulness could hardly have been sustained without an overriding optimism. Continental abundance fostered the conceit that there would always be "more." Crucial to a developing American tendency to ignore harmful side-effects of resource exploitation was the elaboration of this rustic optimism into a theory of externalities which emphasized only the beneficial spillover consequences of economic growth.[11]

The energy policies to be considered in this section fit into three general categories. First are those policies that, while not explicitly energy policies, have significantly shaped our energy situation. Second are those pre-1974 policies that responded to the particular problems and circumstances surrounding various forms of energy and energy resources. The third category covers the attempts during the past few years to respond to the changing perceptions of energy abundance and formulate a comprehensive energy policy for the United States.

Of the nonenergy policies that have had the greatest impact on the energy situation have been those that fostered the rise of the automobile and the shift of population from the cities to the suburbs. The introduction of private automobiles at prices that made them accessible to most American families allowed substantial portions of the urban populace to move to the newly emerging suburbs. While the process of suburbanization had begun earlier, it was in the post-World War II period that the greatest suburban exodus occurred. Government encouraged this movement through the construction of massive highway systems that linked the suburbs to the job markets in the cities. The specific policies of import here were the establishment of the Highway Trust Fund, which ensured that substantial funds would always be available for road construction without having to compete with other public needs, and the inauguration in the 1950s of the Inter-State Highway System, which further encouraged the use of private automobiles and made previously inaccessible rural areas subject to suburban development. The government also encouraged the suburban development by making low-cost home mortgages easily available for veterans and others.

The government perceived that these actions were spurring economic growth and expanding the opportunities available to citizens, and the families settling in the suburbs believed they were securing a more attractive and healthful environment in which to raise their children, but the unforeseen energy impacts of these developments were extraordinary. The individual single-family homes were built with little concern for energy efficiency (as builders sought to keep initial costs low and building codes rarely were drafted with energy efficiency in mind). In addition the various federal mortgage guarantee and public housing programs established cost ceilings on dwelling units. This emphasis on initial cost, rather than lifetime cost, of the structure encouraged the use of the least expensive (and least energy-efficient) heating systems and largely precluded the use of substantial insulation, solar collectors, and heat pumps. The shift to highly inefficient electric heating accelerated in the 1960s as electric rates declined. From 1966 to 1971 the percentage of new homes built with electric heating systems nearly doubled and this increase continued, at a somewhat slower pace, until the end of 1973.[12]

An even greater energy impact involved the energy costs of transportation between the suburbs and the cities. The land use patterns in the suburbs are often such that dependence on the private automobile is almost unavoidable. Given the dispersion of the population, it is extremely difficult to construct mass transit systems that can provide less energy-intensive service at acceptable levels

of convenience for the suburbanites. One indication of the relative
insignificance of mass transit is the finding that 85 percent of employed heads of households use private transportation to their place
of work, and of these, 83 percent travel alone. Between 1960 and
1970 the total municipal highway mileage in the United States increased
by 31 percent while the total municipal mass transit (subways, trol-
leys, railroads) mileage decreased by 40 percent.[13]

A trivial, but illustrative, example of the long-term impacts
of minor decisions is provided by the routing of the Northern State
Parkway built on Long Island by the State of New York in the 1930s.
A short detour from the proposed route, which was necessitated by
an arrangement with the wealthy and powerful residents of the Wheat-
ley and Dix Hills areas, has meant that every commuter using that
route on a daily basis drives an additional 5,500 miles per year.[14]
By the middle of the 1960s, this meant that the country was consuming
an additional 7.9 million gallons of gasoline per year because of a
private arrangement made some 30 years earlier.* Further, unless
a decision is made to relocate this section of the highway (through an
area that has been extensively developed since the 1930s), this waste
of gasoline will continue as long as Americans use gasoline-powered
automobiles.

Another energy-related policy area that illustrates Garvey's
view of American attitudes involves the recycling of materials. The
United States clearly leads the world in the production of solid waste.
Recycling generally uses less energy than does the production of virgin materials, as well as reducing the quantity of solid waste that
must be disposed, preserving resources for future generations, and
avoiding the environmental disruption that often accompanies resource
extraction. Yet American attitudes toward growth and resources
have been such that not only had there been no attempt to promote
recycling, but a series of policies actually discouraged it.

> Federal tax and transportation policies in particular
> create strong disincentives for industry to substitute
> recycled (or "secondary") materials for virgin resources.
> Virgin materials, including timber, coal, and iron ore,
> are treated under federal tax law as capital assets and
> taxed as capital gains rather than as ordinary income
> when they are sold for profit; the effect is to encourage

---

*The estimate assumes 21,500 daily commuters living east of
the detour and an average of 15 miles per gallon.

sales of these rather than the substitution of secondary
materials more harshly taxed. Then, too, the standard
depletion allowance has been granted producers of virgin
materials as another incentive for their use. Finally,
the Interstate Commerce Commission's national rate
structure favors the transportation of virgin over recycled materials.[15]

A final policy area that was not viewed as being related to energy supply but that has significantly shaped energy use patterns is
the price regulation of electric utilities. In order to avoid the costly
and disruptive practice of competing electricity supply companies
running redundant distribution systems, electric utilities have been
established as regulated monopolies. (In some cities the government
has undertaken the role of electricity distribution.) The profits of
electric utilities are set by a governmental body, which has responsibility for setting rates, as a fixed percentage of capital investment.
Thus to increase its profits a utility must increase its investment by
expanding its electrical generation facilities.[16] Utilities were able
to increase demand (and thereby justify increased expenditures) by
offering lower unit costs for electricity to consumers of large quantities of power and by aggressive advertising campaigns that encourage consumers to use more energy-intensive appliances. Thus we
find the anomalous situation in which electric utilities, despite their
government-protected monopolies, were spending $8.00 on advertising
for each dollar spent on research.[17]
By the mid-1970s most such advertising had been stopped,
either voluntarily or by government action, and efforts are under way
in many states to revise rate structures that encourage lavish and
inefficient use of electricity by industry. However, the patterns
established by the use of these practices in the past are not so easily
changed. Consumers who have purchased additional appliances are
not likely to forego their use simply because advertisements are no
longer encouraging them, and industrial processes that are dependent
upon extensive energy inputs cannot be readily converted without
enormous costs. As we have seen in other cases, past policies promote current behaviors. These practices were encouraged or permitted by the government in the belief that they contributed to economic growth. While this belief may have been correct, the practices
also led to the inefficient use of energy and substantial waste.
Coal, the first commercially significant fuel source in the
United States, illustrates the governmental posture toward energy.
As the industry flourished in the latter half of the nineteenth century,
the federal government avoided regulation and left control to the
states. The states, in turn, generally sought to help the industry by

giving it a free hand, as is illustrated by the following description of the situation in Kentucky:

> In the early days of the coal boom, agents of the opera-
> tors systematically crisscrossed the highlands purchasing
> the mineral rights. For a few dollars per acre the native
> mountaineers sold thousands of dollars worth of buried
> coal. Worse yet, they conveyed to the companies the
> right to excavate, to build roads and structures, to use
> the timber, to pollute the water, and to cover the land
> with spoil. The farmers retained the right to plant a lit-
> tle corn and pay taxes. Disastrous as this conveyance
> was, it caused comparatively little trouble as long as
> mining was subsurface. But once an operator chose to
> strip mine, the destruction was complete. Bulldozers
> and shovels would move in, chewing up timber, crops,
> and even the family cemetery. If after a passing rain the
> ravaged mountain slid down into the valley destroying all
> the houses, as in a 1949 case before the Kentucky Court
> of Appeals, that was an Act of God for which the coal
> company could not be held responsible. A few years later
> when the court at first decided in favor of a farmer, the
> coal companies were able to bring enough pressure to
> bear that the court reversed itself on a rehearing. This
> decision guaranteed the rights of strip mining for another
> prosperous decade. Kentucky politicians, like those in
> other states, were reluctant to come down hard on the
> coal companies because they believed that what was good
> for the operators was good for the state. . . . [18]

The one area of the coal industry that required early government intervention was labor relations. Coal first became a concern
for policy makers and a political issue with the emergence of the
United Mine Workers (UMW) and their efforts to organize and represent the nation's miners. The initial response of state governments
was to side with mine operators in their effort to block unionization.
The Commonwealth of Pennsylvania officially sanctioned the notorious
Coal and Mine Police, a group recruited by the mine owners to sup-
press the UMW. Despite their efforts and other actions, a 1902 UMW
strike in western Pennsylvania disrupted the nation's supply of coal,
causing shortages and substantially increasing consumer prices. In
a marked departure from previous governmental behavior, President
Theodore Roosevelt directly intervened in the strike, meeting with
both parties and eventually establishing an arbitration panel.

The reluctance of the federal government to involve itself in the coal industry is illustrated by its record on the health and safety of miners. When the federal Bureau of Mines (BOM) was created in 1911, the agency did not even have the power to enter mines to conduct safety inspections. State governments, often heavily influenced by coal interests, rarely augmented these minimal federal efforts. Even with the passage of the Federal Coal and Mine Safety Act of 1941, BOM had limited authority and even more limited resources to protect the miners. It was not until the 1970s that more far-reaching federal safety regulations were enacted, and today the United States still trails most developed countries in the safety of its coal mines. Similarly, federal efforts to control Black Lung Disease (a debilitating and sometimes fatal respiratory ailment affecting miners) and to compensate its victims came in the 1970s, far later than in many other countries.[19]

A number of factors contributed to the government's lack of involvement in the coal industry. At the time of the coal industry's emergence, the government posture toward all industry was closer to laissez faire than would be true subsequently. Also the coal industry was politically powerful, particularly at the state level, and was able to block government actions it opposed, while securing those governmental benefits (for example, roads, tax advantages, and so forth) it did desire. In addition, the frontier ethic, with its faith in the unqualified benefits of rapid resource exploitation, discouraged governmental regulation. Lastly, coal was not yet viewed as part of a comprehensive system of energy sources and uses.[20]

In the emerging oil industry, government, particularly at the state level, was forced to play a more active role than it had with respect to coal, due to differences in the nature of the extractive process. Ironically, when viewed from the perspective of the 1970s, the principal dilemmas in the oil industry were recurrent gluts of supply and declines in prices that threatened the economic viability of the industry.

One of the important factors fueling the overproduction of petroleum and precipitating calls for government intervention was the "rule of capture."[21] This legal doctrine established that well owners had the right to any petroleum they could pump from their wells. Since subterranean pockets of petroleum did not often correspond to the legal divisions of the surface land, wells on several different tracts of land might be all drawing from the same finite pool. In such circumstances, each well operator would try to pump as much oil as he could in order that the other drillers not deplete the pool first. This led to highly wasteful practices and periods of gross overproduction. The oil industry was also faced with erratic fluctuations in supply and price when new oil fields were discovered and rapidly developed.

To control these boom and bust cycles, the governments of Oklahoma and Texas established first voluntary and then mandatory procedures for limiting production within their borders in order to maintain stable prices. When the federal government balked at establishing national production quotas to supplement these state efforts, the oil-producing states entered into an interstate compact that was ratified by Congress. Through this means and voluntary coordination among the major oil producers (which received tacit federal approval), the most erratic variations in price and supply were controlled.

The next major threat to the U.S. oil industry that brought forth governmental intervention emerged in the 1950s. While the United States had been a net exporter of petroleum through World War II, in the early postwar years less expensive foreign petroleum (first from Venezuela, then from the Middle East) began to capture an increasing share of the market in the eastern United States. Voluntary limitations on imports, adopted in 1954, failed to stem this growing use of imported oil, and in 1959 the Eisenhower administration imposed import quotas. While the ostensible justification for these quotas was the protection of national security, the most obvious beneficiaries were the U.S. oil producers who were spared foreign competition. Among the effects of the policy of "Drain America First" were higher fuel prices for consumers in the eastern states, and the high degree of dependence on foreign petroleum sources that currently leaves the United States vulnerable to price manipulations and interruptions of supply. [22]

In the early oil-drilling operations, natural gas was also detected and released. Owing to its physical properties and the state of technology, there was no large-scale use to which the valuable fuel could be put, and much of it was simply burned (flared) at the drilling site. In addition to fouling the air and irretrievably wasting the resource, this practice reduced the pressure on the oil, adversely affecting the efficiency of the oil wells. However, by the 1920s, technology had devised a pipeline that was suitable for the long-distance transmission of natural gas. A market already existed in the many cities that were using the less desirable coal gas for street lighting and other purposes. With future developments in supply and pricing, the uses for natural gas expanded dramatically.

The physical properties of natural gas required that government play a significant role in the use of the resource. The natural gas distribution system in a city was capital intensive and eventually required that a single company be given a distribution monopoly. Two or more companies installing parallel competing pipelines along every street would have been costly and disruptive. Thus, cities granted authority to monopolistic utility companies (or managed the distribution through municipal agencies) and regulated the prices that were charged. While the same natural monopoly argument applied to the interstate pipelines that brought the fuel from the natural gas fields to the cities, it took several decades and a series of political battles before the Federal Power Commission (FPC) was given juris- diction over the routes and rates for the interstate systems. It was not until the 1960s that FPC authority was further extended to the price of gas entering the interstate system.

These increasing government regulations led to a series of anomalies in natural gas price and supply. In order to keep down costs to the consumer, while encouraging exploration for more natural gas, the FPC allowed higher prices for "new" gas than for "old" gas (that which came from existing wells). Also the FPC did not have authority over natural gas that remained within intrastate systems, and the price for such gas rose considerably above the FPC-mandated interstate price. The combined effect of these measures distorted the supply and availability of natural gas. In addition, because natural gas prices were government controlled, natural gas was "artificially" cheaper than competing energy resources. This led to a rapid expan- sion in the use of natural gas, often as a substitute for coal or petro- leum products, and encouraged the wasteful use of the resource.[23]

The history of civilian nuclear power development followed a unique course, the contours of which were shaped by its physical characteristics and the circumstances of its birth. The highly sophis- ticated, capital-intensive nuclear power plants are an outgrowth of the feverish weapons research undertaken during World War II. The wartime pattern of government control, tight security, and ample federal funding was extended to the peaceful uses of the binding energy of the atomic nucleus.

The Atomic Energy Act of 1946 established the Atomic Energy Commission (AEC) as a civilian agency that would direct the "develop- ment, growth and control" of nuclear power. The tensions among these (partially conflicting) responsibilities would leave the AEC open to charges of slighting control in the pursuit of development and growth and would lead, in the 1970s, to the splitting of these two func- tions into two distinct agencies. The act also created the Joint Com- mittee on Atomic Energy (JCAE) to provide congressional oversight of the AEC. For the first 25 years of the atomic age, the AEC and JCAE exercised nearly total dominance over U.S. nuclear policies, with a minimum of accountability to the public or other governmental bodies.[24]

In 1954 a revised version of the act was adopted that would allow private industry to build and operate nuclear power plants under the supervision of the AEC. (The refining and ownership of nuclear fuel was retained by the AEC, which supplied the fuel to the private companies that operated the plants.) While the Democrats generally opposed the arrangement as providing a windfall for private industry who would profit from tax-supported research and development, the Republican Congress and White House successfully pushed for this change in the structure of the nuclear power field. Several years after this change a number of the vendors of nuclear power systems would join the AEC and JCAE in the domination of nuclear policies.

Despite the passage of the modified act and the interest of many public utilities to benefit from the energy source that was being touted as "clean, virtually limitless, and too cheap to meter," the development of nuclear power plants was delayed. Insurance companies were unwilling to insure this new technology and utilities were not willing to build and operate power plants without liability protection. The federal government, which was already spending billions of dollars on research and development, further subsidized the nascent industry with the passage of the Price-Anderson Nuclear Indemnity Act. This law required that the utilities secure private coverage for the first $60 million in damages. The federal government would then cover damages in excess of that sum up to a limit of $560 million. While $560 million may appear to be a more than adequate sum, government studies placed the possible property damage from a serious nuclear accident at $6-17 billion. Thus, even at the lowest damage estimate, total compensation (private and governmental) would equal less than ten cents on the dollar, and under the provisions of the law, that would be the absolute limit of liability. In other words, the Price-Anderson Act transferred liability from the utility first to the federal government (that is, the taxpayers), and then to the victims, who would be barred from recovering more than a small fraction of their losses. Testifying in 1975 in support of a renewal of the Price-Anderson Act, representatives of the Edison Electric Institute said, "In 1957 when Price-Anderson was first passed . . . no utility would undertake construction of a nuclear plant without the protection this statute afforded."[25]

The extraordinary role played by the U.S. government in fostering commercial nuclear development is noted by John O'Leary, who has served with both the AEC and Department of Energy:

From the late 1950s onward, an enormous promotional effort was funded by the government and executed by the AEC. It took the form of soft-sell campaigns aimed at all levels of our population, from school children through the President and in some cases, particularly where direct financial interest was involved, of more or less hard-sell efforts aimed at the utilities industry. In many aspects the AEC and the individual members thereof acted as an inside lobby, using a broad range of governmental powers in order to launch a new industry.

It is clear that much of the promotional effort had at its roots the almost evangelical fervor with which the commission sought the commercialization of nuclear power. The AEC was morally convinced that the prospect of beating nuclear swords into plowshares justified extraordinary intervention in market decisions. The result was a one-sided presentation of nuclear energy, publicly highlighting its virtues, and particularly from 1965 onward, deliberately downplaying its weaknesses. . . . The commission during this period treated nuclear energy as an energy source that could be used wherever coal or natural gas or oil could be used, without any public acknowledgment of the potential damage to the public health and safety should a poorly located nuclear plant experience a serious accident.[26]

Despite the efforts of the AEC, nuclear development is lagging far behind expectations. In the 1960s, the AEC forecast 200 reactors in the United States by 1980 and perhaps 1,000 twenty years later. As recently as 1972 the AEC projected 150 reactors on-line by 1980. However, by 1975, 55 nuclear power plants were in operation with an additional 62 in various stages of design or construction.[27] By 1979, only 12 percent of the nation's electricity—or about 4 percent of its overall energy—was being supplied by the 72 reactors that were on-line. Perhaps more significantly, the number of new orders for nuclear plants has been falling in the 1970s as a result of serious operating problems with existing plants,[28] an intensifying debate over the safety of nuclear power,[29] and staggering inflation in nuclear construction costs.[30] The failure of nuclear power to fulfill the role expected by its champions has contributed to the precarious energy situation that confronts the United States in the 1970s and beyond.

In addition to the specific intrusions into the energy areas mentioned above, the federal government, beginning with the New Deal, became a producer of electrical power (with the establishment of the Tennessee Valley Authority) and a sponsor of electrical transmissions systems (with the Rural Electrification Administration). Government was also involved in the leasing of public lands and the outer continental shelf to private industry for energy development. The effects of these myriad governmental actions were complex, fragmented, and lacking in coherence or focus. Prior to 1973, federal responsibility for energy was vested in more than 30 different offices in many departments throughout the executive branch. Similarly, more than 30 congressional committees and subcommittees had jurisdiction over energy-related matters.[31] Despite some attempts at reorganization in the postembargo period, it is easy to understand how Jimmy Carter could claim in his presidential campaign that the country did not have an energy policy.

Perhaps no better illustration of this lack of an overall coordinated perspective is in the total absence of governmental efforts to practice energy conservation prior to 1973. If any coherent thread linked the various governmental activities in the energy field it was the assumption that more is better. Nowhere in government was there an attempt to assess the net energy impact or energy efficiency of proposed actions—either public or private. As noted above, through various pricing mechanisms the government sanctioned incentives for ever-increasing energy consumption. Even as the U.S. reserves of petroleum were peaking, and beginning to decline, and dependence on foreign sources was growing, no clear perception of the situation emerged in the government until the events of 1973-74.[32] Compounding the lack of a single agency charged with assessing U.S. energy needs and supplies was the optimism to which Garvey referred, that nuclear energy would meet our needs before fossil fuel shortages became significant. Typical of the overly optimistic assumptions of those charged with overseeing energy policy was the following conclusion of a 1966 government study: "The nation's total energy resources seem adequate to satisfy expected energy requirements through the remainder of this century at costs near present levels. . . ."[33]

It was against this backdrop of traditional perceptions and technological optimism that the events of 1973-74 had their startling impact on the United States. The initiation of the Arab oil boycott following the Yom Kippur War and the substantial petroleum price increases adopted by the members of OPEC during this same period focused U.S. attention on energy policy. While the boycott was the more dramatic of the events and produced quickly visible domestic repercussions, the price increases had the more substantial long-term impact on U.S. interests and U.S. energy policy concerns. Between September and December 1973, the posted price for Saudi Arabian crude, for instance, increased from $3.01 per barrel to $11.65 per barrel.[34] The effects of this and subsequent price increases were felt throughout the U.S. economy as marginal industries were pushed irrevocably into the red and the prices of nearly all capital and consumer goods increased, reflecting the petroleum inflation in construction, production, and transportation. According to then-Secretary of State Henry Kissinger, the OPEC actions

cost Americans half a million jobs and over one percent
of national output; it added at least five percentage points
to the price index, contributing to our worst inflation
since World War II; it set the stage for a serious recession and it expanded the oil income of the OPEC nations
from $23 billion in 1973 to a current annual rate of $110
billion, thereby effecting one of the greatest and most
sudden transfers of wealth in history.[35]

In the absence of bold government action, the OPEC price increases produced shifts in U.S. energy use patterns. In a marked
divergence from the exponential growth in energy usage, total U.S.
energy consumption declined from 1973 to 1974 and again from 1974
to 1975. While government appeals for voluntary conservation may
have had some marginal impact, market response to higher fuel costs
was the primary cause for this decline. A Bureau of Mines study
also ascribed the 1975 drop to the economic recession and the relatively mild winter of that year.[36] Since 1975, energy consumption
in the United States has again been rising, but at rates substantially
below those experienced in the pre-1973 period. One of the major
factors underlying this slower growth rate has been cost-induced
energy conservation by industry. In many instances, U.S. industries
have borrowed techniques and technologies that had been developed
in other industrial countries where high-cost energy had been the
norm. These industrial conservation efforts include the purchase of
new (more energy-efficient) machinery, retrofitting existing machinery
with energy-saving devices, introduction of more energy-efficient
industrial processes (in glass and aluminum production, for instance),
and cogeneration. Between 1972 and 1976, energy efficiency increases
ranging from 12.3 percent (transportation equipment) to 3.8 percent
(primary metals) were achieved by various sectors of U.S. industry.[37]
The federal government's lack of preparedness for coping with
energy as a coherent policy field became clear in the months following the boycott. Not only did the government lack appropriate organization and powers to deal with the situation, it did not even have an
adequate data base on which to make decisions. The government
found itself dependent upon the oil companies for assessments of U.S.
domestic petroleum reserves and production capacity. Initial efforts
were undertaken to prepare the federal government for its new role
as national energy policy manager, as Congress adopted legislation
expanding the government's role in energy price regulation, energy
distribution, and data acquisition. In addition, under Presidents
Nixon and Ford steps were begun to reorganize the government's
energy activities, which reached their culmination in the creation of
the Department of Energy in 1977 under President Carter. While

these steps were being taken, the United States, for the first time, began to assess its energy supply and demand in a comprehensive fashion.

Supply and demand are the two key elements in a national energy program. Owing to the pre-1974 assumptions of abundance, the United States was not accustomed to making long-term projections of energy resource supply. In the period since the oil embargo, many agencies, groups, and individuals have sought to forecast U.S. energy supplies. Such forecasts are necessarily difficult because they are dependent upon a number of critical unknowns: How much of the resource(s) exists in areas that have not yet been explored? How will developments in exploration and extraction technologies affect the available supply? What will be the future price of the resource(s)? (Oil resources that may not be economically attractive to recover when the market price is $3 per barrel may be quite attractive when the price rises to $15 per barrel.)

Some analysts are very optimistic about the future availability of adequate energy resources.

> Allowing for the growth of energy demand . . . we conclude that the proven reserves of these five major fossil fuels (oil, natural gas, coal, shale oil, and tar sands) alone could provide the world's total energy requirements for about 100 years, and only one-fifth of the estimated potential resources could provide for more than 200 years of the projected energy needs. [38]

Critics of this optimistic assessment argue that the existence of vast undiscovered resources is not scientifically supported, that higher prices do not guarantee more resources, and that the environmental impact of accelerated exploration and recovery (in which a higher proportion of the ore is waste) could be devastating. [39] Others note that rather than considering the dollar cost of new energy resources we must consider the net energy to be gained. While a particular energy supply may be economically attractive, the recovery, processing, and transportation of the resource consumes energy. Thus, the net energy gain is not the energy value of the resource, but that value minus all the energy needed to make the resource available. As we move to increasingly exotic sites and sources for our energy supplies, the net energy gains are likely to decrease and approach zero. [40]

It is generally agreed that petroleum and natural gas production in the United States has peaked or will soon peak. On a global scale it is assumed that the peak should occur during the next 15-40 years. The United States does have vast coal reserves (perhaps enough to last 200 years), but coal in its mined form is not readily usable in transportation and other areas and it has significant environmental drawbacks (which will be discussed later). Uranium reserves to fuel the projected number of nuclear reactors are probably adequate for the next 40 years without spent fuel reprocessing or the breeder reactor.[41] Whether uranium reserves would extend beyond that point is difficult to assess given the relative infancy of the uranium mining industry.

Sources of energy other than fossil fuels or nuclear power represent a tiny fraction of current U.S. energy supply. Hydroelectric and geothermal energy have been and continue to be used, but their expansion to major energy sources is most unlikely owing to the availability of suitable sites. Generally the best sites for such power sources were developed earlier in this century.

The most neglected energy source in the United States is solar energy. At present it is economically attractive only for supplementary hot water heating in new homes.

> Energy prices will eventually rise until solar energy is
> cost competitive with nuclear and fossil energy for many
> applications. In general, however, solar energy is not
> yet competitive. This is partially a fact of nature, but
> it is also a consequence of many other factors, including
> past research priorities, past environmental policies,
> and economic policies such as price regulation and deple-
> tion allowances. Simply stated, it has been easier and
> cheaper (although perhaps not smarter in the long run)
> to live off capital (fossil fuels) than income (renewable
> energy resources).[42]

Over the past several years interest in solar energy has increased substantially in the United States. The federal government has significantly increased funding of solar research, although it still lags far behind nuclear research. Industry and government agencies are exploring a host of solar energy applications, ranging from massive orbiting solar collectors to household-sized units. Architects are incorporating more solar components in homes and increasingly designing homes to take advantage of solar heating without the use of any devices (known as passive solar).[43] While the technology exists to produce electricity from solar energy, the process is currently too expensive for widespread use. There is considerable debate over the proportion of U.S. energy needs that solar energy can eventually supply.

One of the most difficult aspects of formulating a national energy policy lies in the forecasting of demand. Such demand

projections must cover a period of at least 30–40 years because of
the long lead times required for the implementation of most energy
policy options. For existing technologies (light-water nuclear reactors, offshore oil wells, and so forth) the period between the time of
the decision to begin the project and the time at which the system is
on-line, producing energy, may be as long as seven to ten years.
For the commercial development of new technologies (nuclear breeder
reactors, fusion systems, and the like), the lead time is unknown,
but it may well extend over three or more decades.

All forecasting methods essentially rely on an examination
of historical trends and extrapolation into the future. The extrapolation may be done with varying degrees of sophistication, and the analyst may attempt to adjust the curve in anticipation of possible future
developments. Some forecasters may disaggregate energy demand
into sectors, examine historical patterns and extrapolate for each,
and then reaggregate these sectoral demands to arrive at a total demand figure. Despite the sophistication and rigor of some of these
methods, the accuracy of an energy forecast is dependent upon the
accuracy of its assumptions about future events and conditions. Virtually all of the forecasts made prior to the OPEC price increases
have had to be revised substantially downward to reflect changes that
were unanticipated at the time the forecasts were made.

One recent study analyzed a number of energy forecasts and
identified the assumptions that were common to all. The assumptions
were:

- Real GNP will grow at 4.0 percent annually.
- Population will grow at 1.7 percent annually (range 1.2–2.8
  percent).
- Prices of fuel will remain constant relative to each other
  and to goods and services in general.
- Supplies of all fuels will be adequate.
- The United States will have no import problem.
- There will be no important environmental restrictions that
  are not offset by technological improvements.
- No "revolutionary" technical change will take place, only
  "evolutionary" change; but nuclear power will become impor-
  tant after 1975 or 1980.
- Business cycle swings will be minor.
- No major wars will occur.
- Energy use per capita will increase at 1.3 percent annually.[44]

Relatively small miscalculations in these ten areas can have
substantial effects on the long-term demand projections. If these
assumptions are correct, total U.S. energy demand by the year 2000

would be 170 quads (x × 10^{15}Btus). If, however, population and per capita energy use were each to grow at 2.5 percent rather than the assumed value, demand in 2000 would be 300 quads. Similarly, if these assumptions are unduly optimistic (in terms of increasing demand) as the post-1974 perspective would suggest, then estimates for the year 2000 would be scaled down, perhaps to the 125-150 quad range.

Complicating further the projection process is the realization that many of the factors influencing the demand level (the subject of the ten assumptions listed above) are capable of being influenced by deliberate actions by the society. Thus energy demand is not a goal established for man by some unalterable forces, but rather is produced, in large measure, by the interplay of a complex series of decisions. Thus, those who prefer a slow rate of growth in energy consumption (or no growth at all) will favor one set of policies designed to dampen demand, while those favoring rapid growth in energy consumption will favor those policies that will fuel energy demand.[45]

By the beginning of 1979 the United States had considered and debated various energy proposals and programs, but a clear commitment to a particular energy future had not yet emerged. Nearly all of the programs offered call for some acceleration in energy production and some efforts to increase energy conservation. Despite this general similarity, the degree of emphasis to be placed on each of these components remains a major point of dispute. While the energy programs proposed by Presidents Nixon and Ford each called for some energy conservation, the primary strategy of each was for the United States to "produce its way out of the energy shortage." This was to be accomplished by reducing obstacles to increased production from conventional sources (by deregulating energy prices and relaxing environmental regulations) and the accelerated development of new high-technology sources (such as the breeder reactor, nuclear fusion, oil shale extraction, the liquefaction and gasification of coal). In contrast, the program President Carter presented in April 1977 placed primary emphasis on conservation of energy, including a reduction in the annual rate of growth of energy consumption from 4.6 percent to less than 2 percent and an absolute reduction of 10 percent in gasoline consumption. The Carter program did not oppose the development of high-technology sources (although he was against the continued development of the Clinch River breeder reactor project), but it put increased emphasis on the utilization of renewable energy resources such as solar power.

In addressing the American people, Carter called for a national effort to overcome the nation's energy problems that would be the "moral equivalent of war." While public opinion poll data indicate that Americans regard energy as a serious matter, in the absence of acute direct effects on their lives (for example, the gasoline shortage of 1974 or the natural gas shortage of 1977) they appear unwilling to make the sort of commitment the president desired. The president's perhaps unrealistic hope of a unified nation rising to adopt a comprehensive energy plan for the future was soon dashed as his energy proposals became subject to the pulls and tugs of politics as usual.

Even before Carter's presidency, partisan considerations had significantly marked efforts to formulate a national energy plan. The energy programs proposed by Republican presidents Nixon and Ford received generally favorable reactions from Republican legislators and criticism from the Democrats. When Democrat Carter presented his energy plan, the pattern of legislative response was largely reversed. Of course, congressional reaction was not purely partisan, with some of Carter's strongest opposition coming from members of his own party.[46] However, since energy had been a major issue in the 1978 presidential election and figured to be significant in future campaigns, politicians were mindful of who would receive credit or blame for particular actions.

Among the most controversial components of Carter's plan were a number of new taxes that were designed to raise the consumer price of energy (and thereby dampen demand) while limiting the profits accruing to the energy producers. These tax proposals led to a protracted fight in Congress, which illustrated the range of interests and factions with stakes in the outcome of the energy plan.

> The deadlock over the fiscal provisions of the energy bills revealed that a new "war between the states" had seized hold of the Congress, if not the country. Dixiecrat and Sunbelt politicians, from both parties in a dozen states, insisted that the oil and gas industries needed more cash—not less—to expand their drilling and refining operations and their capital base. Labor and political groups from the sixteen snowbelt states of the Northeast and Midwest dissented, while the Western states divided between the two factions. If the conservation of supplies was to be effected by raising market prices, the consumer groups proclaimed, it would penalize the least affluent since their energy demands are the least elastic when prices rise. The virtual doubling of energy prices, the liberals warned, would provide the major oil companies with "war profiteering" and "rip off" revenues; and it would penalize the aging industries of the North while bringing new capital to the conservative rimlands of the South.[47]

While the considerations discussed above have all been important in delaying a decisive U.S. response to the energy situation,
perhaps the greatest factor has been the lack of agreement on basic
issues of fact and value. While the disputes are complex and the
divisions multifaceted, two simplified positions can be gleaned from
the debate.[48] The first argues that continued and increasing high
levels of energy consumption, with increasing centralization and electrification, are necessary if the United States is to sustain and enhance its quality of life and standard of living. They claim that an
energy-poor economy would stagnate or decline and that while all
Americans would suffer, the primary victims would be the poor who
would be locked forever into their deprived status. Proponents of
this position see the primary causes of the energy shortage as inadequate economic incentives (owing to excessive governmental price
regulation) and excessive restrictions (primarily environmental
quality and occupational safety regulations). While they recognize
that there are theoretical limits to our resources, they believe that
human ingenuity is limitless and, under proper conditions, can produce technological solutions to energy problems. While they recognize that exponential growth in energy consumption cannot continue
forever, they tend to view a plateau, or leveling in energy consumption, as coming in the distant future. This position is summarized
by an energy company executive:

> At some point in time, well before [power] plants crowd
> us into the ocean or we change the climate of the earth
> through the injection of heat into the atmosphere we will
> reach an energy plateau, a level at which our technology
> will have placed us through the super-efficient utilization
> of power. In other words, we will have learned to do so
> much more with so much less power that more power
> plants as we know them today probably would be super-
> fluous.[49]

Those who oppose this position feel that the time for the energy
plateau has already arrived and they disagree with each of the asser-
tions above. They argue that the relationship between jobs and energy
is not the simple positive correlation suggested above. Rather, they
contend, cheap abundant energy often displaces workers and results
in fewer jobs. In any event, they argue that we can use energy much
more efficiently and provide as much real benefit at much lower levels of energy consumption. In their view, a less-energy-intensive
society could provide a more satisfying lifestyle and abundant opportunities for employment and advancement. Further, they argue that
continued and accelerated reliance on centralized electrical systems

is wasteful of energy and threatens a dangerous social and political centralization of power. Finally, they suggest that it is virtually impossible to sustain high rates of growth in energy consumption as the limits of resources and the absorbtive capacity of ecosystems are reached.

The Congress struggled with Carter's energy bills for 19 months and not surprisingly the product of this battle involving competing industry and consumer interests, state and regional concerns, and fundamental views of energy was indecisive. Many of Carter's more stringent conservation measures were dropped or softened during the legislative journey. A protracted fight over natural gas price deregulation resulted in a policy of gradual deregulation. This compromise was supported by those who favor higher prices to spur conservation and those who favor higher profits to encourage exploration and development. The bill mandated continued improvements in automobile operating efficiency and provided tax incentives for residential, commercial, and industrial investments in energy conservation.

While the adoption of the National Energy Plan[*] in November 1978 represented at most a first tentative step toward a comprehensive response to the nation's energy problems, it appeared that Congress and the public were tired from the ordeal and hoped to forget about energy problems at least for a while. Energy consumption was increasing, energy supplies seemed secure for the next few years, and energy price increases had moderated since the price explosions of 1973-74.

This complacency was shattered quickly by a rapid series of unexpected events in late 1978 and early 1979. A number of small demonstrations against the shah in Iran quickly escalated into a strike that shut down the important Iranian oil industry, and ultimately led to the ouster of the shah. The temporary cutoff of Iranian oil produced a minor petroleum shortage and forced some companies to secure supplies on the more costly "spot" market. The new government in Iran appeared to be a far less secure energy supplier than was the shah. In addition, the new government's links to the Palestinian movement suggested that should there be a future oil boycott

---

[*]The plan actually includes five separate acts: Public Utility Regulation Act (PL 95-617), Energy Tax Act (PL 95-618), National Energy Conservation Policy Act (PL 95-619), Powerplant and Industrial Fuel Use Act (PL 95-620), and Natural Gas Policy Act (PL 95-621).

by the opponents of Israel, Iran would participate in the boycott, unlike in 1973-74. The possibility of such a boycott was increased by the role of President Carter in fostering the peace accord between Israel and Egypt. Moderation in oil price increases ended with the decision by OPEC to increase prices drastically and with President Carter's decision to advance the date of domestic petroleum decontrol, all presaging rapid future increases in gasoline prices.

On March 28, 1979, a number of design and operator errors combined to produce America's most serious nuclear power plant accident. For a week the eyes of much of the country (and the world) were focused on the Three Mile Island reactor near Harrisburg, Pennsylvania, as technicians sought to control the release of radioactive gases and to avert a catastrophic meltdown accident. While the horrors of a meltdown were prevented, the Harrisburg accident is likely to have a significant impact on future use of nuclear power. For the first time the sort of accident the AEC had assured was unthinkable became credible to the American people. Polls showed that Americans' faith in nuclear power was severely shaken and on May 6 over 100,000 demonstrators came to Washington to demand an end to nuclear power.

While the Three Mile Island incident is not likely to lead to an immediate shut-down of all nuclear plants, as some demonstrators demanded, it may well cripple an already weakened nuclear industry. The minimal reforms that are likely after Three Mile Island (the requirement for more safety systems, the up-grading of operating personnel and regulation, and the siting of any future plants in isolated areas distant from population) will further erode the economic viability of nuclear power.[50] By the middle of 1979, the future growth of the nuclear industry in the United States is very much in doubt.

The energy uncertainty resulting from the new questions about nuclear power was reinforced in late April 1979 as gasoline shortages began to appear. In May, California, where the most acute shortages had occurred, was forced to institute alternate-day rationing in order to reduce long lines at service stations. With warnings that the shortages might spread across the country and that gasoline supplies would be short during the heavy-consumption summer months, it began to appear that the winter of 1974 was being repeated.

The incapacity of the government to cope with the problem was made clear in May when the Congress rejected President Carter's proposal for an emergency stand-by rationing program. Representatives apparently voted in terms of the interests of their immediate constituents rather than the overall national interest and were more concerned that their constituents get an abundant supply of gasoline than that the country be able to cope with a future diminution of gasoline supply. Congress, it appeared, reflected a public perception that the shortage was contrived and that it therefore did not warrant sacrifice.

Thus, by the summer of 1979 the United States was only slightly better prepared to deal with energy problems than it had been in 1973-74. Although energy bureaucracies are in place at the federal and state level, and some minor efforts to bring about conservation have begun, energy consumption continues to increase, albeit more slowly than prior to 1974. Despite four years of energy policy debate, the ultimate direction of U.S. energy policy remains unclear and many key issues remain unresolved.

Despite some requirements for improved operating efficiency in appliances and a number of tax incentives, the thrust of current conservation policies rests on increasing the price of energy. While there is evidence that higher prices do dampen demand, Energy Secretary James Schlesinger indicated that gasoline prices would have to surpass $2.00 a gallon to significantly alter driving behavior. In addition, rationing by price impacts inequitably on the poor. Low-income households use only one-half as much energy as do upper-middle-class households[51] and are less able to reduce their energy consumption or afford expensive energy-saving devices or materials. Consumers at all income levels resent higher energy prices, particularly because they do not perceive a genuine crisis requiring such sacrifice and they believe that the energy companies are making excessive profits. Polls show that when forced to choose, Americans prefer rationing to higher prices as a conservation technique, but policy makers (unable to agree on a "fair" rationing program and fearful of the bureaucratic morass of rationing) have avoided formulating or implementing rationing plans. Despite their preference, policy makers may be forced to move to rationing if the energy supply situation worsens, but based on past behavior it will be only as a last resort.

Even before Three Mile Island the future role of nuclear power was controversial and uncertain. Proponents of nuclear power argued that the depletion of petroleum and natural gas reserves necessitates the rapid acceleration of nuclear power use. They claimed that greater use of nuclear power would free the United States from dependence on unreliable and costly energy imports and avert the pollution problems associated with expanded coal consumption. To achieve these ends, nuclear advocates called for the streamlining of the nuclear plant licensing procedures* and increased research

---

*A Carter administration proposal to streamline the nuclear licensing process, which can take up to ten years, did not even get out of the committee in 1978.

and development on nuclear fuel reprocessing, the breeder reactor, and nuclear fusion. Opponents argued that nuclear power is irrelevant to oil imports because nuclear plants can be used only to produce electricity and only 10 percent of our petroleum consumption goes to electricity generation. Nuclear power cannot be used in the gas tanks of our cars or in our oil or gas furnaces. They also argue that the slowdown in orders for nuclear plants reflects the unreliability and excessive cost of nuclear power, and that coal-generated electricity, even with the most advanced pollution abatement systems, will prove less expensive than nuclear power.[52] Lastly, they conclude that the safety hazards posed by the nuclear fuel cycle outweigh any real benefits. A further question that has received renewed attention since Three Mile Island is what would happen if the United States became more dependent on nuclear power and then at some future time a catastrophic accident killing thousands of people were to occur. Would the United States be able to respond to the likely demands of most of the public to shut down our nuclear plants? While the more moderate nuclear opponents call for a ban on new nuclear plants and a tightening of safety and operating procedures at existing plants, other critics are calling for the shutting down of all nuclear reactors.

Another critical policy matter that remains on the agenda is the appropriate future role for renewable energy resources. In part the debate involves questions of technical feasibility, with proponents[53] arguing that solar and allied sources can provide the bulk of our future energy needs and skeptics[54] countering that such an expectation is unfounded. The debate also involves governmental energy research and development (R&D) priorities. Some call for a fundamental reversal of existing R&D priorities from nuclear and fossil fuels to solar energy. Among those supporting a greater reliance on solar energy there is disagreement as to the desirable extent of government involvement. Government could place solar energy in an even position to compete with other energy sources by eliminating the various subsidies and tax advantages currently enjoyed by the other energy sources, or it could take a more active role and adopt policies to subsidize and otherwise encourage conversion to solar energy.

Deciding on the appropriate balance between energy supply and environmental quality will be a continuing concern, particularly if the anticipated increased reliance on coal is to occur. Many people view energy and environmental quality as competing objectives and Congress reflected this perspective in the adoption of the Energy Supply and Environmental Coordination Act of 1974,[55] which permitted delays in achieving previously mandated environmental standards so as to improve the energy availability situation. An alternative perspective views solar energy and conservation as measures that deal with energy needs while having less impact on the environment than do our present energy sources.

INTERNATIONAL ENERGY POLICY

That the international dimensions of energy policy have become significant to the United States was dramatically demonstrated by the events of 1973–74. Before 1950 the United States had been a net exporter of petroleum; by 1970 it had become a net importer. At the time of the oil embargo, the United States relied on foreign sources for about one-third of its petroleum needs. Despite a subsequent reduction in petroleum demand, reflecting the price increases, during the winter of 1976 the United States, for the first time in its history, imported more petroleum than it produced domestically. Accompanying this increased dependence on foreign sources has been a concentration of these sources. Increasingly, U.S. petroleum imports have been coming from the Middle East, as the flow from traditional suppliers, such as Canada and Venezuela, has decreased.

These changes in the structure of international petroleum supply and demand pose a series of threats and problems for the United States and other importing states. Oil-importing nations have become vulnerable to a new oil embargo, which, given increasing dependence, could have even more devastating consequences. In establishing the market price for the petroleum they must import, they are largely at the mercy of OPEC. Given such vulnerabilities, these nations believe that their economic and national security has been threatened and their autonomy in the international system reduced. In addition, increased imports (and greatly increased prices) have a strongly adverse impact on national balance of payments and, consequently, on the domestic economy. While the United States paid $4.6 billion for its petroleum imports in 1972, it is estimated that the cost for 1981 will be $80 billion—a nearly 2,000 percent increase over a nine-year period.[56] Another concern, although its impact is less immediate, is the realization that growing worldwide demand for petroleum will eventually lead to the depletion of accessible global reserves. Some experts foresee this occurring within the present century, while others believe that supplies will last for several decades beyond. There is a shared realization, however, that petroleum cannot be relied upon as the primary fuel for the long-term future.

In response to these concerns the United States convened a conference of oil-importing states in February 1974. The 19 participant nations established the International Energy Agency (IEA) in order to:

- promote secure oil supplies;
- develop an emergency self-sufficiency in oil supplies, restraining demand and allocating available oil among member countries on an equitable basis;
- promote cooperative relations with oil-producing countries and with other oil-consuming countries, including those of the developing world; and
- reduce member-country dependence on imported oil by undertaking long-term cooperative efforts and accelerating energy R&D programs.[57]

It is too early to evaluate the effectiveness of IEA. The emergency plans have been drafted but their utility will remain unknown until tested. The United States has moved ahead with international energy research and development activities. By September 1975 the United States was engaged in 56 separate bilateral R&D projects—involving Soviet bloc and less developed states, as well as members of IEA—and by a year later it was participating in 11 multilateral energy R&D projects.[58] The technological, as opposed to diplomatic, payoffs from these cooperative efforts are problematical.

In addition to these research efforts and the diplomatic activities of IEA, the United States has dealt directly with the major oil-exporting countries. The primary U.S. objective, in which they have had some success, has been the tempering of OPEC price increases. Somewhat less publicly the United States has also urged the OPEC states to maintain stable prices because the development of U.S. alternative domestic energy sources requires assurance that OPEC petroleum will not drop in price and undermine the economic viability of these U.S. sources.

An increasingly important international energy concern, which has become a major policy priority of the Carter administration, is the attempt to control the spread of certain aspects of nuclear technology (and particularly the establishment of a global plutonium economy), because of the associated threat of nuclear weapons proliferation. The seriousness of this matter was demonstrated in May 1974, when India detonated its "peaceful" nuclear bomb. In developing this weapon, India had utilized nuclear material and technology that had been provided by Canada and the United States for the production of energy. In response to the Indian action, Canada and the United States each suspended its nuclear aid to India. Despite additional efforts at control by the International Atomic Energy Agency (IAEA) and the informal association of nuclear-exporting nations, the hazard of nuclear proliferation as an outgrowth of nuclear power plant export remains unresolved. By 1984 it is projected that more than 33 countries will have nuclear power facilities.[59] President Carter has had limited success in convincing other nations to forego the development of the breeder reactor (one of the by-products of which can be fashioned into a nuclear weapon) or to prevent the spread of nuclear fuel reprocessing plants (which can also be used for weapons development). Other countries sharing a concern for nonproliferation but lacking the vast U.S. reserves of coal (as well as reserves of oil and natural gas) see nuclear energy as their only reliable source of power in the immediate future. In addition they favor the development of the breeder or reprocessing as protection against future dependence on the handful of noncommunist uranium-exporting countries (including the United States). [60] They do not wish to substitute dependence on OPEC for dependence on a uranium cartel. President Carter's strategy to date has been to offer guaranteed supplies of uranium fuel and to store nuclear wastes for those countries who are willing to forego their own breeder and reprocessing systems. In so doing, the United States is subsidizing the nuclear programs in other countries in order to enhance national and global security.

ENERGY AND THE ENVIRONMENT

   The linkages between energy and the environment are fundamental to an understanding of either. According to the Second Law of Thermodynamics, all conversions of energy, under thermodynamically optimal conditions, can do no better than produce the amount of energy they consume. In the nontheoretical world, where friction and other factors exist, the consequence of this law is that all energy conversions are to some extent inefficient. That is, less usable work emerges from the process than is consumed. The energy that does not emerge as usable is discharged into the environment as thermal pollution. When energy conversions are relatively few and fuel is abundant, neither of these consequences is regarded as significant. However, when thermal pollution is being discharged in large quantities from many sources, the cumulative effects may overwhelm the coping capacity of biologic and climatic systems. In addition, when fuel supplies can no longer be taken for granted, the losses in conversion may significantly affect the quantities of energy available to do work.
   Most of the energy used in the United States is provided by the burning of fossil fuels. The environmental impacts of such conversion go beyond thermal pollution, as the gaseous and particulate products of combustion (and incomplete combustion) have adverse impacts on various organisms and ecosystems. The release of the binding energy of the atomic nucleus in a reactor (an alternative to fossil fuel combustion) does not produce the gaseous and particulate

pollutants associated with fossil fuels but presents potential hazards of a more virulent nature. The bulk of air pollution, a portion of water pollution, and the major radioactive risks in the United States are the direct by-products of energy production and use.

In a somewhat less direct manner, the production and consumption of energy places additional demands on the environment in terms of the consequences of resource extraction and the industrial/consumer waste products that would not occur save for the ready availability of energy. The extraction, processing, storage, and use of energy resources require large quantities of land and water. Increasingly, as land and water become scarce, the demands of energy production reduce the availability of these resources for other purposes. This issue is most clear at present in the Western states where farmers and energy producers are competing for limited water supplies. The proliferation of nonbiodegradable products and packaging in the United States is a direct result of the increased use of petroleum, as the petrochemical industry needed to develop uses for the by-products of the refining process.

The combustion of fossil fuels in electric power plants and transportation (principally automobiles) is the primary source of air pollution in the United States. The generation of electricity produces 78 percent of sulfur oxides, 44 percent of nitrogen oxides, and 27 percent of particulates. Transportation combustion accounts for 74 percent of the carbon monoxide, 53 percent of the hydrocarbons, and 47 percent of the nitrogen oxides. Attempts to reduce these pollutants often have other consequences, including the production of new pollutants or in greatly increasing costs. Legislation in the 1960s and 1970s that sought to reduce the sulfur oxide emissions in the United States placed a premium on relatively scarce, low-sulfur fuels. An Environmental Protection Agency (EPA) study estimated that the direct cost of air pollution in the United States is $16.1 billion per year, or over $80 per person.[61] This figure does not include indirect costs or the substantial effects on human and other organisms (see Table 1.3).

While some of these air pollutants and other energy by-products (such as mining runoff) find their way into the nation's waters, energy's greatest environmental impact on water is in terms of thermal pollution. It has been estimated that by 1990 electrical generators in the United States will require 200 billion gallons of fresh water per day for cooling purposes.[62] This quantity is the equivalent of one-sixth of the total average fresh water runoff. While water used for cooling can be used for other purposes as well, its usefulness is reduced. Water coming from an electrical generating plant may contain chemicals used for retarding the growth of organisms and other substances. The water, as a function of its higher temperature,

TABLE 1.3

Major Air Pollution Sources in the United States (units are 10^{16} tons per year)

Source	Particulates	Sulfur Oxides	Nitrogen Oxides	Hydrocarbons	Carbon Monoxide
Nature	?	4.2	?	30.7	?
Stationary sources	7.1	22.1	11.0	0.4	1.1
Transportation	0.8	1.1	11.2	19.8	111.5
Industrial processes	14.4	7.5	0.2	5.5	12.0
Miscellaneous	12.8	0.4	2.4	11.2	26.1
Total	35.1	35.3	24.8	67.6	150.7

*Source: U.S. Department of Energy and U.S. Environmental Protection Agency, Energy/Environment Factbook (Washington, D.C.: U.S. Government Printing Office, March 1978).

has diminished capacity to hold dissolved oxygen, which is critical
to aquatic life. Warmer water is also more conducive to the growth
of foul-smelling, oxygen-consuming algae. These reductions in
available oxygen also reduce the capacity of water to process sewage.
Fluctuations in water temperature also disrupt the cycles of aquatic
life. There have been cases in which fish remained in northern
waters during the winter months, because the thermal pollution
masked the seasonal change. When the power plant was shut down
for maintenance and the water returned to its normal winter temperature, large numbers of fish died.

Of the fossil fuels, coal is the dirtiest. In addition to the air
and water pollution effects associated with its combustion, the extraction of coal from the ground carries a significant environmental
cost. The mineral debris of underground mines and the overburden
from surface mines are often left in huge unstable mounds that are
subject to erosion and occasionally tragic landslides. Water seeping
into abandoned mine sites creates sulfuric acid, which increases the
acidity of the soil to a point threatening vegetation and may leech
into waterways. Thousands of miles of waterways in the United
States have already been made highly acidic in this manner. Soil
subsidence in areas above underground mines is another adverse
environmental impact. Underground mining is also quite hazardous
for the workers in the industry. Despite recently stiffened safety
and health legislation for the mining industry, accident rates remain
high and 100,000 miners are afflicted with the serious respiratory
disease known as "black lung."

Surface, or "strip," mining, which is much safer than underground mining, has the potential for permanent disruptions of local
ecosystems. Past strip mining activity has left large areas devoid
of vegetation and subject to serious erosion. After years of debate
and a 1976 veto by President Ford, in the summer of 1977 the United
States adopted national strip mine legislation that requires the restoration of stripped land to its original contours, the creation of a fund
(supported by a coal tax) to restore previously stripped land, and a
prohibition on the strip mining of farmland except when there is the
"technological capability" to restore the land to its original productivity. The coal industry claims this law will greatly increase costs
of production and the long-term effectiveness of the reclamation efforts is still unknown. Particular concern involves the arid Western
lands whose fragile ecosystems may be difficult or impossible to
restore at reasonable cost.

Oil refineries produce both air and water pollution. In general, oil refineries are now in compliance with air quality standards,
but the more costly compliance with water pollution discharge limits
is substantially short of completion. The major environmental hazard

posed by petroleum (other than the combustion effects discussed above) is oil spills. Depending on the particular environment and the type and quantity of the spilled material, the effects may include death, disruption of physiological and behavioral activity, disruption of food chains, and habitat change.[63]

Natural gas is among the more environmentally benign fossil fuels, but the increasing use of liquefied natural gas (LNG) has raised a furor over safety. In 1944 a Cleveland, Ohio, LNG tank cracked and as its spreading contents ignited, a massive fireball incinerated 300 acres of homes, factories, and businesses and killed 128 people.[64] While some experts claim that improved technology would not allow a repetition of the 1944 catastrophe, many groups, particularly those living near proposed LNG storage sites, have been opposing the use of LNG.

The environmental and safety questions associated with light-water nuclear reactors constitute a central issue in the U.S. energy debate. Among the major questions are the effects of routine low-level radiation emissions, the likelihood of catastrophic nuclear accidents, the possibility of nuclear theft, sabotage, and terrorism, and the feasibility of safe and secure disposal of nuclear wastes.

That radiation at particular levels is harmful and lethal to human life is undisputed. In its normal operations a light-water reactor's metal and concrete shielding deflects or absorbs nearly all the radiation produced. The low-level emissions that do escape are below the level considered dangerous by the Nuclear Regulatory Commission (NRC) and less than the radiation from natural sources. There are, however, some scientists who question whether any radiation above the natural levels is safe, and they suggest that there may be long-term effects from the normal operation of nuclear power plants.

A more serious dispute involves the possibility of major accidents at nuclear plants.[65] Owing to differences in design and operation, nuclear power plants cannot explode in the manner of nuclear bombs. Experts, however, can envision a number of eventualities that could result in the deaths of tens of thousands, widespread injuries, property damage in the billions of dollars, and the radioactive contamination of an area as large as Pennsylvania. The major debate has involved two basic issues: What is the likelihood of such a catastrophic accident? and What level of risk is tolerable? The nuclear establishment has argued that the risk of serious nuclear accident is negligible (one study compares it to the risk of being hit by a falling meteorite) and that we tolerate much higher risks in other areas of our lives. Nuclear opponents argue that a major nuclear accident in the next two or three decades is likely and that the risks posed by nuclear power are unique and unnecessary.[66]

Associated with the risk of accident is the question of willful and intentional disruption of nuclear power. In recent years the AEC and NRC have grudgingly responded to criticism of the laxity in nuclear security and several actual cases of nuclear material theft or diversion by increasing some nuclear security.[67] Many experts, however, question the adequacy of current safeguards and wonder if nuclear material can ever be fully secured in a free society.

Periodically the fuel rods in a nuclear reactor have to be replaced. The used rods contain a number of highly radioactive and lethal elements, some of which have half-lives in excess of 24,000 years. To date, no method has been devised for the long-term, safe storage of this nuclear waste. Nuclear advocates claim that this is a relatively simple technical problem and will be resolved shortly. Critics are skeptical of an easy solution and argue that we cannot continue accumulating quantities of this lethal material in temporary storage with only a promise that some final solution will be developed.

FUTURE PROSPECTS

The long-term energy crisis poses a challenge to the United States and the nature of the response has yet to be shaped. Barring unforeseen occurrences (a new oil embargo, for instance) the United States may survive the 1980s without an acute energy shortage. During the next decade energy will have major domestic impacts (higher prices, economic dislocations, and environmental stresses) and remain a major foreign policy concern. Neither the people nor the institutions of the nation have yet regarded the energy situation as one requiring the "moral equivalent of war." Judging from past behaviors such a perception is not likely until the energy crisis becomes acute and immediate. While such a situation may not prevail in the 1980s, the energy policy decisions made during this period will determine, in large measure, the future capacity of the United States to cope with increasing disparities between energy supply and demand. When energy shortages are real and severe there are no acceptable short-term solutions.

The highest energy priority for the United States must be an increase in the efficiency of energy use. It has been estimated that as much as one-half of the U.S. energy consumption is wasted; that is, it provides no benefit and performs no task. Such waste may have been acceptable in a time of energy abundance, but it has become a dangerous anachronism. While some of the waste results from individual behaviors (which might be modified by education, incentives, and sanctions), much is a function of existing infrastructure that can be altered only marginally or at great expense. This underscores the significance of long lead times in energy strategies.

A viable energy strategy for the future must view the exponential growth in consumption over the past several decades as a historical anomaly rather than as an imperative that must be sustained ad infinitum. The logical absurdity of the latter is suggested by the fact that if recent growth rates were sustained for the next 200 years, the land area required for electrical generating plants would exceed the total land area of the United States.

However, even if energy demand can be stabilized or reduced, new energy sources will be needed to fill the gap as supplies of petroleum and natural gas diminish. While various esoteric sources may become significant in the next century, national survival cannot be predicated on a hope, and interim energy sources would be necessary in any event. The relative roles for solar power, coal, and nuclear fission over the next 30-40 years are still subject to heated debate, but affirmative decisions must be made shortly. While solar power is clearly the most benign environmentally, as well as inexhaustible, its ability to satisfy a major portion of future demand remains to be proven. Coal, while plentiful, poses serious (and perhaps insurmountable) environmental problems and substantial increases in coal production may be difficult to achieve. Nuclear fission, which has been viewed for 20 years as the ultimate successor to fossil fuels, is plagued by operating disappointments and escalating arguments over its safety.

Thus, the United States faces a choice between two uncertain energy futures. It may opt for the hard energy path of increasing production and reliance on nuclear power and other high technologies to generate increasing quantities of electricity. In so doing, however, the United States risks substantial damage, accelerated resources depletion, and larger future energy shortages if new technologies do not meet the expectations of their proponents. If, on the other hand, the United States opts for the soft energy path with its reliance on conservation and solar energy, there is a possibility of energy supplies falling below perceived needs and causing socioeconomic dislocations.

As the government continues its debate between these paths, other developments seem to be moving the country closer to the soft path. Energy conservation programs in the manufacturing, commercial, and residential energy sectors are accelerating in response to rising energy costs. Electric utilities are being forced to adjust their long-term demand projections downward, as demand has slackened. Construction and new orders for nuclear power plants are falling far below the level necessary to achieve the hard-path objectives, as utilities are reconsidering the economic, operational, and safety characteristics of commercial nuclear power. Despite these trends, the U.S. energy future is still unclear, but the time for decision has arrived.

NOTES

1. For a fuller description of the history of U.S. energy supply, see John M. Fowler, Energy and the Environment (New York: McGraw-Hill, 1975), pp. 67–80.

2. Energy Policy Project of the Ford Foundation, Exploring Energy Choices: A Preliminary Report (Washington, D.C.: Energy Policy Project, 1974), p. 74.

3. Marc Messing et al., Centralized Power: The Politics of Scale in Electric Power Generation (Cambridge, Mass.: O, G & H Publishers, 1979).

4. For a fuller description of the history of U.S. energy use, see Fowler, Energy and the Environment, pp. 80–94; David Howard Davis, Energy Politics (New York: St. Martin's, 1974); Richard B. Mancke, The Failure of U.S. Energy Policy (New York: Columbia University Press, 1974); Leonard Mosley, Power Play (New York: Random House, 1974).

5. Schipper and A. J. Lichtenberg, "Efficient Energy Use and Well-Being: The Swedish Example," Science, December 3, 1976, pp. 1001–13.

6. Joel Darmstadter, Joy Dunkerly, and Jack Alterman, "International Variations in Energy Use: Findings From a Comparative Study," in Jack M. Hollander, Melvin K. Simmons, and David O. Wood, eds., Annual Review of Energy, Vol. 3 (Palo Alto, Calif.: Annual Reviews, 1978), p. 183.

7. Ibid.

8. Wilson Clark, Energy for Survival (Garden City, N.Y.: Anchor, 1975).

9. Richard Grossman and Gail Daneker, "A Guide to Jobs and Energy" (Washington, D.C.: Environmentalists for Full Employment, 1977).

10. U.S. Senate, Committee on Government Operations (93rd Congress), Confidence and Concern: Citizens View American Government, A Survey of Public Attitudes, Parts 1 and 2 (Washington, D.C.: U.S. Government Printing Office, December 3, 1973).

11. Gerald Garvey, Energy, Ecology, Economy (New York: W. W. Norton, 1972), p. 25.

12. Dorothy K. Newman and Dawn Day, The American Energy Consumer (Cambridge, Mass.: Ballinger, 1975), pp. 43-44.

13. Ibid., p. 72.

14. Robert Caro, The Power Broker (New York: Alfred A. Knopf, 1974), pp. 301-03.

15. Walter A. Rosenbaum, The Politics of Environmental Concern, 2d ed. (New York: Praeger, 1977), p. 269.

16. Herman E. Daly, "Electric Power, Employment, and Economic Growth: A Case Study in Growthmania," in Daly, ed., Toward a Steady-State Economy (San Francisco: W. H. Freeman, 1973), pp. 253-54.

17. Neil Fabricant and Robert Hallman, Toward a Rational Power Policy: Energy, Politics, Pollution (New York: Braziller, 1971), p. 7.

18. Davis, Energy Politics, p. 31.

19. Lynton K. Caldwell, Lynton R. Hayes, and Isabel M. MacWhirter, Citizens and the Environment: Case Studies in Popular Action (Bloomington: Indiana University Press, 1976), pp. 304-08.

20. For a fuller discussion of the coal industry, see Henry M. Caudill, Night Comes to the Cumberlands (Boston: Little, Brown, 1963); Morton Baratz, The Union and the Coal Industry (New Haven, Conn.: Yale University Press, 1955); Davis, Energy Politics, pp. 19-47.

21. Robert Engler, The Politics of Oil (New York: Macmillan, 1961), pp. 132-33.

22. For a fuller discussion of the history of oil in the United States, see Mancke, The Failure of U.S. Energy Policy, pp. 86-106; Davis, Energy Politics, pp. 48-109; Engler, The Politics of Oil.

23. For a fuller discussion of the history of natural gas policy, see Mancke, The Failure of U.S. Energy Policy, pp. 106-21; Davis, Energy Politics, pp. 110-40.

24. Common Cause, "Stacking The Deck: A Case Study of Procedural Abuses by the Joint Committee on Atomic Energy (Washington, D.C.: Common Cause, 1976).

25. Carolina Environmental Study Group V. AEC, 9 ERS 1964.

26. John F. O'Leary, "Nuclear Energy and The Public Inter-est," in Robert J. Kalter and William A. Vogely, eds., Energy Supply and Government Policy (Ithaca, N.Y.: Cornell University Press, 1976), pp. 239-40.

27. Nuclear Reactors Built, Being Built or Planned in the United States as of June 30, 1975 (Washington, D.C.: U.S. Energy Research and Development Administration, 1975) (TID-8200-R32).

28. David Bird, "Con Edison Scores Makers of Nuclear Gen-erating Plants for Glorious Promises," New York Times, November 19, 1972; Thomas Enrich, "Atomic Lemons: Breakdowns and Errors in Operations Plague Power Plants," The Wall Street Journal, May 3, 1973.

29. For a fuller discussion of the nuclear power industry in the United States, see Peter Metzger, The Atomic Establishment (New York: Simon & Schuster, 1972); Spurgeon M. Keeny, Jr. et al., Nuclear Power Issues and Choices (Cambridge, Mass.: Ballinger, 1977); John G. Fuller, We Almost Lost Detroit (New York: Ballan-tine, 1975).

30. Anthony J. Parisi, "Nuclear Power: The Bottom Line Gets Fuzzier," New York Times, April 8, 1979, Section 3, pp. 1, 4; Charles Komanoff, "Doing Without Nuclear Power," New York Review of Books, May 17, 1979, p. 14.

31. J. Herbert Holloman and Michael Grenon, Energy Research and Development (Cambridge, Mass.: Ballinger, 1975), pp. 55-62.

32. George A. Lincoln, "Background to the U.S. Energy Revolution," in Joseph S. Szyliowicz and Bard E. O'Neill, eds., The Energy Crisis and U.S. Foreign Policy (New York: Praeger, 1975), p. 25.

33. Interdepartmental Study of Energy Research and Develop-ment, "Energy," in Roger Revelle, Ashok Khosla, and Maris Vinov-skis, eds., The Survival Equation (Boston: Houghton Mifflin, 1971), p. 212.

34. Robert B. Krueger, The United States and International Oil (New York: Praeger, 1975), p. 68.

35. Henry A. Kissinger, "Energy: The Necessity For Decision," Address before the National Press Club, Washington, D.C., February 3, 1975.

36. Federal Energy Administration, Energy Reporter (Washington, D.C., June 1976), p. 2.

37. Macauly Whiting, "Industry Saves Energy: Progress Report, 1977," in Jack M. Hollander, Melvin K. Simmons, and David O. Woods, eds., Annual Review of Energy, Vol. 3 (Palo Alto, Calif.: Annual Reviews, 1978), p. 183.

38. Herman Kahn, The Next 200 Years (New York: William Morrow, 1976), p. 64.

39. Preston Cloud, "Mineral Resources in Fact and Fancy," in Daly, Toward a Steady-State Economy, pp. 50-75.

40. For a discussion of net energy and the implications of entropy, see Clark, Energy for Survival, pp. xiv-xv, 7; Nicholas Georgescu-Roegen, "The Entropy Law and the Economic Problem," in Daly, Toward a Steady-State Economy, pp. 37-49.

41. Keeny, Nuclear Power Issues, pp. 71-94.

42. Division of Solar Energy, U.S. Energy Research and Development Administration, Solar Energy in America's Future: A Preliminary Assessment, 2d ed. (Washington, D.C., March 1977), p. 71.

43. For an elaboration of possible solar energy applications, see Clark, Energy for Survival, pp. 355-512.

44. Lee Erickson, "A Review of Forecasts For U.S. Energy Consumption in 1980 and 2000," in Barry Commoner et al., eds., Energy and Human Welfare—A Critical Analysis (New York: Macmillan, 1975), pp. 5-6.

45. The following studies are a representative sample of the projections of future U.S. energy supply and demand: Federal Energy Administration, Project Independence Blueprint (Washington, D.C.: U.S. Government Printing Office, 1974); Project Independence, U.S. and World Energy Outlook Through 1990 (Washington, D.C.: U.S. Government Printing Office, 1977); Department of Commerce,

Forecast of Likely U.S. Energy Supply/Demand Balances for 1985 and 2000 and Implications for U.S. Energy Policy (Springfield, Va.: NTIS, 1977) (NTIS PB 266 240); Executive Office of the President, The National Energy Plan (Washington, D.C.: U.S. Government Printing Office, 1977); H. Franssen, U.S. Energy Demand and Supply, 1976-1985, Limited Options, Unlimited Constraints (Washington, D.C.: U.S. Government Printing Office, 1977); Projections of Energy Supply and Demand and Their Impacts (Washington, D.C.: Department of Energy, 1978).

46. For a discussion of the impact of legislative reforms and lack of party cohesion on energy legislation, see Bruce I. Oppenheimer, "Policy Effects of House Reform: Energy Legislation, 1975-1977," paper presented at 1978 Annual Meeting of the American Political Science Association, New York.

47. Walter Goldstein, "The Political Failure of U.S. Energy Policy," Bulletin of Atomic Scientists, November 1978, p. 19.

48. For extensive presentations of these two perspectives, see Alternative Long-Range Energy Strategies, Joint Hearings before the Select Committee on Small Business and the Committee on Interior and Insular Affairs, U.S. Senate, 94th Cong., 2d sess., December 9, 1976, and Additional Appendices, 1976 (Interior Committee Serial No. (94-47) (92-137).

49. John W. Simpson, President, Westinghouse Power Systems Company, in Congressional Record, March 8, 1971, p. E-1566.

50. Parisi, "Nuclear Power."

51. Newman and Day, The American Energy Consumer, p. 88.

52. Komanoff, "Doing Without Nuclear Power."

53. Amory B. Lovins, Soft Energy Paths (Cambridge, Mass.: Ballinger, 1977).

54. Charles B. Yulish, ed., Soft vs. Hard Energy Paths: 10 Critical Essays on Amory Lovins' "Energy Strategy: The Road Not Taken?" (New York: Charles Yulish Associates, 1977).

55. 42 U.S.C. 1857 (f).

56. Hobart Rowen, "Needed: An Energy Program That Really Hurts," Washington Post, June 16, 1977.

57. U.S. Department of State, Compendium of U.S. Multilateral Energy R & D Agreements (Washington, D.C., September 1976), p. 4.

58. U.S. Department of State, Compendium of U.S. Bi-lateral Energy R & D Agreements (Washington, D.C., September 1975).

59. Stockholm International Peace Research Institute, World Armaments and Disarmament: SIPRI Yearbook, 1977 (Cambridge, Mass.: MIT Press, 1977), pp. 38–39.

60. Robert L. Gallucci, "Fission in Atlantic Politics: Non-Proliferation, Nuclear Energy, and the Europeans," paper presented at the 1978 Annual Meeting of the International Studies Association, Washington, D.C., February 1978.

61. L. Barret and N. Waddel, "Cost of Air Pollution Damage: A Status Report," U.S. Environmental Protection Agency, February 1973.

62. Fowler, Energy and the Environment, p. 175.

63. Interagency Energy-Environment Research and Development Program, Accidents and Unscheduled Events Associated with Non-Nuclear Energy Resources and Technology (EPA-600/7-77-016) (Washington, D.C.: U.S. Environmental Protection Agency, February 1977), p. 95.

64. Sidney Wolf, "Liquified Natural Gas," Bulletin of Atomic Scientists, December 1978, pp. 20–25.

65. For a discussion of possible nuclear accidents, see Richard E. Webb, The Accident Hazards of Nuclear Power Plants (Amherst: University of Massachusetts Press, 1976).

66. For a presentation of the two general perspectives, see U.S. Nuclear Regulatory Commission, "Reactor Safety Study—An Assessment of Accident Risks in U.S. Commercial Power Plants" (Washington, D.C.: October 1975) (WASH-1400); and "The Risks of Nuclear Power Reactions" (Cambridge, Mass.: Union of Concerned Scientists, August 1977).

67. Mason Willrich and Theodore B. Taylor, Nuclear Theft: Risks and Safeguards (Cambridge, Mass.: Ballinger, 1974).

2

THE SOVIET UNION
Donald R. Kelley

THE PATTERN OF ENERGY USE

Viewed in historical perspective, the energy consumption of
the Soviet Union has been shaped both by the regime's single-minded
attention to rapid industrial growth, with primary emphasis on heavy
and defense industries at the expense of consumer production, and
by the predictable shift from reliance on coal and lignite toward the
greater utilization of oil and natural gas and the development of
nuclear energy. In the early years of Soviet industrialization, which
began in earnest with the first five-year plan in 1929, the growing
industries were fueled by the intensive development of the indigenous
fuel resources of European Russia; in part because of their ready
availability, and in part because the Soviet economy initially experienced a period of the disproportionate development of key sectors
that did not require the creation of a comprehensive and diversified
national energy system, little attention was given in the early years
either to the conservation of energy resources or to the evolution of
a balanced energy profile. While Soviet planners were unquestionably
aware of the future limits of the indigenous fuel supplies of the rapidly
industrializing European sector and of the potential for the development of far greater—and admittedly more costly—energy resources
east of the Urals, the pressures for short-term developmental payoffs
and hence the need for cheap and readily available energy seductively
drew their attention away from the long-term development of the more
extensive resources in West Siberia and beyond.

With the postwar reconstruction of the Soviet economy largely
completed by the 1950s, Soviet planners initiated two policies of long-
term significance to the future energy profile of the nation. First,
they began the gradual transformation of the pattern of domestic consumption, shifting toward a more diversified fuel mix increasingly

dependent upon oil and natural gas. Second, they began ambitious developmental programs to tap the fuel resources east of the Urals, initially in West Siberia and Soviet Central Asia, and then increasingly in the more remote and inhospitable reaches of East Siberia. Both policies were viewed as absolute necessities if the economy were to advance much beyond prewar levels; the shift to oil and gas was regarded as closely tied to the further technological modernization of industry, and the increasing reliance on non-European resources was seen as an open acknowledgment that existing sources in the European sector were thought to be close to exhaustion or economically unprofitable. Both commitments also entailed massive outlays of capital and manpower either to accomplish extensive conversions in industry and in heat and electric power generation or to locate and develop the new fuel supplies east of the Urals.

Patterns of Soviet fuel consumption from 1950 onward clearly show the impact of the shift to oil and gas (see Table 2.1). In 1950 and 1955, the production of coal and lignite constituted 59.1 and 60.1 percent of total fuel production (205.7 and 310.8 million metric tons of standard fuel equivalent, or s.f.e.); peat contributed another 4 percent of the national total, and imports exceeded exports by a substantial margin (9.1 versus 1.7 million metric tons s.f.e. in 1950 and 9.1 versus 5.8 in 1955). Beginning in 1960, the relative contribution of coal and lignite began to register a marked decline, dropping to 50.8 percent that year, and then to 40.6 percent in 1965, 33.8 percent in 1970, and 30.2 percent in 1975, even though in absolute figures the production of coal increased from 1960 to 1975 from 373.1 to 490.4 million metric tons s.f.e. The contribution of peat also predictably slipped from 2.8 percent of the total national fuel supply in 1960 to 1 percent 15 years later. The import-export trend had also reversed by 1960, when exports numbered 14.7 million metric tons s.f.e., and imports dropped to 5.4 million metric tons s.f.e.; 15 years later the ratio remained roughly the same, with 30 million metric tons s.f.e. being exported while only 10.7 million metric tons s.f.e. entered the country.[1]

Equally dramatic was the rapid increase in the nation's oil production and consumption. In 1950, Soviet sources produced 54.2 million metric tons s.f.e. of domestic crude oil, or 15.6 percent of the total national energy consumption, which was supplemented by an additional 4 million metric tons s.f.e. of imported oil (more than double the export total of 1.9 million metric tons). By 1955, domestic production had climbed slightly to 19.6 percent of the total energy consumption, and exports now exceeded imports in the measure of 13.3 to 7.5 million metric tons s.f.e. Five years later domestic production had climbed to 28.8 percent of the national total, a figure that increased to 34.1 percent in 1965, 39.3 percent in 1970, and

TABLE 2.1

Soviet Energy Production and Trade, 1950-75
(in million metric tons s.f.e.)

	1950	1955	1960	1965	1970	1975
Coal, lignite, peat						
Production of coal and lignite	205.7	310.8	373.1	412.5	432.7	490.4
Production of peat	14.8	4.0	20.4	17.0	17.7	16.9
Exports	1.7	5.8	14.7	25.8	28.4	30.0
Imports	9.1	9.1	5.4	7.4	7.8	10.7
Oil						
Production of crude oil	54.2	101.2	211.4	346.4	502.5	701.8
Production of shale oil	1.3	3.3	4.8	7.4	8.8	11.7
Exports (incl. petroleum products)	1.9	13.3	53.7	102.3	151.2	204.6
Imports	4.0	7.5	7.4	2.8	6.8	9.9
Natural gas						
Production	7.3	11.4	54.4	149.8	233.5	345.7
Exports	—	—	—	—	3.9	23.3
Imports	0.06	0.2	0.3	0.5	4.2	14.9
Primary electric power						
Hydropower	1.56	2.84	6.25	10.0	15.3	15.5
Nuclear power	—	—	—	0.5	1.2	7.0
Exports from all sources	—	—	—	0.2	0.7	1.4
Imports	—	—	—	—	—	0.2
Fuel wood						
Officially sold	27.9	32.4	28.7	33.5	26.6	23.8
Privately collected (estimated)	35.0	35.0	35.0	40.0	40.0	40.0

Source: Adapted from Herbert Block, "Energy Syndrome, Soviet Version," in Annual Review of Energy, Vol. 2, ed. Jack M. Hollander (Palo Alto, Calif.: Annual Reviews, 1977), pp. 462-63.

43.3 percent in 1975, representing 701.8 million metric tons s.f.e. Over the same period from 1960 to 1975, exports of crude oil and petroleum products had increased from 53.7 to 204.6 million metric tons s.f.e., while imports had inched up from 7.4 to a mere 9.9 million metric tons. The production of crude oil from shale remained fairly constant for the entire postwar period, contributing only slightly less than 1 percent of the national total at any point in time.[2]

The production of natural gas also registered sharp increases over the same two and one-half decades. In 1950, only 7.3 million metric tons s.f.e. were produced, representing 2.1 percent of the nation's total energy production. By 1955, output had jumped to 11.4 million metric tons s.f.e. (but still only 2.2 percent of the national total), and by 1960, production had increased to 54.4 million metric tons s.f.e., or 7.4 percent of total energy production. Within the next five-year period, however, production increased threefold to 149.8 million metric tons s.f.e. (14.7 percent), and then grew to 233.5 million metric tons (18.3 percent) in 1970 and 345.7 million metric tons (19.5 percent) in 1975. Until 1970, imports and exports remained negligible; but from 1970 to 1975, exports rose from 3.9 to 23.3 million metric tons s.f.e., while imports increased from 4.2 to 14.9 million tons, with the former shipped largely to East European allies or other European consumers and the latter imported from Iran and Afghanistan to satisfy energy shortfalls in the Caucasus and the southernmost territories of European Russia.[3]

Hydroelectric power has also proven attractive to Soviet planners, who were motivated both by the pragmatic desire to harness untapped rivers as a source of electrical power and by the grandiose revolutionary symbolism of commanding nature itself in the interest of building socialism. Despite the public fanfare attached to the building of gigantic hydroelectric dams—some even commemorated in poetry, as was the case at Bratsk—the actual contribution of such facilities has remained relatively limited; in 1950 they contributed only 0.4 percent of the national fuel balance, a figure that would inch upward to only 1.2 percent in 1970, and then drop once again to 0.9 percent five years later. Moreover, further potential for hydroelectric development is severely limited; the major waterways in European Russia have already been utilized virtually to full capacity, and development in the north or in Siberia, while feasible, entails both high cost and construction difficulties.[4]

Nuclear power has also contributed only a small share of the total national energy package, increasing from 0.1 to 0.4 percent of the total from 1970 to 1975. While significantly greater development is slated for the next several decades, especially in the European areas where the energy shortfall is the greatest, the contribution of nuclear energy will still trail behind more conventional sources, particularly with the rapid development of West Siberian oil and gas reserves and the new-found emphasis on the greater utilization of coal.[5]

The use of wood as an industrial and domestic fuel has predictably declined over the last two and one-half decades, although it remains important, especially for domestic use, in some of the more remote regions. In 1950, both the officially sold and estimated privately collected stock of firewood comprised 18.1 percent of the total fuel balance, a figure that would drop to 13 percent in 1955, 8.7 percent in 1960, 7.2 percent in 1965, 5.2 percent in 1970, and 3.9 percent in 1975. Despite this precipitous decline, it should be pointed out that as of 1975, in terms of the total national fuel balance, firewood consumption still exceeded the contribution of nuclear power by almost a factor of ten and hydroelectric power by just over a factor of four.[6]

Overall statistics on national energy production and consumption conceal a considerable regional imbalance between the energy-rich territories of West and East Siberia and high level of consumption in the more densely populated and industrially developed regions of European Russia. Almost 80 percent of all Soviet energy is consumed in the European sector, including the Urals and the Caucasus, with some 65 percent utilized west of the Urals alone. Even with new initiatives to shift the most energy-intensive industries to the east into the energy-producing territories, the European sector is still expected to consume at least 70 percent of total national energy output 15 years in the future. For years, this region has experienced increasingly severe energy shortfalls in terms of indigenous production. It contains no more than 12 percent of the nation's total energy resources, and by 1980 it is expected to provide no more than 60 percent of its own energy needs, a figure that will drop even further to 40 percent a decade later. Current programs to expand production of European resources—once regarded as too costly, to utilize less desirable local coal, and to accelerate nuclear development in the region—will do little to alleviate the overall problem. In the 1976-80 five-year plan, resources west of the Urals were to account for less than 10 percent of increased coal production, to register no gains whatsoever in terms of the output of natural gas, and to suffer a 5 to 10 percent decline in crude oil production. Virtually all of the yearly increment in energy consumption, which Western observers estimate to be approximately 5 percent per year through the 1980s, will have to be provided by the rapid development of the resources of Central Asia and West Siberia, with increasing reliance on the even more difficult to obtain resources of East Siberia as time progresses. The development of these resources—examined at length in the energy policy section below—will provide the key to the continued energy self-sufficiency of the Soviet Union and has become the focus of ambitious and far-reaching development programs (and no small degree of regional and interbureaucratic conflict) and of growing concern on the part of Soviet leaders that lagging exploration and construction, a lack of sophisticated technology, and overly optimistic projections concerning the size of untapped Siberian resources may cumulatively place the nation in the throes of a self-made energy crisis within the next decade. [7]

The actual energy consumption patterns in the Soviet Union provide an instructive profile of the contemporary economy. Industry ranks as the top consumer, taking 72.3 percent of all fossil fuel, 71.9 percent of electrical energy, and 71.9 percent of thermal energy in 1975, excluding consumption as raw materials. Within industry, the greatest consumption is by electric power stations and the iron and steel industry, with the extractive industries for coal, oil, and gas following in close order. Fuel expenditures for the latter category are expected to rise dramatically in the coming years because of the massive expenditures of energy required to develop the fuel industries of Siberia and the problem of transporting these energy resources to the European sector. Construction follows in second place, consuming 2.4 percent of all fossil fuel, 2.1 percent of electrical energy, and 3.1 percent of thermal energy. Agriculture takes 6.3, 5.3, and 1.9 percent of the same fuel resources, a slight increase over the previous decade because of increasing mechanization in the countryside. Transportation claims only 7.9 percent of the nation's fossil fuel and 7.2 percent of its electrical energy, a much smaller proportion than in most other industrialized nations because of the extensive development of public transportation and the relative scarcity of private autos (although demand for fuel for the latter has grown in recent years, frequently producing a profitable black market for gasoline in the major cities). Housing and consumer services consume 7.2 percent of all fossil fuels, 8.5 percent of electrical energy, and 15.7 percent of thermal energy. In the major and medium-sized cities, thermal energy is now largely supplied by centralized heat and power stations that provide both heat and electricity, thereby considerably increasing the efficiency of the energy expended. The poorly developed state of consumer services per se undoubtedly contributes to their limited consumption of energy compared with other industrialized nations, although demands in this area can be expected to grow in the coming years. [8]

DOMESTIC ENERGY POLICY

Soviet leaders have unambiguously endorsed a policy of maximum economic growth virtually since the beginning of forced-pace

industrialization in 1929. In the minds of Soviet political leaders
and economic planners—and unquestionably in the mind of the man in
the street himself—rapid industrial growth has always been envisioned
as the key to the growth of domestic economic strength and the protection of the socialist motherland. Whether the task was to rebuild
an economy virtually destroyed by two world wars and a prolonged
civil war or to narrow the gap between the USSR and the economic
accomplishments of the capitalist world, the answer was always the
same: growth for the sake of growth, for everything depends on it.
Even the recurring political struggles over the allocation of always
scarce resources between the production and consumer sectors have
taken place within a larger framework of mutual agreement on the
need for maximum overall growth, and the Western discussion of the
limits to growth animated by the Club of Rome's dire projections has
attracted only fleeting attention even from Soviet environmentalists
and scientists, to say nothing of the rejection and scorn of official
commentators, who label such problems as indigenous only to the
capitalist world.[9]

Despite their seemingly universal consensus that maximum
growth is a desirable national goal and their wide-ranging ability,
through centralized planning, to manipulate economic activity along
desired paths, Soviet planners have in recent years become increasingly concerned about the prospects for continued growth. In part their
concern stems from a long-term slowing of the growth rate, common
as any economy reaches a level of industrial maturity; but also in part
it is derived from well-founded concern that a number of approaching
economic and social problems will retard the growth rate in the 1980s.
Having shifted from "extensive" to "intensive" forms of development,
the Soviet economy will become increasingly dependent for further
growth on the skillful mastery of high-level technology, whether
domestically developed or imported from the West. Yet in this area
the performance record to date has been woefully inadequate; Soviet
research and development efforts lag behind their Western counterparts in virtually all important growth-producing areas, and industry
has shown great reluctance to introduce even modest technological
advances into actual production. While Western technology has provided some assistance, the results have been disappointing from the
Soviet perspective, in part for political reasons but also in large part
because the intended spillover effect has been minimal. Manpower
problems will also grow more intense in the coming decade, confronting Soviet leaders with a shrinking labor force and the need to stress
the importance of increased productivity from each worker, a measure
itself reliant in large part on the technological upgrading of industry.
While programs to encourage extended work years beyond normal re-
tirement or to foster full or part-time employment among women

(already a disproportionately large segment of the work force) will provide some relief, the problem is expected to emerge as a real threat to further growth.

In numerical terms, the growth rate of the Soviet GNP has dropped considerably in the postwar years. During the period from 1955 to 1969, it averaged only 5.7 percent annually, a sizable decline from the high level recorded before the war or in the period of war-time reconstruction. During the early 1970s, the GNP fluctuated erratically, averaging somewhere between 4 and 5 percent annually, depending upon the method of calculation. Soviet and optimistic Western commentators project a similar growth rate for at least the 1980s, although more pessimistic outside observers predict a de-creasing rate that could bottom out at around 2 percent by the end of the decade if domestic and international conditions prove unfavorable.[10]

Even more significantly, the energy/GNP ratio, which had begun to decline in the late 1960s and early 1970s to register higher GNP gains per unit of energy input, has now once again begun to inch upward. In 1970 it reached a ten-year low of 2.94 kilograms of fuel input for each ruble of GNP. The following year it rose to 3.10 kilo-grams for each ruble, and then registered 3.02, 3.02, and 3.10 for the years from 1973 to 1975.[11] In real terms, this means that the aggregate growth of energy consumption, which had slowed in the 1960s in response to the shift to oil and gas and other technological improvements in energy utilization, has also begun to climb. From 1970 to 1975, the USSR averaged a 5.1 percent increase in energy consumption annually, and the projections of the five-year plan for 1976-80 suggest that the rate of growth will be roughly the same. More stringent conservation efforts, begun within the past few years, show some hope of ameliorating this trend, especially in the critical area of oil consumption, where conservation measures have reduced annual increases in consumption from 8.3 percent in 1972 to a low of 3.9 percent in 1977.[12]

While projections of Soviet energy needs through the next sev-eral decades will be discussed below, it must be noted that increasing concern has been voiced both by Soviet planners and Western observers that future growth rates for the economy as a whole may be closely tied to perhaps overly optimistic projections concerning the produc-tion of domestic energy sources. In part because the development of the more plentiful resources of West and East Siberia has met with delays, and in part because short-term but highly visible and annoy-ing energy shortfalls, especially in the winter months, have height-ened Soviet consciousness of the potential for an energy crunch, Soviet political leaders and economic planners have shown greater concern about instituting meaningful conservation programs and mo-bilizing even greater efforts to increase domestic production, es-pecially in West Siberia.

Little of this new concern can be attributed to the impact of the 1973 energy crisis per se. The 1973 embargo itself had no perceptible impact upon the Soviet energy scene. To the extent there was any initial reaction at all, it was principally at the rhetorical and ideological level—a combination of "that's what you get for siding with the Israelis" and the predictable lecture about the inability of a capitalist system to plan for and regulate the production of basic commodities such as fuel. It did not take long, however, for Soviet leaders to perceive the new market and hard currency earning potential for fuel exports from their own seemingly plentiful stocks. In the short run, Soviet foreign trade officials began to market existing stockpiles abroad, earning capitalist-like windfall profits in the process. Negotiations were also begun concerning long-term commitments with other major industrial consumers, some envisioning complex arrangements for the utilization of comparatively advanced Western technology in areas such as gas liquification in exchange for fuel and others entailing extensive investments of foreign capital to develop as yet untapped energy sources in East Siberia. The initial flush of optimism on the part both of Soviet officials, who saw such measures as a quick fix for their own technological limitations and a source of hard currency, and of energy-hungry Western and Japanese officials, who viewed the Soviet initiatives as a possible, albeit expensive, alternative to Middle Eastern suppliers, quickly turned to a sense of growing pessimism. While Soviet energy exports did grow appreciably after the 1973 embargo, coming to constitute the largest hard currency earning item in the foreign trade balance, the earlier optimistic projections were not met. In part as a consequence of the political complexities of the trade and development commitments (see below), and in part because Soviet planners increasingly perceived the real and potential energy shortfalls within their own nation and acknowledged their continuing commitment to supply energy (although at more profitable rates) to their East European allies, expectations were lowered across the board, and attention shifted to short- and long-term measures to supply the Soviet Union itself with adequate energy supplies over the next several decades.[13]

That the USSR was itself beginning to face the possibility of serious energy shortfalls was only slowly realized by Soviet officials, who had been lulled into a sense of complacency by the enormous energy potential of domestic resources and the continuing self-deception that a centrally planned socialist system was immune to such events. That self-deception came to a painful end in the early and mid-1970s as a growing series of energy shortages in major industrial regions such as Sverdlovsk, Chelyabinsk, and Perm, and then serious shortages of fuel for domestic consumption during the winter, culminating in a serious heating crisis in most major cities during the winter of 1978-79, pressed upon the consciousness of Soviet leaders. Also no less important was the apparent slowing of a number of ambitious efforts to tap new and less accessible energy sources in West Siberia; as the Soviet press reported with increasing frequency in the early and mid-1970s instances of construction delays at new mine or well sites, and as it became apparent that efforts to force greater production from existing sources would be only marginally successful, Soviet leaders acknowledged that the rapidly growing energy demands of the economy and the limited potential for increasing domestic production were converging toward a potential crisis.[14]

The result of this realization—and of the political implications of accepting the potential consequences of present policy and the costs involved in changing it for an inherently conservative regime, delicately balanced in terms of major institutional power blocs, and on the verge of the Brezhnev succession—was to produce a mélange of short-term policy initiatives and reassessments of long-term options. Into this growing debate have entered not only the advocates of various technical options but also (and predictably in the context of Soviet politics) the advocates of various institutional and regional interests. Reviewing recent commentary among Soviet energy experts, Leslie Dienes observes that "the controversy and lobbying seem more intense than at any other time in the post-Stalin years."[15] This is hardly surprising, both in light of the political style of the Brezhnev regime and its unwillingness to enforce its will on a host of competing bureaucratic and regional power centers, thereby tacitly encouraging an intensification of the game of bureaucratic politics, and in light of the stakes involved in the conflict. At issue, in the final analysis, are nothing less than the profile of industrial development over the next several critical decades, when the Soviet economy is expected to pass through a phase of technological modernization similar to what Western analysts have termed the "second industrial revolution," and the fate of major institutional and regional interests always jealously defensive of their bailiwicks and prerogatives.

Before any assessment can be offered either of short-term conservation and energy production programs or of plans for long-term development, some attempt must be made to project Soviet energy needs over the next several decades. The task is far from easy. The growth rate of the economy is itself uncertain, as are trends concerning the energy intensiveness of the most rapid growth sectors. Assuming that overall growth until 1990 will remain in the 4 to 5 percent range, and further assuming that shifts toward greater energy intensiveness and the high fuel demands necessary to tap new energy sources in remote regions will only marginally affect overall national needs, one can estimate that the Soviet Union will require somewhere between 2.9 to 3 billion tons of standard fuel equivalent by 1990 for domestic purposes alone. Export requirements will add another 280 million tons s.f.e., assuming that gas exports will continue to grow rapidly, coal exports will hold at the present level or expand only slowly, and that oil exports to Eastern Europe will remain roughly at their present level and shipments to noncommunist nations will remain in the range of 20 to 25 million tons a year. As Dienes points out, these totals are far lower than earlier Soviet and Western estimates, signaling both the recognition of the potential for energy shortfalls and the tacit acknowledgment that Siberian resources will be brought on-line more slowly than originally anticipated.[16]

The most attractive short-term options available to Soviet planners to ease growing fuel shortages, particularly in the European sector, lie in a judicious combination of better conservation efforts and the greater utilization of lower-grade fuels, with immediate attention to the further exploitation of coal mined in the Urals or the European territories from deposits that were once thought to be too close to exhaustion, too expensive, or too poor in quality for further production. Official interest in such a policy first came to light in 1974 when high-ranking officials of the State Planning Committee (Gosplan) recommended that lower-grade local fuels be utilized whenever possible. Noting that the inferior coal and lignite of the Moscow basin and the Urals and the extensive peat and oil shale deposits found elsewhere in the European sector were of little use to industry or transportation, planners argued that they were ideal for consumption in thermal generating plants. A new interpretation emerged in the discussion to explain earlier conversions to oil and gas in such facilities: such changeovers had occurred not for industrial or environmental reasons, but rather because of the retarded (or neglected) development of the coal industry itself, a shortcoming that could be easily overcome with increased investment in already proven technology. Now potentially scarce gas and petroleum products, so the argument went, "must first and foremost be used in industry as valuable raw materials or as a precious fuel that makes it possible to employ highly efficient technological processes."[17] Support for the exploitation of European coal reserves grew throughout 1974, with laudatory comments spreading from the pages of economic journals such as Planovoe Khozyaistvo (Planned Economy) to the Communist Party's theoretical journal Kommunist, signaling that important political backing now stood behind the pro-coal forces. The tenth five-year plan for the period 1976 to 1980 also reflected these new priorities, although it did not make the argument for coal with quite the force of earlier pronouncements. It did, however, clearly call for "the wider use of cheap solid fuel in the production of electric power," although accelerated development of oil and gas reserves, nuclear power, and hydroelectric generating capacity was also promised.[18]

The new emphasis on the production of low-grade local coal in the European sector is not without its problems. Even setting aside the potential deterioration of air quality in the major cities—a consideration that, after long delays, has recently become important to Soviet planners—both technical and political issues have arisen. At the purely technical level, there is considerable question about whether extensive conversions of gas- or oil-fired generating plans are possible; while the technology is theoretically available, the conversion costs would be enormous, especially in light of the long-term prospects that such coal-burning installations would eventually be forced to consume coal shipped from distant Siberian mines. The construction of new facilities that would utilize coal is much less problematical, although their environmental impact and long-term fuel sources remain issues of concern.

Other difficulties plague the further development of European coal resources. With few exceptions, most coal production in the region comes from deep shaft mines and narrow seams. The level of mechanization remains woefully inadequate, in terms of both the quality and quantity of available technology.

Even in those areas such as the Moscow basin or the more industrialized sections of the Ukraine, where one would expect the new pro-coal policy to be most vigorously carried out, complaints about inadequate machinery and support from higher officials have remained frequent, and production targets are often unmet. Economic problems have also beset the mining industry. Coal production remains unprofitable in the European sector, and unit costs are far higher than for other fuel sources. Even some of the better-grade Siberian coals prove to be economically competitive despite the transportation costs, and the economic indexes will undoubtedly improve as the technology of long-distance power transmission is improved, permitting mine-mouth generating stations in the East to supply power directly to consumers in the European sector (although industrial growth in the East itself is expected to consume an increasing share of this energy). Thus while the long-term prospects for the development of coal as an important fuel source remain bright, its significance as a local fuel within the European sector, and especially its ability quickly to fill the gap between production and consumption, is questionable.[19]

The political dimensions of local coal and lignite development in European Russia cannot be overlooked. There is mounting evidence that local industrial and party officials see the pro-European coal commitment both as a new lease on life for a declining local industry—a "second wind," as one put it—and as yet another opportunity to assert their region's merit in the competitive battle for resources. It must be recalled that Soviet long-term developmental plans (and a good deal of current investments) have focused almost exclusively on

the further industrial growth and mineral riches of the non-European
territories to the east. While the European sector will unquestionably
remain important both as an economic and population center, the further improvement of its economy is seen more in terms of the modernization and increasing technological sophistication of existing industries, tasks that have proven difficult even under the best of circumstances and that do not provide the excitement—or the challenges
and motivation—of opening up and mastering the wilderness of Siberia.
One does not have to probe too deeply into the open regional competition for resources or the implicit psychology of those involved to
recognize that the spokesmen of the European sector are waging a
struggle not only on the usual institutional and bureaucratic battle-
grounds but also, in a sense, against time, as national priorities
inevitably shift eastward. To be sure, the battle has not been joined
in either the terms or to the degree that it was a decade ago when the
debate was about whether to develop Siberian resources; that question
is now irrevocably resolved in the affirmative, and the issues at
hand are the pace of such development (and thus implicitly the extent
to which resources must be shifted from other economic goals and
regions) and the extent to which the European territories will become
increasingly dependent on Eastern suppliers. In political terms, the
industrial leaders and regional party secretaries in the European
sector who have joined in the discussion to press for local advantages
already possess considerable political clout. Strongly represented
in higher party circles, including both the central party bureaucracy
and the Central Committee itself, these individuals are well situated
to press their case. Moreover, the real possibility of open political
conflict over the selection of a long-term successor to Brezhnev at
least tentatively opens the door to raising the regional issue once
again in highly political form; and while it may be objectively clear
that the future of Soviet energy production lies in the less densely
settled and marginally industrialized East, the political reality is
that the majority of the important institutional, bureaucratic, and
regional vested interests whose interaction will shape the succession
have their principal power bases in European Russia.

Greater emphasis on energy conservation programs has also
been seen as another important stop-gap measure, and Soviet leaders
have shown increasing recognition of the long-term need to curb
wasteful consumption. Any comprehension of both the energy-related
and environmental aspects of such programs must begin with an understanding of the very different nature of conservation problems in the
USSR. Unlike most other highly developed industrial states, in the
Soviet Union the efforts of the average citizen-consumer can account
for only marginal energy savings even under the best of conditions.
The typical consumer, in fact, has comparatively little direct control

over the most energy-consuming aspects of his or her life, especially in the major cities. The use of public transportation is virtually an economic necessity for most residents, and little control can be exercised over the heating of residential facilities and the workplace, which are supplied from centralized thermal generating plants. To be sure, these measures are themselves important and, on the whole, successful conservation efforts, but they involve few direct decisions on the part of the consumer. Yet paradoxically the thrust of the public conservation campaigns has until recently been directed at the individual consumer rather than at the considerably more wasteful practices of industry and transportation. At this level, conservation programs have had notable impact; Russians do seem conscious of the small things: turning out lights or doing their best to fill cracks around windows in the winter.[20]

In recent years, as the seriousness of energy shortfalls has become more apparent, attention has shifted to the more critical problem of energy waste by industry, transportation, and fuel producers themselves. The 1976-80 five-year plan established norms for the reduction of fuel consumption; furnace fuel usage was to be cut by 3 to 4 percent, electrical and thermal power by 5 percent, and gasoline and diesel fuel for transportation by 8 percent.[21] Beginning in 1977, even more stringent demands were heard concerning waste in industry, and the media began an extensive campaign to uncover abuses. Judging from the continued references to such shortcomings over the next several years, attempts to limit wasteful consumption in industry have been less than successful, although efforts to recover thermal energy for other uses have been widely attempted. The transportation industry also came under closer scrutiny at this time, especially in terms of the considerable loss of coal and lignite that occurs in shipment. The fuel industry itself has also been reprimanded for its wasteful practices. For economic as well as technical reasons, significant quantities of fuel are left untapped; as late as 1978, Soviet commentators reported that as much as 30 percent or more of the coal, 65 to 70 percent of the petroleum, and inestimable quantities of casinghead gas were left in the ground or simply burned off at existing mine and well sites, resulting in the loss of billions of rubles.[22] While efforts have been undertaken to increase these yields or to harness wasted secondary fuels, the programs have not moved much beyond the experimental or test-site stage.

At the root of many of these problems is a basic feature of the Soviet economy itself: both the pricing and incentive systems have done little until recently to foster conservation. Regarding the former, prices for all commodities are set by central planners to reflect what are regarded as national priorities; the actual prices set for fuel supplies do not necessarily reflect either their scarcity or production costs. While in theory such a system could be manipulated to set fuel prices sufficiently high to bring about conservation measures, in practice this has never effectively been done. Although energy prices have been increased over the last few years, they still do not represent true scarcity values. More recent efforts directed at industry have utilized fines for excessive consumption, the threat of maximum-use quotas, or outright cutoffs rather than more subtle pricing levers. Further diminishing the economic impact of these measures have been the traditional priorities and success criteria operative in industry. While whole industries and individual factories are theoretically responsible for fulfilling an assortment of norms dealing with all aspects of production, knowledgeable production officials have always understood that their overall performance would be evaluated in terms of their ability to meet certain critical criteria such as gross output or profitability. In either case, adherence to energy-consumption norms did not bulk large in the evaluation of overall performance. Moreover, other potentially conservation-relevant measures such as better insulation or the auxiliary use of thermal energy from industrial processes were classified as "non-productive" investments and given scant attention until recently, although they are now being stressed.

Economic constraints have also affected the performance of the energy-producing industries. The cost structure for an individual mine or well site is not related to the real scarcity value of the resources or to actual production costs. In the past, financial pressures resulting from this situation have led to the premature abandoning of still potentially productive sites. Recent increases in fuel prices and the rental fees imposed on the extractive industries were designed in part to rectify the situation, but it is as yet too early to pass final judgment on their effectiveness.[23]

Apparently the existence of conflicting legal norms concerning energy usage in industry and the decentralization of enforcement powers have also contributed to a lessening of the impact of well-intended conservation measures. Speaking before a joint meeting of the Standing Committee on Industry of the two houses of the Supreme Soviet in the summer of 1977, energy officials complained about conflicting regulations and out-of-date technological indexes. The continued decentralization of responsibility for conservation efforts concerning oil, petroleum products, and coal also drew their fire, as did the lax enforcement of existing norms. New regulations were enacted by the Supreme Soviet the following month, but their obvious intent was to instruct both industry and local officials to take coordinated measures to prepare for the winter of 1979.[24] By the spring of 1978, calls for conservation measures had grown more strident, and factory and local party committees were summoned to increase their supervision of energy use and to intervene to correct blatant abuses.

Uncommonly cold, even by Russian standards, the winter of 1978-79 proved to be a particularly sharp test of both domestic production and conservation efforts. A combination of lagging production of domestic coal and oil and a reduction of gas supplies from Iran, where political turmoil disrupted production, resulted in major energy shortfalls in the major cities of European Russia as well as the south Ukraine and the Caucasus. Apparently fuel supplied for domestic consumption was the first to suffer, leading the Moscow correspondent of the New York Times to describe the city in January as a place "where everything freezes but the rumors."[25] Efforts were made to fill the gap with more readily available fuels and to force greater conservation in industry, but apparently with only partial success. Despite official attempts to play down the significance of the shortages and blame the situation on the uncommonly cold winter, it is obvious that Soviet leaders were shaken by the experience; local party committees apparently assumed the principal burden of dealing with the shortages as best they could as well as the task of bearing the brunt of public discontent.[26]

Undoubtedly as a consequence of the severe shortages of the previous winter, Soviet officials issued in June 1979 new and much strengthened directives on fuel production and conservation for the coming winter. While such ritual exhortations to prepare for the winter have always been offered, the 1979 pronouncements represented a significant up-grading of concern. The overall tone of the joint resolution by the Communist Party Central Committee and the Council of Ministers was far more strident than before, and the measures required of both producers and consumers were far more specific and demanding. The People's Control Commission was instructed to intensify its inspection efforts, the media were told to conduct an extensive campaign to educate the public about the problem and suggest conservation efforts, and the party was enjoined at all levels to take the situation under "its unremitting control."[27]

In addition to greater emphasis on the development of coal resources and conservation efforts, Soviet officials have also accelerated the development of their most productive oil-producing areas, most importantly the Samotlor and nearby fields in Tyumen district. Because of the rapid depletion of other fields and disappointing results from the search for new large deposits to the east, attention has been focused in desperation on the better-known fields of West Siberia, which will unquestionably maintain their lead in production well into the 1980s. Western specialists suggest, however, that the Samotlor region's production will peak much earlier—or may have already peaked, if the more pessimistic assessments of the CIA are accepted at face value—despite the accelerated pace of sinking both production and wildcat exploratory wells and the disappointing attempts to develop nearby sites and improve recovery techniques.[28]

While a continued commitment to conservation measures is now spoken of as an important element in maintaining energy self-sufficiency, long-term future developments will be dominated by increased coal production, especially in West and East Siberia, increasing reliance on Siberian oil and gas, and greater stress on nuclear and hydroelectric power. The prospects seem bright, even allowing for the possibility of an energy squeeze in the late 1980s or 1990s because of lagging energy growth rates.

In addition to viewing the production of lower-grade local coal and lignite as a stop-gap partial solution to European energy short-falls, Soviet planners envision extensive development of the nation's estimated 6.8 trillion tons of coal as an important source of future energy supplies. Increasingly the center of production is to shift eastward; the 1976-80 plan called for only marginal growth in the production of the Donets basin in the Ukraine (up from 221 to 231-233 million tons) but for much larger increases from the Karaganda basin in Kazakhstan (from 92 to 127 million tons), the Kuznets basin in West Siberia (from 137 to 161 million tons), and the more distant Kansk-Achinsk basin in East Siberia (more modestly up from 25 to 40 million tons). Overall coal production was predicted to rise to somewhere between 790 and 810 million tons by 1980, a figure slightly lower than earlier official estimates. Such growth envisioned an annual increment of somewhere between 2.4 to 2.9 percent over the five-year period, a level that early reports suggest will not be met. From 1975 to 1976, coal production increased only slightly over 10,000 tons, or 1.49 percent, and in the following year (the first of the current five-year plan and the only one for which figures are available), the growth rate actually slipped to 1.46 percent. Adding to the problem is the progressively lower calorific value of increased production. Over the last two decades, the standard fuel equivalent of the raw tonnage mined has dropped from 80 to 70 percent, and further declines are inevitable because of the relatively poorer quality of Siberian coal.

Over the next decade, Soviet planners will probably concentrate on the further development of the Ekibastuz fields, north of Lake Balkhash, the production of high-ash coal from which could reach 100 million tons by 1990, and on surface mining in the Kuznets basin, where costs are constantly rising. In the late 1980s or early 1990s, the center of production would inevitably shift further eastward to the Kansk-Achinsk basin, which could produce enormous quantities of low-quality fuel at relatively cheap costs. The problem is that coal from the latter is not economically competitive west of the Urals, does not lend itself to enrichment and briquetting, and is prone to self-combustion. Moreover, efforts to develop alternative arrangements to consume Kansk-Achinsk coal in other regions, freeing their higher-quality fuel for use in the more densely populated areas, or to develop mine-mouth thermal electric power stations or coal slurries have thus far fallen victim to repeated construction delays or the low level of Soviet technology in these areas. Research on long-distance direct current lines has now become a high priority, as has the development of slurry lines, but their profitable commercial utilization is years in the future. Moreover, the development of the Kansk-Achinsk complex is apparently far behind schedule; a _Pravda_ correspondent who visited the region in March 1979 noted that all that existed of the first power generating station, scheduled to go on line in four or five years, was a sign at the construction site reading "Site of the future Berezovskoe State Regional Power Station No. 1," and that the understaffed and ill-equipped local construction agency would need at least ten years to complete the limited number of projects already begun, to say nothing of those still on the drawing board.[29]

Of all of the potential fuel supplies available, oil presents the most immediate problems for Soviet planners. While oil production has expanded rapidly over the last decade, in response to both the greater utilization of crude oil as an industrial raw material and the growing need for oil products as high-energy fuels, the oil industry has followed the short-sighted practice of sacrificing exploratory drilling for the sake of maximizing yearly production. Although total reserves for the nation as a whole are estimated at 175 billion tons of standard fuel equivalent, the figures for explored and proven reserves drop to 20 billion and 4.9 million tons, giving the Soviet oil industry a highly unfavorable ratio of proven reserves to output of 9.5:1.0. This situation is further complicated by the shift of oil production to the east, where prospecting and test drilling are far more costly and the required lead time far longer.

Given both the approaching depletion of the existing production centers west of the Urals and the continuing controversy about the future prospects of West Siberian sites in the Tyumen district, there is real concern that oil shortages may soon be a reality. A pessimistic assessment issued by the U.S. Central Intelligence Agency in 1977 indicated that by the mid-1980s oil production would fall to a point where it critically endangered the growth rate of the domestic economy and virtually eliminated hard-currency exports to the West and politically important shipments to East European allies, although other Western observers have offered a more sanguine view. Official Soviet plans have called for crude oil production to reach from 620 to 640 million tons s.f.e. by 1980, a figure that most analysts thought overly optimistic, especially in light of the rather substantial under-fulfillment of the growth plan in 1975. Since that time, however, the growth rates for 1976 and 1977 have risen once again to 5.0 and 5.8 percent, respectively; assuming that the annual increment can be held to at least 5 percent, production in 1980 should reach the 632 million metric ton level, safely within planned guidelines. Much of this increment is being purchased through a combination of neglecting prospecting activities and the development of deposits outside of West Siberia and of draining the Samotlor and other Tyumen district sites with all possible haste. A substantial down-turn in production from these West Siberian sources—an event that may already have begun or is certainly inevitable within the next five years barring new discoveries or dramatically improved recovery technology—would significantly worsen the overall national fuel situation. CIA estimates hold that such a decrease has already begun, with total oil production peaking in April 1979 at 11.3 million barrels a day, and even the initially more optimistic Western commentators such as the prestigious Oil and Gas Journal have noted that annual production increases have started downward once again, with the 1978 increment the lowest since the 1960s and the projected 1979 figures lagging even further behind. While Soviet planners have themselves shown greater realism concerning production figures in recent years—the lowered 1979 goals are now below initial optimistic estimates—they nonetheless express the view that accelerated production from West Siberia and greater reliance on rapid increases in the production of natural gas resources will meet overall national needs.

Even if the accelerated pace of exploration in other regions such as East Siberia discovers sizable new oil fields—and the results thus far are not encouraging—meaningful production is not likely before the late 1980s at the earliest. Thus the greatest immediate pressure for increased output will continue to fall on West Siberian sources such as Samotlor and surrounding fields, where the increased output of recent years has apparently been purchased at the expense of considerable damage in the form of dropping well pressure and water incursion. Until other options become available—and perhaps until the political issue of regional priorities is settled—there seems to be no other alternative; greater conservation efforts will yield some effect, as will conversions from oil to other fuels for industrial and domestic purposes, but the end result may well simply be to postpone the projected oil shortages until the 1990s.[30]

Technological problems and construction delays have also taken their toll of efforts to increase oil production. Soviet drilling technology has proven inadequate to the task of sinking deep wells, and domestically developed turbodrills have reached their maximum potential. Robert W. Campbell attributes the declining rig productivity

for the industry as a whole mainly to such technical problems, even
in areas of relatively shallow wells such as Samotlor. While better
foreign technology could provide a quick solution to this dilemma,
Soviet purchases have been subject to Western political restrictions
in a number of cases. Soviet offshore drilling rigs are regarded as
even more primitive by Western standards, and while adequate for
the shallow water drilling in the Caspian, they are not up to the task
of exploratory drilling in much deeper water where potential offshore
deposits are thought to exist. Recent purchases of foreign-built jack-
up rigs and pumping equipment, plus greater emphasis on improving
Soviet-built equipment, will certainly improve the situation, but it
will be years before offshore operations contribute significantly to
meeting the nation's energy needs. Pipeline technology historically
has been a major problem area, although the Soviets have recently
launched a major effort both to lay new lines and improve the technological level. Construction delays and other problems of the support infrastructure are cited with predictable regularity in the Soviet
press, and even seemingly well developed areas such as Samotlor
lack adequate road networks, frequently suffer from power shortages,
and must bear with inadequate housing and domestic services for their
workers.[31]

While the Soviet Union undoubtedly enjoys the world's largest
reserves of natural gas, with proven and probable reserves set at
22.4 trillion cubic meters and potential reserves possibly as high as
100 trillion cubic meters, their development over the last two decades
has been retarded both by the distant location of the most plentiful
supplies and continuing problems with inadequate extractive and pipeline technology. As with virtually all other important fuel sources,
the most extensive reserves lie to the east of the Urals; European
sources that provided the bulk of gas supplies in the 1950s and 1960s
are now receding in importance, and few additional important deposits
have been discovered in the region. The most important production
centers are now located in the West Siberian lowland and to the north,
with the most potentially productive future sites within the Arctic
Circle. Because of these difficult conditions, Soviet policy in the
1960s and early 1970s deemphasized these regions and sought instead
to increase production at the better-developed European sites and
further to the south in the Siberian lowland; the results were less than
hoped, and the rate of growth for the industry as a whole declined for
a decade and a half. Now attention has once again returned principally to the north Siberian fields and, to a lesser degree, to the more
limited fields of Central Asia. The rapid development of these deposits closely parallels the tacit Soviet acknowledgment that gas production will soon have to compensate for decreases in oil production,
both for domestic consumption and in the export market. Western

critics point out that recent production increases were the result of bringing extensive new fields on line and are thus likely to be the exception rather than the rule unless new deposits are found, which has not been the case despite renewed exploratory efforts. Greater attention has also been given to improving the nation's pipeline network, an important consideration concerning the remote location of the major production sites.

According to official Soviet estimates, natural gas production is to reach 400 to 435 billion cubic meters by 1980, thus increasing by from 6.7 to 8.5 percent annually during the 1976–80 five-year plan. While yearly shortfalls below planned quotas have been commonplace for the industry for over the last decade, the first few years of the current five-year plan show promising results: The overall growth rate from 1975 to 1976 was 10.9 percent, while from 1976 to 1977 it dropped to 7.8 percent. Assuming that it held somewhere in the 7 percent range, production by 1980 would rise to well over 400 billion cubic feet.

By 1990, however, the problem could grow far more serious. The depletion of older fields and the increasing difficulty of drilling producing wells further to the north would inevitably take its toll. Adding to the difficulty would be the enormous requirements for better technology for pipeline construction and possible gas liquification and the sheer physical commitment of manpower and resources to construct the drilling facilities and pipeline network. In the future, the real problems besetting the further development of the gas industry seem to lie more in these areas than in the inherent availability of potential gas supplies.[32]

At present only about 13 percent of the USSR's hydroelectric power potential is being utilized. However, further development of hydropower will entail high construction and transmission costs. Only 18 percent of the hydropower potential is located in the European sector, the greater portion of which has already been tapped. Only one-third of the 14,000 megawatts of new capacity to be constructed in the 1976–80 plan is located in this region, with 48 percent in Siberia and the Soviet Far East and 19 percent in Central Asia. The great hope lies in the extensive development of Siberian and Far Eastern waterways. In Siberia, the installed capacity in the Angara-Yenisey region is now just under 13,000 megawatts and is expected to reach nearly 20,000 megawatts by 1985 and possibly 60,000 by the end of the century. Other hydroelectric installations are planned for the region in connection with the construction of the second major rail link to the east, the Baikal-Amur mainline, and the extensive development of high-energy-consuming local industries. The hydropower potential of West Siberia is considerably more limited, as are the possibilities for development in the Far East. The actual contribution

that the hydropower potential of Siberia will make to national consumption patterns will depend directly on the success of experimental
work now underway on several long-distance direct current transmission lines and the further integration of a truly national power
grid.[33]

While Soviet planners have spoken optimistically in recent
years about the ability of nuclear power to make an important contribution to energy shortfalls in the European sector, overall nuclear
capacity is still quite limited, supplying only 0.4 percent of total
power in 1975. Even with accelerated development during the current
five-year plan, by 1980 it will account for only 6 percent of all total
electrical power production, compared with 2 percent in 1975. Total
generating capacity is slated to reach 20,000 megawatts by the end
of 1980, a goal that probably will not be met because of construction
delays and problems with on-line operations that have held operational
reactors below their rated capacity. Optimistic Soviet projections
of increasing generating capacity six or eight times by 1990 are also
unquestionably realistic, but even if successful, would constitute
only 4 to 5 percent of total energy output. Soviet scientists have also
devoted considerable attention to breeder reactors, submitting their
first commercial station to testing in 1972; results have been disappointing, and the reactor has operated at only 30 percent of rated
capacity. These difficulties have done little to dissuade the advocates
of even further nuclear development, who have argued for future
widespread dispersion of reactors (some small enough to supply energy to individual factories or residence complexes) throughout the
European sector as a partial answer to the region's energy shortfall.[34]

While increased attention has been given in recent years to
experimentation with esoteric energy sources such as wind, solar,
geothermal, and tidal power, their contribution to the overall energy
picture will be negligible in the near future. According to Soviet
estimates, the potential for wind power utilization in agriculture
alone could reduce the need for organic fuels by roughly 2.5 million
tons of s.f.e. each year, and solar energy for domestic heating in
the southern part of the country would conserve another 4.0 to 4.5
million tons. The development of these esoteric sources has been
retarded for a number of reasons, not the least of which has been
the tendency until recently to assign research and development tasks
to energy-related ministries whose principal responsibility was the
production of other fuel sources. In some instances, such as wind
power research, this dependency link has now been broken, and an
independent, although poorly funded, research institute has been
established outside of Moscow. In most other cases, the poor-cousin
status continues, occasionally prompting public complaints from the

advocates of solar or geothermal power that their projects are not being taken seriously.[35]

Wind-powered electrical generating facilities have been created for use in agriculture or for light household equipment in remote areas. Somewhat larger generating facilities are now planned for the Far North to supply small communities where great distances make the transportation of conventional fuels prohibitively expensive. By the end of the 1980s it is estimated that some 100 megawatts will be generated by wind power in this region, and smaller installations are planned for sparsely settled areas of Central Asia. Cost projections indicate that wind-generated power will be competitive with hydro-electric power and considerably less expensive than nuclear power, at least in the Far North.

Solar energy has received limited development in Soviet Central Asia, primarily for agricultural uses, the heating of domestic housing, the powering of household appliances (a special factory is to be estab-lished by 1980 exclusively devoted to their further development), and the elevation and desalination of ground water. Several research institutes are at work on future applications, which may now be taken more seriously because of the real prospect of decreased gas imports from Iran.

Geothermal power has also received limited attention in the Caucasus, some parts of the Urals, Central Asia, Siberia, and the Far East. Present applications are focused mainly on heating hot-houses for agricultural production, space heating, and, to a lesser extent, the generation of electricity. The brightest prospects for future development lie in those remote areas such as Siberia or the Far East where geothermal applications are economically competitive with other fuels.

While tidal power has been commercially harnessed since the commissioning of the 800-kilowatt Kislaya Bay generating station in 1969, there is little to indicate that the additional facilities planned or now under construction will have anything more than local impact. Two additional stations are under construction on the White Sea: the Lumbovka plant, which will generate 300 megawatts, and the Mezen River installation, which will eventually produce six gigawatts, and additional facilities are planned for the Sea of Okhotsk in the Far East.[36]

INTERNATIONAL ENERGY POLICY

As a major energy exporter, the Soviet Union is in a position to enjoy unique economic and political advantages, at least in the short run. Hard-currency earnings from fuel exports have become an important element in the balance-of-payments situation and figure critically in the reduction of Soviet indebtedness. Moreover, long-term trade arrangements linking energy exports to Western and Japanese consumers and the influx of foreign capital and superior technology were seen by Soviet planners in the mid-1970s as an easy solution to their own research and development problems, and although the most optimistic of these ventures have fallen victim to a combination of Soviet reluctance to make large export commitments in light of possible domestic shortfalls and foreign reservations about long-term commitments to ambitious developmental schemes or objections to other aspects of Soviet policy such as emigration, expanded exports to fuel-hungry nations will still remain an important element of the total trade balance. Soviet fuel supplies to Eastern Europe will also continue to have both economic and political significance. Even though Soviet trade officials began cautioning their East European allies to search for alternative sources for a part of their energy needs several years before the 1973 oil crisis, and even in light of the substantial increase in Soviet prices to East European consumers in the mid-1970s (although still below world market levels), there is little to indicate that the USSR will drastically reduce energy exports to the region unless absolutely forced to do so by domestic shortfalls. The political value of these trade ties is considerable, in terms of both Soviet-led attempts at bloc economic integration and outright political leverage in a potentially unstable buffer zone.[37]

Soviet trade in oil and refined petroleum products with its East European allies has grown rapidly since the war, rising from 2.2 to 63.3 million metric tons s.f.e. from 1955 to 1975. Paradoxically the Soviet Union also imports limited quantities of Middle Eastern oil for eventual shipment to its East European allies; from 1970 to 1973, the amount soared from 5 to 19 million tons s.f.e., only to drop once again in 1974 and 1975 to 6 and 9 million tons, respectively. Such imports have always been seen as the safety valve permitting the USSR to meet its short-term obligations in Eastern Europe and will doubtlessly decline in the future unless severe domestic oil shortages arise. The overall trend in Soviet oil exports to the region is clearly toward diminished Soviet responsibilities, with the projected increase for 1976-80 set at 3.4 percent annually as opposed to the much larger annual increase for slightly over 10 percent for 1970-75. Continued reductions are likely in the future, a result both of greater consumption at home and increasing shipments of Soviet gas to the region, thereby freeing oil products for use elsewhere.

Soviet oil exports to hard-currency areas have risen rapidly in recent years, growing from 3.2 million metric tons s.f.e. in 1955 to 35.5 million in 1965 and 52.6 million in 1975. Western Europe is the principal consumer, drawing just over 7 percent of its entire oil

consumption from the USSR. Significantly, however, oil exports to
the West are expected to decline during the present five-year plan,
dropping to 65 million tons s.f.e., considerably lower than the 1975
high of 75 million tons and just a shade above the 64 million ton figure
set in 1970. In terms of hard-currency earning potential, these
losses will be offset by the sizable increases planned for natural gas
exports, which will become the major growth area in the energy
export market for the next decade.[38]

Gas exports are expected to triple from 1975 to 1980, with supplies provided to East European allies increasing from 14 to 40 million tons s.f.e. and shipments to nonbloc nations rising even more
quickly from 9 to 31 million tons. Gas imports, which were scheduled
for only moderate growth, may stagnate or even decline because of
the changing situation in Iran. The most propitious hard-currency
export market for the near future remains Western Europe, which is
within reach of the pipeline network extending from Soviet production
areas into and through Eastern Europe and which can supply badly
needed technology and equipment in exchange for gas. More questionable are the once-touted plans to pipe Soviet gas either to Mur-
mansk or to Nakhodka for liquification and subsequent shipment to
the United States and Japan. While Soviet efforts to construct the
necessary pipelines have gone ahead, other developments have cast
considerable doubt on the projects' continued viability; both U.S. and
Japanese interests have had second thoughts about the enormous
investments needed to develop liquification facilities and the long
delays until the beginning of deliveries in the late 1980s, doubts that
have been strengthened by the prospects of shortages in the USSR
itself. Political considerations have also taken their toll, both in
terms of a progressive deterioration of Soviet-U.S. relations, especially in the area of trade, and a cooling of Soviet-Japanese friendship because of improved Sino-Japanese ties.

The Soviet Union also exports a small amount of coal. In 1975,
about 26 million tons s.f.e. were shipped abroad, two-thirds of which
went to other bloc nations and one-third to hard-currency areas. In
terms of earnings, exports to the West and Japan contributed 5 percent of the total hard-currency earnings for that year. The Japanese
have been particularly eager consumers and have entered into a long-
term agreement to assist Soviet authorities in the development of
coal reserves in Southern Yakutia, a more realistic project that has
survived the economic and political strains between Moscow and Tokyo.
Coal imports into the USSR, on the other hand, amount to no more
than just under 11 million tons s.f.e., with most in the form of high-
grade coal or coke from Poland. Imports have inched up only slightly
over the last decade, however, and are unlikely to grow appreciably
in the future.

The Soviet Union also exports about 1 percent of its electrical output, or about 11.9 billion kilowatt hours in 1975. Principal consumers are Finland and the bloc nations of Hungary, Bulgaria, and Czechoslovakia, which are fed through the "Mir" grid system, which interlinks East European nations and the Soviet Union.[39]

ENERGY AND THE ENVIRONMENT

Any understanding of the interface between energy development and environmental quality in the Soviet Union must begin with a basic comprehension of the political and institutional milieu in which such policies are made. Setting the stage for such interaction are a host of long-standing national priorities that dictate maximum economic growth at virtually any cost. While questions of environmental quality have risen to increased national prominence in the last decade and important measures have been undertaken to alleviate the worst abuses, especially in high-priority areas such as the major cities or important waterways, it is clear that Soviet leaders have not lost their devotion to economic expansion or promoted environmental concerns to a coequal national priority; the costly economic tradeoffs that adequate measures for across-the-board environmental protection measures would entail are painfully apparent to Kremlin leaders, and although they have shown a willingness to take forceful action in certain cases—to improve air quality in the major cities, to tackle water pollution problems in the Volga-Ural basins, or to deal with a pollution threat to Lake Baikal, for example—they have largely avoided irrevocable, sustained commitments to extensive national programs.[40]

The institutional arrangements for environmental protection tellingly suggest both the level of political commitment and the manifest problems involved in making and enforcing abatement measures. Unlike most industrial nations, the Soviet Union has not established a single environmental protection agency with wide-ranging authority. Despite an apparent campaign by proenvironmental forces for further centralization, authority has remained divided among a frequently complex mélange of user agencies, the Ministry of Land Reclamation and Water Resources, the Sanitary-Epidemiological Service of the Ministry of Health, the State Committee on Hydrometeorology and the Environment (recently up-graded and retitled to include limited environmental functions), a number of water basin inspectorates, and the local soviets. Coordination is supposed to be provided by a special Division for the Protection of Nature within the State Planning Committee, which was intended to reconcile environmental imperatives with other interests in the formulation of annual and long-range plans; evidence suggests that this neophyte agency has been a weak lobbyist for environmental interests, its failings stemming primarily from its lack of bureaucratic clout in comparison with industrial interests. These efforts aside, there is little attempt to provide effective coordination of environmental programs. The Ministry for Land Reclamation and Water Resources is the lead agency on questions of water pollution, although it shares some responsibilities with the water basin directorates; in both cases, the agencies in question are simply mandated to honor other higher priorities first (the ministry is predictably more concerned with land reclamation, and the directorates are heavily involved with regional development), and they are both chronically understaffed and poorly armed with effective legal sanctions to enforce protection measures. The State Committee on Hydrometeorology and the Environment takes the lead on questions of air pollution and in monitoring and research tasks, but its prospects as an effective policeman are limited for very similar reasons.[41]

Environmental lobbies can play at best a very limited role in the Soviet Union, although there is considerable evidence that ecological consciousness has increased dramatically within the educated elite and, to a lesser degree, among the average citizenry. The problem is essentially institutional: any public organization must possess party approval before it can function, and Kremlin leaders have been understandably unwilling to open that potential Pandora's box by sanctioning the creation of environmental lobbies. The existing organizations such as the All-Russian Society for the Conservation of Nature are principally conservation and wildlife oriented and less concerned with problems of urban industrial pollution. Even at that, their style has been exceptionally timid and self-effacing, and it is unlikely that they will ever play an aggressive role. More concerned with purely pollution questions have been the relatively few ad hoc environmental coalitions that have formed around specific pollution problems such as the fate of Lake Baikal or the potential for serious damage to the fragile Siberian environment in connection with energy development schemes and the settlement and industrialization of the region. However, as loose, transitory coalitions dependent upon the personal prestige or unchallengeable expertise of their most visible spokesmen or on the cooperation of sympathetic media, these groups have not been able to generate an on-going, national organizational base, although they have had impact on specific, localized problems.[42]

On the whole, air quality in the Soviet Union has been the unintended beneficiary both of the decision in the 1950s to convert large portions of industry and thermal electric generating facilities to cleaner fuels and of the limited number of private autos. Particularly in the major cities of European Russia, air quality has been greatly improved from the early 1970s onward, in part because of the conversions noted above but also in large measure because central authorities placed high priority on maintaining or improving air quality in these areas and were willing to make the financial contributions necessary for pollution control devices or to accept the costs of closing or relocating particularly dirty industries. This is not to suggest, however, that anywhere near national pollution standards were articulated and then stringently enforced; quite the opposite is the case, and the less important urban areas in the European sector and the major industrial centers in the Urals and elsewhere still suffer from considerable air pollution.

The principal energy-related threat to air quality obviously lies in the short-term emphasis now being placed on Moscow basin and other European coal and lignite deposits and on the long-term implications of the utilization of considerably dirtier Siberian coal. In the short run, actual conversions of existing heat- and power-generating facilities in major urban areas will probably be limited by technological and cost considerations, although the June 1979 directives on preparations for the coming winter again stressed the preparation of oil- and gas-burning installations for emergency conversion to other fuels. The real environmental danger comes with the increasing consumption of Siberian coal; even if consumed locally in mine-mouth generating stations or shipped to areas like the Kuznets basin to permit the utilization of its higher-grade coal in European Russia, the local impact will be enormous, and it is unlikely that Soviet leaders will allocate the funds necessary for adquate emission controls in such low priority areas.

The other major threat to air quality is the growing number of fleet and private autos. In the early 1970s, auto exhausts constituted the fourth most potent national pollution hazard, accounting for over 13 percent of total air pollutants. Expanded production of autos and particularly greater sales to private consumers, whose maintenance of engine emission standards is less likely than for fleet vehicles, are the principal causes of concern. Adding to the difficulty is the high concentration of these vehicles in urban areas, where air pollution problems are already the greatest. Soviet authorities have promised the construction of special inspection stations in these urban clusters, but it is unlikely that they will be able to police the growing auto population in any comprehensive fashion.[43]

More than any other environmental threat, water pollution has always been taken far more seriously by Soviet leaders. Given the particular technological configuration of Soviet industry and the comparatively greater impact of agricultural and mine wastes, water pollution has unquestionably emerged as a more serious problem than air pollution. By the late 1960s, virtually all important rivers that flowed through industrial or densely populated areas were badly polluted, and major bodies of water such as the Caspian, Azov, and

Baltic seas and Lake Baikal were already dead or seriously threatened. Adding to Soviet recognition of the problem was the growing concern with the availability of adequate fresh water supplies for industry, agriculture, and human consumption. Rapidly rising demand for fresh water, coupled with an increasing pollution threat and significantly lowered expectations concerning their ability to divert northward flowing rivers to provide new sources for deficit areas, led Soviet leaders in the early 1970s to launch important new initiatives. In 1970 a national water code came into effect, and special decrees followed rapidly concerning water quality in key areas such as the Volga-Ural basins, the Caspian, Azov, and Baltic seas, and Lake Baikal. Institutional arrangements were strengthened, although subsequent developments have suggested that even the lead agency, the Ministry of Land Reclamation and Water Resources, places other priorities before water pollution. Funding levels were clearly increased; in 1973 annual investment rose fivefold from 300 to 1,500 million rubles, and in the spring of 1976, Brezhnev announced a five-year, 11 billion ruble environmental program, most of devoted to water pollution.

Despite the fanfare and additional investment, it is apparent that Soviet leaders are following roughly the same strategy as with air pollution, albeit at a higher level of commitment. The greatest progress has been made where special enactments have singled out particularly acute problems and mobilized both public attention and government resources. Even in these instances, however, problems remain: The construction of waste purification facilities has lagged years behind schedule even in key areas such as the Volga-Ural basins simply because the few construction agencies that specialize in this technology are swamped with projects, and the technological level of emission control devices for industrial use is generally low.

The greatest potential for greater water pollution associated with current energy development programs comes from the mining industry. Mine run-off has always been a problem in the European sector and the Urals, and the almost complete shift to strip mining practices for the new Siberian deposits will intensify the problem. While land restoration is theoretically possible and in fact mandated by several laws on the topic, Soviet practice to date has set an unenviable record. The problem will grow even more complex in the future because of the fragile nature of the tundra environment where the most massive strip mining is planned. [44]

The further development of nuclear energy represents several threats to the environment, especially in the European sector, where nuclear installations will be most numerous. Adding to the normal low-intensity pollution potential of any functioning reactor are the catastrophic threats of reactor malfunction and the disposal of nuclear wastes. On the whole, Soviet authorities seem relatively unconcerned with the problems, and there has been no discernible pressure from the scientific community or citizens' groups for better safety procedures or a moratorium on construction. Official reaction to the near disaster at Three Mile Island produced only vague reassurances about the better quality of Soviet technology and the predictable lecture that one could expect as much from privately owned utilities interested only in profits. This optimism has led Soviet authorities to take dangerous shortcuts in developing existing reactors. Burned reactors lack back-up cooling systems to prevent core overheating, and siting regulations, although recently strengthened, still permit reactors to be located in or near population centers; indeed, current discussions of the greater potential for small reactors in providing heat and power at factory and residential sites seem to suggest that there will be an even greater proximity between the population and reactor sites in the future. According to the testimony of Western scientists who have examined Soviet breeder reactor technology, safety concerns have been deemphasized in the interest of rapid development. Soviet reactors also seem to be as prone as any others to "bugs" and on-line problems, and many have been compelled to operate below rated capacity or have experienced small accidents or near-misses, although the latter point is difficult to verify. The problem of nuclear wastes has been handled either by dumping them in the Black Sea—apparently with some impact on the local environment—or by burial in underground vaults. According to the dissident scientist Zhores Medvedev, improper underground facilities caused an accidental vault rupture in the late 1950s, creating a nuclear cloud that drifted several hundred miles. While Soviet scientists now speak of using improved vaults, Western experience with similar programs suggests that problems will remain.[45]

The development of hydroelectric power also entails potential environmental costs both in terms of the disruption of existing land and water ecosystems—and the bulk of new construction will be in West Siberia and the Far East, where delicately balanced ecological systems are easily susceptible to extensive damage—and the flooding of potentially productive farm land, forests, or possible mining sites. Since the per capita amount of arable land has constantly declined over the last decade, Soviet officials have become particularly concerned with further losses and have instituted new guidelines for the evaluation of the alternative uses of land. These new pricing guidelines do not, however, represent actual opportunity cost calculations, nor do they accurately evaluate the potential value of subterranean resources. As in the past, the ultimate decisions on land use questions will be resolved more in terms of the political influence of the respective ministries involved, a terrain on which hydropower developers have usually had considerably more power than the spokes-men of agricultural, mining, or forestry interests. The Soviet track record has been better concerning the environmental consequences of hydropower dams that have disrupted aquatic ecosystems; especially after they realized that the fishing industry was being adversely af-fected, dam builders installed by-pass devices such as fish ladders or took other measures to lessen the impact on feeding and spawning areas, sometimes even restocking depleted areas. [46]

Quality, like beauty, lies ultimately in the eye of the beholder, and this is certainly true for a citizen of the Soviet Union seeking to assess his or her own priorities in terms of energy, environmental, and consumption tradeoffs. There is indeed a paradox in Soviet thinking about the complex interrelationship of these themes. On the one hand, there is strong evidence to suggest that Russians have maintained a close psychological tie with nature; even hardened ur-banites typically seek out rented summer dachas in the countryside or government resorts closer to nature for their annual vacation, and in the public discussion of the new 1977 Constitution the theme of environmental protection emerged as an important "motherhood" issue. On the other hand, the average Russian consumer has not shown any inclination to reduce his or her demands for improved housing, better clothes, a more diversified diet, or a host of con-sumer goods and appliances solely for the sake of energy conservation or environmental protection. Perhaps understandably in a nation where economic growth has until recently been channeled primarily into heavy industry and the military, and the consumer was instructed to wait for a better day, the sentiment to postpone even further the joys of consumerism is not strong. Soviet leaders have shown in-creasing concern with satisfying the growing consumption demands of the population and not insubstantial nervousness about the prospects that consumer frustration could spill over into anomic political op-position, as has happened several times in Eastern Europe. The situation may become particularly acute in the 1980s, as stagnant or even declining growth rates for the economy as a whole come into headlong conflict with the need for substantially greater investments both to modernize industry itself and to develop and transport distant energy resources, to say nothing about the continued heavy commit-ment to defense spending. In such a crunch, the odds are that a vast majority of the nation would choose growth and consumerism above alternative environmental or conservationist goals. Given the rela-tive weakness of proenvironmental forces and the absence of a single centralized and institutionalized proenvironmental force within gov-ernment itself (perhaps not an unintentional omission), it is unlikely that any effective political opposition could be mobilized in defense of the environment.

FUTURE PROSPECTS

While other major industrial nations are struggling to compre-
hend the resource, environmental, and social implications of the new
"age of limits," Soviet leaders still view the world in terms of virtually limitless horizons of upward growth. To be sure, they acknowledge and even grudgingly accept that future growth rates will shrink
and perhaps stagnate, but this is seen as a product of a number of
converging factors affecting the normal growth rate of any advanced
industrial nation rather than any fundamental downturn impelling a
rethinking of their basic weltanschauung. Even within the scientific
community, the "age of limits" argument finds few adherents willing
to state publicly that there is something deeply flawed with the assumptions and aspirations of modern Soviet society.

It is no less important to understand that both Soviet leaders
and the average person view the world from a perspective that has
not come to think in terms of tradeoff relationships. In an act of
considerable self-deception, both view the goals of adequate fuel and
resource production, economic growth and a rising standard of living,
and environmental protection as mutually reconcilable goals. To
the extent that there have been problems or that there may be difficult moments ahead—and both parts are now readily acknowledged—
they are seen as the result of inept planning or the inevitable excesses
and errors of a system attempting to compress a century of economic
growth and the modernization of an always lagging economy into a
few decades. Deeply ingrained in Soviet thinking is the assumption
that what they would term "scientific planning" is capable of recon-
ciling all of the conflicting problems that for the capitalist world,
where such rational planning is allegedly impossible, have led to a
seeming cul-de-sac. Virtually as an article of faith, they accept the
notion that only short-term difficulties lie ahead and that the "limits"
imposed on further development lie only in the oversights of policy
makers, the shortcomings of a perfectable technology, and the reluctant malleability of the human spirit.

In more concrete terms, any assessment of probable future
scenarios concerning the energy situation and its environmental impact must begin with the recognition that the Soviet Union is still in
an enviable position in comparison with virtually any other major
industrial state. At the worst, energy shortfalls in the mid-1980s,
as predicted by the CIA, would intensify the problem of maintaining
the growth rate and probably exacerbate a host of institutional and
regional rivalries. The more probable scenario—at least to this
author—is that serious national problems (as opposed to localized
shortages) will be delayed until the late 1980s or the 1990s and their
severity determined more by Soviet haste in developing East Siberian

resources, and the influx of advanced Western technology, than by any shortage of energy resources per se. Unlike most deficit nations, the problem for the USSR is not to develop alternative and as yet experimental technologies, but rather to apply existing advanced technologies to Soviet conditions. Given even a worst-case scenario in which Soviet domestic production fell to a point where hard-currency-earning exports to the West and Japan were curtailed and shipments to Eastern Europe endangered, the Soviet Union would still likely be able to satisfy overall domestic needs and would probably enter into the world market, hopefully on barter terms, to obtain sufficient resources to maintain at least a portion of the politically important shipments to Eastern Europe.

The situation with the environment is far less certain. An outside observer is tempted to lecture Soviet authorities on the overall backwardness of their environmental program. The absence of a centralized environmental agency, of powerful lobbies, and of a meaningful across-the-board commitment to the environment per se are troublesome, and the obvious intent of Soviet leaders to maintain their flexibility to determine which high-priority areas will receive protection and which will be permitted to deteriorate in the national interest strikes a particularly Orwellian ring that some animals are still more equal than others. In objective terms, some aspects of Soviet environmental policy are clearly ill-considered and short-sighted—their handling of the environmental implications of nuclear power is probably the best example where purely technological arguments can be made against current policy. Other aspects of environmental policy strike deeply to the heart of fundamental philosophical issues that seemingly predetermine Soviet responses—as with the question of calculating the true value of natural resources, which any Marxist would reckon differently from a Western economist on philosophical grounds alone. Yet short of a fundamental reorientation of this world view and a thorough reassessment of national priorities, Soviet environmental policy is likely to remain a patchwork of serious commitments and conscious oversights. As Thane Gustafson has pointed out, a hammer and sickle on a field of ecology green is an unlikely banner for this or any probable future generation of Soviet leaders.[47]

NOTES

1. For the best summaries of Soviet energy production, see Iain F. Elliot, The Soviet Energy Balance: Natural Gas, Other Fossil Fuels, and Alternative Power Sources (New York: Praeger, 1974); Marianna Slocum, "Soviet Energy: An Internal Assessment,"

Technology Review, October–November 1974, pp. 17–23; J. H. Chesshire and C. Huggett, "Primary Energy Production in the Soviet Union: Problems and Prospects," Energy Policy 3, no. 3 (1975): 223–41; and Herbert Block, "Energy Syndrome, Soviet Version," in Jack M. Hollander, ed., Annual Review of Energy (Palo Alto, Calif.: Annual Reviews, 1977), pp. 455–97. Annual Soviet statistics are to be found in the yearly series Narodnoe Khozyaistvo v SSSR v 1977 godu (Moscow: Statistika, 1978) and earlier volumes.

2. Block, "Energy Syndrome," pp. 462–63; and Robert W. Campbell, Trends in the Soviet Oil and Gas Industry (Baltimore: Johns Hopkins University Press, 1974), pp. 9–14.

3. Block, "Energy Syndrome," pp. 462–63; Campbell, Trends in the Soviet Oil and Gas Industry, pp. 49–68; and Leslie Dienes, "The Soviet Union: An Energy Crunch Ahead?" Problems of Communism 26, no. 5 (1977):41–60.

4. Block, "Energy Syndrome," pp. 475–76; Emily E. Jack et al., "Outlook for Soviet Energy," in Soviet Economy in a New Perspective: A Compendium of Papers Submitted to the Joint Economic Committee, Congress of the United States, October 14, 1976 (Washington, D.C.: U.S. Government Printing Office, 1976), pp. 466–67; Alan B. Smith, "Soviet Dependence on Siberian Resource Development," in ibid., pp. 485–87; and Violet Conolly, Siberia, Today and Tomorrow (New York: Taplinger, 1975), pp. 61–90.

5. Philip R. Pryde and Lucy T. Pryde, "Soviet Nuclear Power," Environment 16, no. 3 (1974):26–34.

6. Block, "Energy Syndrome."

7. Dienes, "The Soviet Union."

8. Daniel R. Kazmer, "A Comparison of Fossil Fuel Use in the U.S. and U.S.S.R.," in Soviet Economy in a New Perspective, pp. 500–34.

9. Donald R. Kelley, "Economic Growth and Environmental Quality in the USSR: The Soviet Reaction to The Limits to Growth," Canadian Slavonic Papers 28, no. 3 (1976):266–83.

10. Donald W. Green, "The Soviet Union and the World Economy in the 1980s: A Review of Alternatives," and F. Douglas Whitehouse and Daniel R. Kazmer, "Output Trends: Prospects and

Problems," both in Holland Hunter, ed., The Future of the Soviet
Economy: 1978-1985 (Boulder, Colo.: Westview, 1978).

11. Block, "Energy Syndrome," pp. 461-62; and Kazmer, "A
Comparison of Fossil Fuel Use."

12. Kazmer, "A Comparison of Fossil Fuel Use"; "Soviets
Mount Drive to Avert Oil Turndown," Oil and Gas Journal, September
18, 1978, p. 62.

13. Donald R. Kelley, "The USSR: The Self-Made Energy
Crisis," in Donald R. Kelley, ed., The Energy Crisis and the Environment: An International Perspective (New York: Praeger, 1977), pp.
41-64.

14. Kelley, "The USSR"; and Dienes, "The Soviet Union."

15. Dienes, "The Soviet Union," p. 57; "Soviet Geologists Rap
Energy Planners," Oil and Gas Journal, February 5, 1979, pp. 38-42.

16. Ibid., p. 44.

17. Planovoe Khozyaistvo, No. 7 (1974).

18. Kommunist, No. 1 (1975); and Pravda, November 28,
1974, November 16, 1975, December 14, 1975, and December 28,
1975.

19. Block, "Energy Syndrome," pp. 466-69; and Dienes, "The
Soviet Union."

20. Kelley, "The USSR" and "Economic Growth."

21. Pravda, November 21, 1977.

22. Voprosy Ekonomiki, No. 6 (1978), pp. 3-14.

23. John Kramer, "Prices and the Conservation of Natural
Resources in the Soviet Union," Soviet Studies 24, no. 3 (1973):364-73;
Marshall I. Goldman, The Spoils of Progress: Environmental Pollution in the Soviet Union (Cambridge, Mass.: MIT Press, 1972),
pp. 43-76.

24. Izvestiya, July 30, 1977.

25. New York Times, January 10, 1979.

26. See, for example, Pravda, January 7, 1979; and Kommunist, January 13, 1979, p. 2.

27. Izvestiya, June 14, 1979.

28. "Soviets Push W. Siberia Oil Production," Oil and Gas Journal, May 15, 1978, pp. 46-48; "Soviets Mount Drive to Avert Oil Turndown," p. 63; "CIA: Soviets Won't Reach Oil Target," Oil and Gas Journal, October 23, 1978, pp. 62-64; "Soviets Confirm Some CIA Assessments," Oil and Gas Journal, January 15, 1979, pp. 46-47; and "Work at Samotlor Bolstering Soviet Oil Production," Oil and Gas Journal, December 4, 1978, p. 33.

29. Jack et al., "Outlook for Soviet Energy," pp. 464-68; Smith, "Soviet Dependence," pp. 487-88; Block, "Energy Syndrome," pp. 466-72; Dienes, "The Soviet Union"; Narodnoe Khozyaistvo v 1977 g. (Moscow: Statistika, 1978), pp. 148-49; Pravda, March 22, 1979.

30. Central Intelligence Agency, Prospects for Soviet Oil Production, Report ER 77-10270, April 1977; for a more optimistic assessment, see Block, "Energy Syndrome." See also Dienes, "The Soviet Union"; Campbell, Trends in the Soviet Oil and Gas Industry, pp. 26-35; and Narodnoe Khozyaistvo; "CIA: Soviets Won't Reach Target," p. 63; "Soviets Confirm Some CIA Assessments," p. 46; "CIA: Global Oil Supply Outlook Poor," Oil and Gas Journal, September 3, 1979, pp. 50-51; "Communist Crude Output Hits 13.755 Million B/D in 1978," Oil and Gas Journal, February 26, 1978, p. 38; and Richard Halloran, "CIA Sees Soviet Importing Oil Soon," New York Times, July 30, 1979, p. D1.

31. Campbell, Trends in the Soviet Oil and Gas Industry, pp. 14-25; John P. Hardt, "West Siberia: The Quest for Energy," Problems of Communism 22, no. 3 (1973):25-36; Pravda, August 10, 1977; Ekonomika i Organizatsiya Promyshlennogo Proizvodstva, No. 6 (1977); Ekonomicheskaya Gazeta, No. 17 (April 1978); Sovetskaya Rossiya, December 8, 1978; Ekonomika i Organizatsiya Promyshlennogo Proizvodstva, No. 2 (1979); and "Soviets See Offshore Program Speedup," Oil and Gas Journal, June 25, 1979, pp. 40-41.

32. Campbell, Trends in the Soviet Oil and Gas Industry, pp. 49-67; Jack et al., "Outlook for Soviet Energy," pp. 463-64; Dienes,

"The Soviet Union"; Izvestiya, February 24, 1978; Pravda, October 30, 1977; Narodnoe Khozyaistvo; "Soviets Press Line Work To Boost Gas Production," Oil and Gas Journal, April 2, 1979, pp. 38-43; and "Soviets Mount Drive to Avert Oil Turndown," p. 62.

33. Block, "Energy Syndrome," pp. 475-76; and Smith, "Soviet Dependence," pp. 485-87.

34. Pryde and Pryde, "Soviet Nuclear Power"; Robert W. Campbell, "Issues in Soviet R&D: The Energy Case," in Soviet Economy, pp. 97-112; Block, "Energy Syndrome," pp. 476-77; Izvestiya, May 25, 1977, and April 11, 1979.

35. Rolf Grunbaum, "Alternative Energy in the USSR," Environment 30, no. 7 (1978):25-30; and Pravda, December 15, 1978.

36. Grunbaum, "Alternative Energy"; Campbell, "Issues in Soviet R&D"; Chesshire and Huggett, "Primary Energy Production"; and Elliot, The Soviet Energy Balance.

37. Jeremy Russell, Energy as a Factor in Soviet Foreign Policy (London: Saxon House-Lexington Books, 1978); Martin J. Kohn, "Developments in Soviet-East European Terms of Trade, 1971-1975," in Soviet Economy, pp. 67-80; John M. Kramer, "Between Scylla and Charybdis: The Politics of East Europe's Energy Problem," Orbis 22, no. 4 (1979):929-50; and Christopher C. Joyner, "The Energy Situation in Eastern Europe: Problems and Prospects," East European Quarterly 10, no. 4 (1978):496-516.

38. Block, "Energy Syndrome," pp. 483-86; Campbell, Trends in the Soviet Oil and Gas Industry, pp. 74-84; Kohn, "Developments in Soviet-East European Terms of Trade"; Smith, "Soviet Dependence," pp. 492-97; Marshall I. Goldman, "Autarchy or Integration—The USSR and the World Economy," in Soviet Economy, pp. 86-96; "Soviet Oil, Gas Exports Value Hits High," Oil and Gas Journal, July 30, 1979, pp. 117-27; "Soviet Bloc Warned of Higher Russian Oil Prices," Oil and Gas Journal, May 21, 1979, p. 106; David A. Andelman, "An Oil Shock of Their Own Hits Members of Comecon," New York Times, August 12, 1979, p. E4; and Halloran, "CIA Sees Soviet Importing Oil Soon."

39. Block, "Energy Syndrome," pp. 490-99; Dienes, "The Soviet Union"; and Russell, Energy as a Factor, pp. 140-206.

40. Donald R. Kelley, Kenneth R. Stunkel, and Richard R. Wescott, The Economic Super Powers and the Environment: The United States, the Soviet Union, and Japan (San Francisco: W.H. Freeman, 1976), pp. 130–35; and Donald R. Kelley, "Environmental Problems as a New Policy Issue," in Karl W. Ryavec, ed., Soviet Society and the Communist Party (Amherst: University of Massachusetts Press, 1978), pp. 88–107.

41. Thane Gustafson, "Environmental Policy under Brezhnev: Do the Soviets Really Mean Business?" in Donald R. Kelley, ed., Soviet Politics in the Brezhnev Era (New York: Praeger, 1980).

42. Donald R. Kelley, "Environmental Policy Making in the USSR: The Role of Industrial and Environmental Interest Groups," Soviet Studies 28, no. 4 (1976):570–89.

43. Gustafson, "Environmental Policy under Brezhnev."

44. Ibid. See also Kelley, "The USSR: The Self-Made Energy Crisis," pp. 55–57.

45. Pryde and Pryde, "Soviet Nuclear Power"; New York Times, November 7, 1976.

46. Kelley, Stunkel, and Wescott, The Economic Super Powers, pp. 84–85. See also Kramer, "Prices and the Conservation of Natural Resources," pp. 364–73; and Phillip R. Pryde, Conservation in the Soviet Union (New York: Cambridge University Press, 1972), p. 115.

47. Gustafson, "Environmental Policy under Brezhnev."

3

JAPAN
Kenneth R. Stunkel

THE PATTERN OF ENERGY USE

Japan, about the size of California, is a small country that manages to use enough energy to make it the third-ranking economy in the world after the United States and the Soviet Union.* As a feudal agricultural society of some 30 million people in 1868, when modernization began, Japan's impact on energy was negligible. As a great economic power (keizai taikoku) of 114 million people in 1978, Japan absorbed roughly 6 percent of global energy supplies with only 3 percent of the world's population. In the shadow of two dramatic facts this was a remarkable achievement. First, at the time of the allied occupation in 1945 Japan was a shattered nation with marginal economic prospects. Second, the Japanese homeland was all but destitute of resources necessary for industrialization.

Japan's energy use in the past 90 years of industrialization has been associated with high economic growth rates. World War II, when seen in perspective, was a traumatic but nevertheless ephemeral interruption of a fairly consistent domestic growth pattern. The trend of economic growth from 1890 to 1941 was 3 to 5 percent of GNP, comparable to any industrial state of the period, and was accomplished despite the great depression and a catastrophic earthquake in 1923.[1] Japan was an energy glutton well before the spectacular economic resurgence of the postwar era. Its economic machine,

---

*Japan has been the third largest economy and second largest market economy since 1965.

the most imposing in Asia, was fueled mainly from coal and hydro-electric power.

Military defeat in 1945 left the economy prostrate and few observers expected more than a halting, modest recovery thereafter. Yet by 1954 the prewar peak for GNP, about $10 billion in 1939, had been exceeded. By 1958 the economy "took off" in a rapid, steep ascent of growth and development unprecedented in modern history.* GNP had reached $39 billion by 1960 and topped $493 billion in 1975, with an average growth rate of better than 10 percent annually, well beyond the performance of any other country. In the past decade GNP has grown some 4.6 times, with an even better showing for exports and imports. Throughout this period of accelerated growth, energy demand persistently outstripped increments of GNP; thus in 1972 real GNP was 5.1 times that of 1955 while energy use expanded 6.3 times. Between 1960 and 1972 energy demand rose 5 percent annually with the result that Japan's share of world energy supplies doubled.[2] In 1976 Japanese demand for oil exceeded that of Africa, China, India, Pakistan, and Eastern Europe combined. With the exception of Americans, never have so few taken so much in so short a time.

Currently the energy structure of Japan is dominated by oil, which in March 1976 was 74 percent of primary energy use. Nearly all of it was imported. The general picture can be studied in Table 3.1. A shade more than 10 percent of the energy used was produced at home, less than 1 percent of which was domestic petroleum. The nearest competing source was hydropower , followed by coal. This dependence on imported oil contrasts sharply with the situation in 1955 when 76 percent of primary energy came from domestic coal and stream flow. Sources other than petroleum will edge up somewhat in the future, but the prognosis is that Japan will go on being more reliant on oil—and generally more reliant on energy imports—than other major industrial states.[3]

The distribution of energy use can be approached by looking at how total oil consumption in 1972 (210 million kiloliters)

---

*This "economic miracle" has yielded satisfactorily to rational explanation. For a good discussion of how it was done, see Edward Denison and William Chung, How Japan's Economy Grew so Fast (Washington, D.C.: Brookings Institution, 1976).

TABLE 3.1

Japan's Long-Term Energy Outlook

	Fiscal Year 1975 (April 1975–March 1976) Actual		FY 1985 Reference Case		FY 1985 Accelerated Case		FY 1990 Accelerated Case	
Energy demand								
Demand before saving	390m Kl (oil equiv.)		740m Kl		740m Kl		916m Kl	
Energy saving ratio			5.5 Percent		10.8 Percent		13.5 Percent	
Demand after saving			700m Kl		660m Kl		792 Kl	
Sources of energy	Amount	Percent	Amount	Percent	Amount	Percent	Amount	Percent
Hydropower	24.9m Kw	5.8	39m Kw	3.3	41m Kw	3.9	51m Kw	3.9
Geothermal	0.05m Kw	0.0	0.5m Kw	0.1	1m Kw	0.3	3m Kw	0.7
Oil and LNG (domestic)	3.5m Kl	0.9	8.0m Kl	1.2	11m Kl	1.7	14m Kl	1.7
Coal (domestic)	18.6m ton	3.3	20m ton	2.0	20m ton	2.1	20m ton	1.8
Nuclear power	6.62m Kw	1.7	26m Kw	5.4	33m Kw	7.4	60m Kw	11.2
LNG (imported)	5.06m ton	1.8	24m ton	4.9	30m ton	6.4	44m ton	7.7
Coal (imported)	62.34m ton	13.1	93m ton	10.7	102m ton	12.4	144m ton	14.1
New energy	—	—	—	—	2.3m Kl	0.4	13m Kl	1.8
Subtotal (oil equiv.)	104m Kl	26.6	195m Kl	27.8	228m Kl	34.5	340m Kl	42.9
Oil imports	288m Kl	73.8	505m Kl	72.2	432m Kl	65.5	425m Kl	57.1
Total (oil equiv.)	392m Kl	100	700m Kl	100	660m Kl	100	765m Kl	100

Source: Japan Ministry of International Trade and Industry. Compare these projections with those contained in Japan's New Energy Policy (Tokyo: Ministry of International Trade and Industry, 1976).

was divided up. Close to 88 percent went to four sectors of the economy:

Mining, manufacturing, and industry	38.3
Vehicles	18.0
Electric power	17.9
Petrochemical industry	13.6

The category of "people's consumption" accounted for 6.1 percent and all other uses, including jet fuel, absorbed the remainder.[4] Energy use in Japan's declining agricultural sector is covered by the first four categories; for example, pesticides and artificial fertilizers are petrochemical products, which Japanese farmers apply to the soil in prodigious quantities. Industry presently accounts for 57 percent of all energy use, with steel taking a large share (20 percent).

Japan may have the most energy-intensive and pollution-prone industrial structure on the planet.[5] It is a creation of the postwar "second industrial revolution" and at its core are the steel, oil refining, petrochemical, aluminum refining, paper-pulp, textile, automobile, and power generation industries. There is a dramatic relationship between the growth of these industries and changes in the pattern of energy demand. Consider a few examples:

Steel production has swollen from 8 million to 102 million tons between 1939 and 1978, though the latter figure is well below capacity. Optimistic planners and forecasters have anticipated production levels by 1985 anywhere from 173 to 193 million tons.[6] Domestic coal reserves were unable to supply the steel industry with coking coal on such a scale, so imports rose from 9 to 57 million tons between 1960 and 1973. Total coal imports in 1973 reached 64 million tons a year, may rise to 100 million tons by 1980, and break into the range of 111 to 121 million tons by 1985—surely a formidable rate of increase.[7]

Oil refineries appeared in the 1950s to process imported crude oil, a development intended to circumvent the higher cost of importing refined petroleum products, and one that has made Japan's refining capacity second in the noncommunist world. Since 1960 an agile transition from coal-based acetylene to oil-based ethylene chemistry has led to a tenfold jump in ethylene production and a tripling of other forms of chemical production in the 1960s. The share of oil in Japan's energy picture rose from less than 38 percent in 1955 to 76 percent in 1975,[*] with a corresponding grip on world petroleum

---

[*]In barrels of oil the increase was 20.4 times.

imports that would stand at 20 percent right now if pre-1973 growth rates had been sustained. [8]

Although a tiny land of 370,000 square kilometers, Japan provides sites for more than 600 paper and pulp mills, ranking only below Canada and the United States. [9] Production has been cut recently because of slipping demand, but the industry is still dependent on heavy imports of wood pulp to compensate for shrinking domestic forest reserves, one of the few genuine resource assets of the past. [10]

The automobile industry burgeoned after the 1965 recession. Some 8 million vehicles of all kinds have become 20 million or more, most of which are supplied domestically. This abrupt glut of vehicles—requiring mountains of processed materials, millions of tons of fuel, and a supporting infrastructure of roads, bridges, and tunnels—has a diversified impact on energy demand.

In such a profligate industrial structure it is understandable that power plants would spread across the land. A journey through the Japanese islands seldom leaves one without a conspicuous view of great towers straddling the landscape with their sagging burden of power lines. Japanese power capacity by 1980 will be nearly triple the European Economic Community figure for 1959 and double Japan's own 1970 level of 300 billion kilowatts. Should growth optimists have their way, power demand in 1985 could be more than triple that of the early 1970s. [11] Even on the assumptions of "stable growth" economics—generally the Liberal Democratic Party (LDP) line since 1973—power generation would reach colossal proportions. [12]

The disadvantages of Japan's industrial structure are not unrecognized. The problem is just what to do about it. The challenge is to reduce or phase out energy-intensive industries in favor of high-technology ones like precision instruments, fine chemicals, and electronics. The magic phrase is "knowledge-intensiveness," a concept that government planners are still trying to define. Unfortunately there are stubborn obstacles in the way of rational change. As a business spokesman says, "highly orchestrated planning is not so easy anymore with size and less freedom of movement." [13] Low corporate profits and uncertain business prospects since 1973 inhibit the investment necessary to change product lines. Very high debt-equity ratios* leave most industries with little room to maneuver.

---

*Most industries have borrowed heavily to capitalize their expansion and must grow in productivity to meet a staggering (from the viewpoint of Western businessmen) debt service.

Companies are trying to reduce bank loans in anticipation of slow growth and a profit squeeze. And, of course, no one wants to be cut back or to go under. The industrial structure division of the Ministry of International Trade and Industry (MITI) sees no drastic changes in either industrial or import structure in the foreseeable future. As of 1977 MITI had come up with no long-range plan for industrial transformation, for "the process of reorganizing the industrial structure would generate problems even in normal times."[14] Indeed, steel will remain intact to supply China, a major new trading partner on its way to becoming the world's leading importer of steel.

In the meantime energy consumption per unit of land area is already 18 times the world average, compared to six times for the United States. The value of GNP per square mile is $4,245, compared to $357 for the United States. What these dry figures imply is that Japan is burning, figuratively, like a roman candle.

DOMESTIC ENERGY POLICY

In July 1970 the Energy Council projected energy needs for 1985 at 1 billion kiloliters of oil equivalent—more than twice the level in 1974—and at 2.5 billion for the turn of the century. After the oil crisis of 1973 planners hastened to revise their forecasts, saying in one instance that "in light of the present situation, these projections seem considerably exaggerated."[15] In 1975 MITI said Japan would be using 560 million kiloliters of oil equivalent by 1985; in 1978 that estimate was scaled down to 420 to 432 million kiloliters.[16] It remains to be seen whether Chinese and Mexican oil reserves can stave off a further reduction of expectations in the short term.

The potential for more growth in Japan rests on production factors (capital, labor, technology), balance of payments (that is, import-export balance), and physical constraints (available resources, land sites, and pollution levels). A realistic assessment of these factors suggests that a lot more growth is possible in the near future.[17] The 1973 embargo did not mandate zero growth, nor would Japan be able to live with such an option without great national stress. A growth rate lower than 6 or 7 percent a year is an unlikely policy choice because of anticipated negative effects on wage and employment levels, exports, public investment (parks, antipollution measures, housing, and the like), and other aspects of national life.[18] Unfortunately the Japanese economy is geared to rapid growth, for a distinct momentum has emerged in the past 20 years. Nevertheless, most analysts, in or out of government, appear to concede the inevitability of a reduced pace,* somewhere in the range of 6 or 7 percent of real growth annually.[19] Even at 7 percent some analysts, like Kanamori,† believe that Japan will crack a trillion-dollar GNP by 1985, which would mean better than a doubling of energy use.[20] In reality the growth rate has been lower than 6 percent since 1973.[21] While Japanese business is not pleased with this, the 5.4 percent rate of growth in 1977-78 was still one of the highest in the world. A slower pace for Japan still means terrific pressure on energy and resources.[22]

One can surmise that government planners are ambivalent on the issue of growth policy. Frequently the right and left hands are working at cross purposes. A 1975 economic white paper, "In Pursuit of a New Stable Course," reads in part:

> this new course [that is, stable growth] will mean saying goodbye to the rapid economic growth of the past. The present is a time which requires that we realize that the postwar growth period is growing to a close and that we set our aims on a truly affluent society while standing up to the trials and tribulations that will come in the dimness prior to the dawning of a new age.[23]

Unsteady, contradictory effusions of this sort symptomize the fact that Japan's growth momentum has been retarded only slightly by tougher circumstances. There is extreme reluctance to act on a conscious choice of fresh priorities, not entirely without reason.

Japan is being prodded to high future growth rates and energy use by internal dynamics of the economy. The debt-service problem has already been mentioned. There is also the need for a continual flow of exports, without which Japan cannot purchase the fuel, food, and other materials necessary to sustain the country. A growth rate of 4 percent or less might entail heavy unemployment. An interim report on the National Development Plan in 1975 claimed that 5 million

---

*MITI saw in 1976 a 6.2 percent growth rate for 1973-85 without energy conservation and 5.3 percent with it, although a genuine conservation policy is plagued with vague ifs. Japan's New Energy Policy (Tokyo: MITI, 1976), pp. 24-26.

†Long associated with the Japan Economic Research Center (Nihon Keizai Kenkyu Senta), an influential think tank, Hsiao Kanamori has been a strong advocate of unimpeded growth toward a trillion-dollar GNP by 1985.

people would be thrown out of work at a 3 percent rate of growth to 1985.[24] Population pressure is likely to discourage any fundamental movement toward a reduction of consumer demand, at least on a voluntary basis. The low rate of Japan's population growth (about 1.2 percent on the average since World War II) is highly deceptive, because it resulted in the addition of 38 million people between 1945 and 1975, a leap from 72 to 110 million people, hardly a negligible increase.* Each year a net increase of some million citizens must be absorbed into the economy. The expected size of Japan's population in 1985 is on the order of 120 million, which is a bit like crowding a Sweden into the country every five years. During the era of high economic growth, population lagged behind both GNP and energy, but the ratio of population growth to per capita consumption rose steadily, as it continues to do.[25] In the tense modern setting of resource and environmental limits, one might suppose that the Japanese would feel uneasy about another 30 million citizens joining consumer ranks in the next half century, at a near replacement level of 1.05 percent annually. The uneasiness is held at bay with the assumption that economic growth will provide for so many people in such a small, resource-poor land.

In summary, Japan's "growth policy" is "to ensure a long-range stable growth of the national economy through effective supply and demand of energy."[26] The attainment of this objective, according to the Industrial Structure Council, requires the promotion of conservation, obtaining a stable supply of petroleum, diversifying energy resources, and pursuing energy research and development. If all goes well, it is hoped that Japan can sustain a comfortable growth rate through 1985, accompanied by a reduction of oil dependence.

The most obvious crimp in this program of stable growth is the handsome cost of oil since the October 1973 Arab price offensive began. Japan is impaled securely on the petroleum hook. Oil imports in the last quarter of 1978 were running 11.7 million barrels a day. More than 99 percent came from abroad and 74 percent of that came from the Middle East. Petroleum that had cost a total of $7

---

*It took a bare 30 years to add more people than the total population of the country in 1872. An increase that once took 2,000 years was achieved in three decades. Density of population has gone from 91 per square kilometer in 1972 to 288 in 1973, compared to 72 per square kilometer in Asia as a whole. Much of Japan's population is jammed into a small area along the Pacific seaboard.

billion in 1972 commanded $24 billion in fiscal year 1978.* While
the immediate effect on Japan's enviable balance of payments was
disconcerting, recovery from the shock was swifter than anyone expected. By 1979 the balance-of-payments situation was good enough
to cause tension between successful Japan and annoyed trading partners like the United States, even though imports and exports were
down considerably, with a good deal of industrial capacity idle. The
near zero growth of GNP in 1974 accelerated to better than 5 percent
in 1976 and has managed to stay at that level through 1979.[27] Thus
the resilience of the Japanese economy seems to have been vindicated
in this particular crisis. In a mood of resignation the future escalation of oil prices was accepted by planning groups. Experts at the
World Petroleum Congress in Tokyo (May 1975) generally agreed
that oil and gas prices would rise steeply, followed by the depletion
of oil and gas reserves toward the end of the century or shortly
thereafter.[†]

In some other respects the Japanese performance was less
impressive. Government and business reactions to the oil crisis
tended to be abrasive, opportunistic, and hysterical. With government encouragement, the trading and oil companies scrambled aggressively for oil and ended by pushing oil prices up in Iran, Libya,
and Nigeria. Other oil users and the poor countries were ignored as
the Japanese pressed frantically to close deals in the Middle East.
There was also domestic hoarding of oil supplies. Altogether Japanese
behavior was less than seemly.

Evidence is that the embargo of 1973 taught Japan only a modest
lesson. Tokyo has responded with a show of caution and much stylized
hand wringing, not with major plans to change economic behavior.
The general strategy is to muddle along from month to month and
hope that a stoppage of oil flow here can be compensated for elsewhere.
When the Iranian crisis broke, petroleum lost to Japan (17 percent of
the total used) was made up from Saudi Arabia, a few other foreign
sources, and from 84-day stockpiles. Japan's relative cheerfulness
in the face of such perilous dependence on a line of oil tankers stretching across the sea must count as one of the great examples of national
faith in modern times.

---

*Add another $7 billion from OPEC hikes in 1979, between $18
and $23.50 a barrel.

†On optimistic assumptions it does not appear that oil production will satisfy demand by 2000 A.D., says Desmond Dewhurst,
president of the Institute of Petroleum. "World Energy Resources—
The Looming Gap," Petroleum Review, May 1976, pp. 277-78.

No doubt many consumers and businesses are willing to try conservation because of high energy costs and the discomforts of economic contraction. On the other hand, there are few guidelines or regulations on a national level and it is not easy to assess how much of the problem has been relieved by voluntary measures.* Probably no more than a fraction of the energy has been saved that ought to be. A sensible energy policy should heed Eiji Ozaki's plea that "the energy program for Japan, if any, must be to reduce the use of energy in Japan."[28] It is understood by planners that conservation measures might save up to 40 million kiloliters of energy in oil equivalent by 1985 (see Table 3.1) and moderate a high ratio of energy to GNP.[29] But the conservation scenario is to be achieved by voluntary compliance and "administrative guidance" (gyosei shido), a process of informal consultation between government and business in which personal relationships, persuasion, and subtle pressure are orchestrated into a consensus. The mandatory approach has been laid aside because "energy conservation largely hinges on the will and efforts of industry and the consumer public," and "a psychological burden would be imposed on the people if energy conservation measures were forced upon them."[30] What, then, may be taken as the government's energy conservation policy? It is a policy of "continuing and expanding the current government guidance for saving."[31] What this seems to mean is that conservation measures will be improvised without a definite plan or legislation.

The potential for conservation in Japan has been lightly scratched. In spite of International Energy Agency recommendations, there has been little effort to promote insulation and district heating. Kerosene is held at an artificially low price, thus stimulating extravagance with light crude petroleum in the private sector. Since 1973 there has been no deliberate reduction of fuel consumption. Visible reductions are due to economic slump rather than to premeditated policy. MITI has admitted that since "it is difficult to accomplish this [that is, conservation] through the market mechanism alone," a variety of measures "should be considered."[32] The aim is a 10.8 percent conservation rate to avert shortages in the mid 1980s. The measures turn out to be conservative rather than seriously conservationist. Since 1973 the biggest energy-consuming industries have been subject to some restrictions. Japan's steel mills, the world's most efficient,

---

*The United States consumes twice the energy per dollar than does Japan, but the contrast is due more to American extravagance and inefficiency than to model Japanese frugality.

have been using gaseous waste to expand power generation. The chemical industry has taken steps to recycle waste heat and to use energy-saving catalysers. Cement manufacturers are replacing oil kilns with types that use half the fuel. Paper-pulp mills are taking on more waste paper materials to trim their consumption of electricity. MITI is urging high-energy manufacturing to relocate abroad or reduce domestic plans for expansion.

The central weakness of official conservation policy, apart from the flaccid voluntary approach, in which everyone is invited to decide for himself how much conservation is in the national interest, is the lack of a comprehensive plan to reduce consumption of energy. One way of doing this without painful inconvenience would be the active promotion of thermodynamic efficiency in domestic and industrial energy systems. The cheapest and easiest way to produce energy is to save it through efficient conversion techniques. A thermodynamic approach to energy conversion wants to extract the maximum amount of work from the minimum amount of fuel, or, phrased another way, to perform a task by means that entail the least amount of work. The kind of efficiency sought is "equal to the ratio of the least available work that could have done the job to the actual available work used to do the job."[33] No awareness of these principles is expressed in a basic document like MITI's Japan's New Energy Policy. While industry is more efficient in Japan than in the United States, even greater efficiencies are to be had with leadership and incentives.

Japan has the option of encouraging substantial investment in total energy systems, which could produce energy for commercial and industrial purposes and then recycle the waste heat for space heating and air conditioning.[34] Total energy efficiencies are in the range of 85 percent, compared to 60-65 percent for conventional power plants; thermal pollution could be reduced by half and combustion-generated pollutants by some 35 percent. The drawback of total energy systems is the need to custom design them for the building or community in which they are to be used. Obviously it is cheaper and simpler in the short run, though not better thermodynamically, to have one local power plant serving everyone in a region or community, but a recycling method of energy utilization is technologically feasible (the United States has about 600 functioning total energy systems) and conforms to the imperatives of thermodynamic efficiency. So far little is being done with this alternative in Japan, where there is enough capital on hand to bring the costs of total energy down over a decade or two.

The Japanese are showing deference of late to the concept of energy self-sufficiency. If one means by "self-sufficiency," however, the economic ability of a nation to survive arbitrary, unexpected energy shutouts by foreign suppliers, Japan has no hope of becoming self-sufficient in the balance of the century. At best the syndrome of dependence can be mitigated somewhat by the judicious development of unimpressive domestic energy resources and a risky commitment to nuclear power. In present circumstances one might suppose that top priority would be given to the achievement of maximum energy independence at the earliest moment through new energy technologies and modification of the old ones. The reality is that national policy is seeking to obtain more oil by all practicable means.[35] There is a rational choice to be made here—rigorous conservation and intensive development of domestic energy alternatives—and Japan has opted to stress the expedient, dangerous course of nailing down more oil abroad.

The explanation for this behavior is twofold. First, it is cheaper in the short term to obtain more fossil fuels than to invest massively in novel energy sources. Second, even a modest transition to greater self-sufficiency would require a radical transformation of Japan's national life, including conservation and recycling practices now barely glimpsed, a reduction of total energy use after conservation, and a firm policy of zero or even negative population growth. With respect to the last point, Ozaki maintains that "the population of Japan should be reduced by half, that is, 50 million."[36] With these introductory reflections out of the way, let us ask about the substance of Japan's "project independence."

Domestic oil and gas production is minuscule and unlikely to become significant in coming years, but resolute efforts are trying to squeeze the last drops and whiffs of each from both land and sea. For the development of oil and gas there have been four five-year plans, the last beginning in 1969. By the close of 1972 Japan's confirmed deposits of crude oil stood at 24 million barrels and deposits of natural gas at 300 billion cubic feet. In relation to current and projected needs, domestic production is clearly marginal (in 1971 Japan used 1.5 million barrels of oil). Dozens of exploratory wells have been sunk in the vicinity of Japan. Offshore areas are being probed near Hokkaido and Honshu, especially in the Sea of Japan. Some oil was found off Akita, Yamagata, and Niigata prefectures, though a 15,000-barrel-a-day yield seems rather poor for the capital and energy invested. The extraction of natural gas from ground water has been slowed because of widespread ground subsidence in heavily industrialized areas.[37]

To cushion the shock of any future embargo the Japanese want a 90-day stockpile of oil. Storage is an expensive option, quite apart from the tense issue of finding suitable space. Japan has paid the price on both counts. Sixty days of consumption at the end of 1974 would have been 42 million cubic meters; a 90-day reserve for 1980 would have to be 76 million, a huge addition of 34 million. The cost of storing one cubic meter (1,000 kiloliters) for a year is about $15 at 1976 prices. Accommodation of a further 34 million cubic meters would require seven more storage bases the size of Kiire (the world's largest) in Kagoshima prefecture. Some 425 tanks of 100,000 cubic meter capacity would sprawl across 4,480 acres of land costing $934.5 million. The tanks would cost another $1.5 billion. The present 84-day stockpile has brought Japan within reach of its goal of a three-month grace period in the event of an oil shutout.[38]

There is promise on the domestic front in hydroelectric and coal power, traditional sources of energy that seem more attractive in the wake of oil politics and a limping nuclear power program.[39] In 1974 Japan had 1,600 hydroelectric plants capable of generating a total of 21.5 million kilowatts, or about 4.5 percent of primary energy. Electric power companies want to discard the older plants— 40 percent of which are more than 45 years old, with an average capacity of 2,700 kilowatts—and build a new line of facilities near key reservoirs of Japan's rivers. The modern plants would generate 100,000 kilowatts each, thus doubling hydroelectric capacity from 21.5 million kilowatts in 1974 to 49 million by 1981. Engineers have been working on an innovative hydraulic power plant capable of pumping surplus water uphill during slack hours so it can be used during periods of high demand. Due for completion in 1979, this pump storage system in Nara prefecture is expected to produce 603,000 kilowatts. An array of such plants could supply 10 percent of Japan's primary energy needs by 1985. The stumbling blocks are insufficient capital for plants and dam building and the perennial threat of earthquakes, for earthquake-proof dams are costly. A reminder is in order as well that hydroelectric power provides only electricity. Apparently the future expense of oil and antipollution equipment has wreathed the hydropower option in an economically palatable glow, but the government would have to come up with a sizable part of the capitalization. Unfortunately the government has not chosen to exploit fully the hydropower potential of Japan, having been lured away by nuclear power. The most power anticipated from stream flow by 1985 is another 7 million kilowatts (28.3 million altogether), about 3.7 percent of primary energy, much less proportionately than in 1974.[40]

Government support is needed also for the expansion of the coal industry. Actually the immediate need is for a revival. In 1960 about 52 million tons of domestic coal provided 35 percent of primary energy; the 21 million tons extracted in 1974 supplied less than 6 percent, and of this amount 11 million tons were poor-quality coking coal. The hope of planners in government is an annual output of 20 million tons through 1985, which would mean a domestic source for 20 percent of all coal used.[41] In pursuit of this sensible objective,

TABLE 3.2

Demand for Coal in Japan, 1960, 1973, 1985
(1 million tons)

		1960	1973	1985
Coking coal	Demand	20	68	99
	Domestic supply	11	11	11
	Import	9	57	88
Steam coal	Demand	41	10	29
	Domestic supply	41	10	9
	Import	0	0	20
Total	Demand	61	78	128
	Domestic supply	52	21	20
	Import	9	57	108

Source: Look Japan, October 10, 1974, p. 20.

MITI has run a survey of the entire country with a view to boosting coal production. In the mid-1970s a number of Japan's 37 mines were expanded and several new ones were opened.[42] There has been pressure for nationalization of the industry to prevent mine closures and reductions in production.* The government-linked Electric Power Development Company has supported the construction of coal-fired plants in Kyushu coal mining areas where fuel could be provided domestically should imports become too costly. A small plant has been built in Kyushu that can process low-grade brown coal (lignite) competitively.

All of this is to the good beneath the certainty that coal demand will rise (see Table 3.2). The impetus toward coal is logical because the Japanese mining companies have superb liquefaction and gasification technologies. Moreover, coal is a better alternative in the short run than Japan's dubious nuclear power program, which has alienated even some officials of the Electric Power Development Company.

___

*On the condition of Japan's coal mines, see Japan Economic Yearbook, 1978-1979 (Tokyo: Oriental Economist, 1979), p. 83.

The drawbacks are mainly environmental. Air pollution is already a national blight. An escalation of coal burning, even with filtering devices, will exact a price in contaminated air. Millions of once quiescent citizens have become restive about their health in congested urban areas where much of the country's industry happens to be located. Other obstacles are rising costs, especially miner's wages, and the fact that only 6 percent of all electric generating plants can be converted to coal power. Serious work is being done by the Mitsui group on coal liquefaction and gasification. By 1980 Mitsui—a leading industrial combine—wants a liquefaction plant in Canada or Australia (200 million-ton capacity) to export sulfur-free fuel to Japan. There exists also a promising agreement between the Americans, Germans, and Japanese to pursue a large-scale coal liquefaction project. But the specter of high costs should not be treated lightly. Investment for coal development is put at $5.5 million for the period 1976 to 1985. In the past decade or so domestic mining costs have risen nearly 70 percent.[*]

Another expensive alternative is liquefied natural gas (LNG), of which Japan is the world's largest importer. In 1975 this energy source figured as a bit more than 1.8 percent of total energy demand. LNG is not cheap, but the Japanese, at least for now, can afford it and LNG imports for 1985 are likely to run between 5 and better than 6 percent of total energy use, with 83 percent of the imports generating 16 percent of the country's electricity needs. Naturally, storage and transportation of the volatile substance will pose all the difficulties for Japan that have been encountered elsewhere.

The erratic career of nuclear power has been a nagging disappointment for Japan. In 1973 it was announced by MITI that nuclear-generated electricity would account for 11 percent of all primary energy by 1985.[43] Because of setbacks and obstacles—steeply rising prices for reactors[†] and nuclear fuel, huge investment needs for fuel enrichment and reprocessing plants, reactor failures and mounting safety problems, the dilemmas of nuclear waste disposal, and determined local opposition to nuclear construction projects—expectations have been shaved down to 5.4 to 7.4 of primary energy by 1985, or less than one-quarter of electric power generation.[44] Earlier

---

[*]All of Japan's coal companies have big deficits and require government subsidy. See Japan Economic Yearbook, 1977-1978 (Tokyo: Oriental Economist, 1978), p. 80.

[†]The light water type will be dominant for some time, with only slight impact from thermal and breeder types by the early 1990s.

predictions that as much as 90 percent of all electricity production
would be nuclear by the end of the century have proved to be fanci-
ful.[45] Japan began its quest for nuclear power in 1954. The first
reactors came on-line in Mihama and Fukushima in 1970. As of 1977
there were 13 reactors operating at a yield of 7,430 megawatts;
another 17 reactors are approved or under construction with an output
of 13,840 megawatts. In 1978 about 7 percent of the country's need
for electricity was supplied by nuclear power, surely a modest result for all the years, expense, uncertainties, and headaches.[46]

Despite many impediments, officials and technicians are
loath to acknowledge a dim future for the nuclear option, or to admit
that criticisms of the program are anything but ill-founded, misguided, or premature. Several troublesome questions refuse to back
away.

First, there is the question of cost. In the early years of
nuclear development the outlays came to $7 billion. Investments for
1975-80 are expected to surpass $19.5 billion. Around $5 billion a
year is being spent on the commercial atom while only a minor fraction of that amount has been assigned to other sources of energy.
Between 1973 and 1978 the costs of nuclear construction rose at twice
the rate of general wholesale costs. Uranium prices jumped eight-
fold and reprocessing charges tenfold between 1975 and 1979. At the
present time the cost of nuclear waste disposal remains a foggy im-
ponderable, but it is not likely to be cheap. Adding to the escalation
of cost is the fact that many of Japan's reactors have had the world's
lowest efficiency rating and that complete shutdowns of problem-
ridden reactors have been frequent.

Second, there is the question of enrichment and reprocessing,
or the maintenance of a viable fuel cycle. Japan has only minute
domestic reserves of uranium ore, predominantly low-quality material amounting to about 8,000 tons.[*] Needs have been estimated at
more than 100,000 tons in 1985 and 190,000-230,000 tons by 1990,
the bulk of which must be imported from suppliers like Canada. Enriched uranium needs are likely to double between 1980 and 1990.[47]
The United States is presently the chief supplier, but a limit to those
shipments is anticipated by 1980, thus spurring the construction of
enrichment facilities in Japan.[48] All but one of Japan's reactors—
whether operational, under construction, or planned—are light water
types fueled by enriched uranium. In a move to reduce nearly total
dependence on U.S. supplies, an enrichment facility on Japanese soil

[*]Some estimates are lower. See Bulletin of IAEA 5 (1973):11

was completed in 1979, at a cost of $2.5 billion. Utilizing a centrifuge enrichment technology,* this pilot plant will produce at most 2.5 percent of the enriched fuel needed, which means that dependence on imports will persist into the 1980s and probably beyond. A reprocessing plant was completed in Ibaraki prefecture in 1977 with a capacity of 210 tons a year, but it is plagued with doubts in the spheres of economics and safety; the plant discharges more radiation in a day than comes from a nuclear plant in a year. The problem of securing enough atomic fuel may be compounded in years to come by a spreading distaste for involvement in the processes of enrichment and reprocessing among current and potential suppliers.[49] Moreover, the countries supplying Japan have their own energy crises and nuclear power ambitions to cope with.

Third, there is the question of safety. The early Mihama plant has a long record of accidents and was shut down four times between 1972 and 1977. One of those accidents threatened the nuclear fuel rods and was concealed from 1973 to 1976, thus feeding public distrust of safety measures and government assurances. A range of such failures in the nuclear industry has tended to break down the distinction between "absolute" and "social" safety in the public mind. Some part of this problem in the safe management of nuclear technology can be explained by the fact that Japan is working with borrowed technology and has a weak foundation of basic research. On the other hand, one should not underestimate intangibles like the "nuclear allergy" that swept Japan after the catastrophes of Hiroshima and Nagasaki, which has been intensified by a rash of technical and structural failures in "nuclear prefectures" like Fukui (north of Kyoto) and Fukushima (north of Tokyo), and more recently by the Three Mile Island incident in the United States. Three Mile Island has stiffened the determination, for example, of an antinuclear group in the town of Kucho on Shikoku, the least developed of Japan's islands. The Shikoku Power Company wants to build a nuclear facility near Kucho. A 66-year-old spokesman for the antinuclear movement, Fusaichi Hirono, said that "the Three Mile Island accident showed that the government and company's view that nuclear power generation is perfectly safe because of double and triple safety checks was a sheer lie." It was said in the Mainichi Shimbun, one of Japan's leading newspapers: "It is as if we were all living in Pennsylvania."

---

*Using 10,000 centrifuges, the technique is expected to be six times more energy efficient than the diffusion method used in the United States.

The accident came like a kamikaze, a "divine wind," the allusion being to a great storm that destroyed an invading Mongol fleet in the thirteenth century.[50]

The investment of $10 million for nuclear safety research has not improved the Japanese safety record, which remains one of the world's least reliable.[51] Yet the builders of nuclear plants—Mitsubishi Heavy Industries, Ishikawajima Heavy Industries, and Kawasaki Heavy Industries—dismiss citizen protests as an ephemeral public relations issue but are willing to confess that local resistance has slowed things down.

Fourth, there is the question of nuclear waste disposal and other forms of pollution. A reactor generating 1 million kilowatts leaves a kilogram of lethal ash each day. The Science and Technology Agency has estimated that fuel levels by 1985 will be 1,600 tons a year in Japan, with an annual waste yield of 45,000 cubic meters, or 225,000 drums. So far the only methods of disposal have been ocean dumping and limited shipments of waste to Britain. In the first instance, a test program between 1957[*] and 1965 saw 1,661 drums (406.8 curies) of radioactive solids dropped to the sea bottom a few thousand meters off Tateyama in Chiba prefecture.[52] In the second instance, citizens of both countries were surprised to learn that nuclear waste had been moving by sea for some time from Japan to Britain.[53] The Japan Atomic Energy Commission has assured the public that low-level wastes are "to be discharged into the environment in such a way as to ensure an adequate degree of safety," while both medium- and high-level wastes "are to be stored carefully for the time being."[54] What these statements mean, of course, is that Japan has no solution to waste disposal.

For meteorological reasons, heat from nuclear plants in Japan cannot be discharged from cooling towers. A large amount of water is needed. If atomic power expanded to the level of 100,000 megawatts by 1990, warm water would be discharged (0.7 tons a second for every kilowatt at a raised temperature of 10 degrees centigrade) at the rate of 250 billion tons a year. This would be thermal pollution on a grand scale, for the total stream flow of Japan is just 400 billion tons.

On first blush the atom seems to be an easy, technologically glamorous escape route from bondage to fossil fuels. The other side of the coin is that nuclear planners in Japan are pushing ahead in reckless disregard of real technical uncertainties and environmental

---

[*]When Japan's first U.S. research reactor went critical.

constraints, quite apart from escalating costs of nuclear fuel and construction.

In 1974 considerable fanfare announced the "Sunshine Project," whose ostensible purpose was to "cope with the burning question of energy crisis" by promoting new energy sources: solar, geothermal, synthetic natural gas (SNG), and hydrogen energy. In less than two years the program dropped quietly out of sight. The reason is not obscure. In the short run it is more convenient to buy additional supplies of fossil fuel and to move along, however ineptly, with nuclear power. This choice inflicts the least amount of immediate discomfort and is perceived to be the line of least resistance. The predictable outcome of this policy is that renewable energy sources, whatever their theoretical prospects, will not be conspicuous in Japan's larger energy picture until the next century, assuming an orderly transition can be made from an economy based on nonrenewable energy to one exploiting renewable energy. Before summarizing the alternatives of interest in Japan, it is worth repeating that productive conservation is the best way to turn up large amounts of energy at low cost. Moreover, conservation as a key element of energy planning would stretch out Japan's fossil fuel subsidy while reducing imports.

Nuclear fusion technology lies 20 to 30 years in the future in the most optimistic assessment. At least $10 billion will be needed for testing facilities and the development of fusion technology. An experimental reactor is the goal for 1995, to which end Japanese plasma physicists are trying to achieve, sustain, and contain the incredible temperatures basic to a fusion reaction. Given the extent of Japan's dependence on fossil fuel imports, the money would be perhaps better spent on sources that can provide immediate relief, such as investment for greater efficiencies from existing energy conversion systems.

The oil equivalent of solar energy available in Japan is estimated at 288 million kiloliters a year, which compares favorably with the 224 million kiloliters of oil imported in 1971. About 2 million solar water heaters have been installed around the country, making Japan a world leader in limited solar application. Tentative long-range plans had included solar technology for industrial purposes—power generation, air conditioning, and furnace operation—by the close of the century. Even before the Sunshine Project went into eclipse, the view prevailed that solar energy would be too expensive and aggravate the land crisis by eating up space for solar collectors. As a result, the outlook for solar energy is that it may provide relief for an indeterminate number of homeowners wise enough to accept it before the bottom falls out of the fossil fuel market. The land issue is not the problem it was made out to be, for it would be worsened only if nuclear and solar plants were built. Solar collection space is no more expansive than the "safety space" surrounding nuclear plants. The point is that a choice is necessary; Japan cannot have it both ways.[55] The real weakness of the solar option for Japan is neither money, space, nor technology, but rather inauspicious conditions of climate in many parts of the country. Maneuvering around this weakness would need more commitment from government and industry than has been forthcoming. If there is a quick way to make photovoltaic cells economical, a reasonably sure bet is that the discipline and resourcefulness of Japanese industry can find it, but the attempt has to be made.

Although Japan seems well suited to geothermal energy, there is uncertainty about its true potential. At the present time six geothermal power stations are operating in Japan. Their combined capacity is around 133,000 kilowatt hours. Estimates suggest that thermal energy from volcanoes and dry-hot rocks might generate in excess of 20 million kilowatts a year. Japan's success with geothermal energy, after 14 years of effort (the first station was built in 1966 at Matsukawa in Iwate prefecture), must be compared with that of other countries: 390,000 kilowatts in Italy, 300,000 in the United States, and 200,000 in New Zealand. Why is Japan's performance so relatively modest? The low profile is explained by a combination of difficulties, including knotty technical problems, siting dilemmas (27 out of 30 prime geothermal areas are in national parks), ambiguous interpretation of survey data, insufficient pilot programs, and a halting political will to act.[56] Environmental objections are a prominent obstacle to geothermal development. The Environment Agency opposes the use of national parks for power plants of any kind, a point of view easy to respect in a crowded nation with little usable land. Furthermore, the cost of antipollution measures, such as preventing the discharge of arsenic in large amounts, has risen to prohibitive levels. Even should the optimism of planners be justified by a geothermal yield of 6 million kilowatts by 1985, this would be no more than a drop in Japan's energy bucket.

So far the Japanese have neglected tidal and wind power, both of which are variants of solar energy. Whatever the potential, neither seems practical or economical in the context of energy needs through the next decade. By way of illustration, a wind power plant designed to yield 1,000 kilowatts might cost $3.5 million, and 4,880 windmills would be needed on offshore platforms to supply the same amount of energy on an annual basis that could be obtained from one 12,000-barrels-a-day oil platform.[57]

Magnetohydrodynamics (MHD), a technique for converting oil and other fuels directly into electrical energy by means of a high-intensity magnetic field, is being explored, although the field of MHD research is in the nascent stage and confronts numerous difficulties still far from solution. It will be a long time before relief comes from that direction. Hydrogen chemical fuel, the basis of a potentially innovative source of energy, capable of fueling vehicles as well as power plants, could provide a minute fraction of the country's energy in 25 years, but no one expects more than that.[58] All listed commercial processes for hydrogen production, except the electrolysis of water, depend on the combustion of fossil fuels and cannot be described as "clean." A hydrogen energy system has yet to be developed that satisfies reasonable criteria of economy. The "hydrogen economy" is, at best, a distant prospect for Japan.

INTERNATIONAL ENERGY POLICY

No country in the world with major economic status is more dependent than Japan on imports of fuel, minerals, and other staples essential to national well-being.[59] Almost 90 percent of the country's energy must be hauled in from the outside. An embargo on foreign oil supplies, for whatever reason, would paralyze the economy within a few months. Without a steady flow of fuel, raw materials, and food, a once self-sufficient Japan would perish.

As noted earlier, the foundation of Japan's energy policy—indeed, the substance of it—is to stabilize a reliable, growing supply of fossil fuel from abroad. This is the central theme of the General Energy Council's report: "Policy for Securing a Stable Supply of Energy in the Decade Beginning 1975."[60] In virtually all assumptions and estimates made since 1973, oil will be king in Japan's economy through the 1980s, and in all calculations about the future, dependence on foreign sources of petroleum is the looming fact of life. A nearly perfect 100 percent of Japan's oil is imported (99.7 percent) and 80 percent[*] of that comes from the Middle East.[61] Until recently the two biggest suppliers were Saudi Arabia and Iran, their shares being about equal in 1975. By 1979 Saudi Arabia was shipping ten times the amount coming from troubled Iran (see Table 3.3).

The oil industry in Japan has virtually no producing sector; the emphasis is on refining crude oil and marketing products. Even

---

[*]The proportion was greater (90 percent) in 1966, but the absolute quantity doubled in six years. Calculated with regard to primary fuel rather than final energy use, the import-dependence factor jumped from 18.8 to 82.3 percent between 1955 and 1970.

TABLE 3.3

Japan's Crude Oil Imports by Supplier, FY 1974

Area	Total (1,000 kl)	Percent
Near and Middle East		
Saudi Arabia	61,506	22.3
Kuwait	24,733	9.0
Neutral Zones	15,539	5.6
Qatar	270	0.1
Abu Dhabi	26,842	9.8
Dubai	1,568	0.6
Oman	6,285	2.3
Iran	73,695	26.8
Iraq	2,572	0.9
Bahrain	79	0.0
Egypt	113	0.0
Subtotal	213,202	77.4
Asia		
Indonesia	37,251	13.5
Malaysia	9,095	3.3
Subtotal	46,346	16.8
Africa		
Libya	4,442	1.6
Nigeria	4,862	1.8
Cabinda	509	0.8
Subtotal	9,813	4.2
Other Areas		
Venezuela	362	0.1
Australia	162	0.1
USSR	232	0.1
People's Republic of China	5,119	1.9
Grand total	275,233	

Note: Quantities are not precise and do not total 100 percent.
Source: Japan Economic Yearbook, 1975–1976 (Tokyo: The Oriental Economist, 1976), p. 79.

141

in the case of refining and marketing companies, ownership is shared with foreign oil interests like Caltex, Exxon, Shell, Mobil, and others. Idemitsu is the only Japanese company that has escaped cross-share-holding deals. There is no vertical system, as in the United States, structured from wellhead to consumer. Petroleum dependence for Japan sits on two legs: Middle East oil fields and the international oil companies. In this dual arrangement the former source has become dominant since 1973 through the assertiveness of OPEC, something of a break for the Japanese because it has given them more flexibility in negotiations than was the case when oil imports were ruled by long-term, restrictive contracts with international companies.[62] The shift of oil power to Arab and Iranian producers has meant that Japan can maneuver freely in the Middle East to consummate advantageous business deals and to improve supply and pricing agreements, all with less danger of provoking the companies to vengeful acts (see Table 3.4).

The issue of Japanese-developed oil since 1973 has become vague and somewhat pointless. Japanese "development" implies sources independent of the oil companies, which have become the agents of OPEC, that can be worked by Japanese personnel, capital, and technology. Even if such consortia were a reality in the Middle East, they could not prevent manipulations of supply and price by Arab owners. On the most optimistic assumptions of a low Japanese growth rate, freedom of Japanese Middle East concessions from production-sharing agreements, and substantial annual production from Arabian Oil, JAPEX Indonesia, and the Siberian Tyumen fields (say 100 million tons altogether), the Middle East share of Japan's oil supply by 1985 would still be over 70 percent.

A further headache for Japan's independent companies is the recent national commitment to buy oil from the People's Republic of China (PRC). Oil imports in the quantities expected by the Chinese could wreck the Japanese independents unless government subsidies come to the rescue. The dilemma is shaped by: oil purchase commitments abroad, other than PRC; the wish to pursue more independence in the energy field; and pressure from oil suppliers and businessmen at home to purchase more oil than is needed to avoid hurting business should the oil purchase agreement with the PRC run into trouble. Another wrinkle is that U.S. oil majors may buy surplus Chinese oil and resell it to Japan, thus striking an indirect blow at feeble Japanese independent companies.[63] These tensions and uncertainties have persuaded MITI to consider allowing the Japan National Oil Corporation (JNOC) to engage directly in oil prospecting and production. Heretofore the law has allowed JNOC to do no more than provide financial assistance to private companies. Now JNOC has been asked by the Chinese to involve itself directly in the

TABLE 3.4

FY 1976 Crude Oil Imports into Japan
by Foreign Suppliers

Supplier	Percent
Caltex	13.9
Exxon	12.3
Mobil	7.9
Gulf	7.1
Shell	10.8
B.P.	3.9
Compagnie Francaise des Pétroles	1.4
Total for majors	57.3
Getty	1.8
Unoco	2.2
Others	1.3
Total for independents	5.3
State-run oil companies	8.2
Oil developed by Japanese-affiliated firms	7.1
Trading firms	16.8
Grand total	100.0

Source: Japan Economic Yearbook, 1978-1979 (Tokyo: The Oriental Economist, 1979), p. 85.

development of oil fields in the Pearl River estuary. There is readiness to accept the invitation because of Japanese oil company weakness compared to international giants like Exxon, Union, and Mobil, which have been approached also by the Chinese. [64]

Japan's failure to create an independent shelter in the world's energy market has influenced policy objectives in two ways: first, in the direction of cultivating good relations with the oil-producing nations and, second, in the direction of diversifying sources where it is practical to do so. In 1976 foreign assets came to $13 billion, most of it scattered geographically but concentrated in the resource and manufacturing industries, an investment pattern designed to speed the flow of raw materials into the country.[65] For example, the greater part of $1.4 billion of investments in Britain goes to British Petroleum's Abu Dhabi concession, in which Japan is a minor shareholder.

Several investment-cooperation trends can be identified as having more than transitory significance. The first is Japan's offensive to squeeze more energy out of nearby Asian sources. In 1977 about 22 percent of crude petroleum imports came from Asia. MITI has set a goal of 30 percent by 1990.[66] With JNOC active, the rate of Japanese-developed oil might rise from 9 percent in 1977 to 20 percent in 1990. There is less emphasis on Southeast Asia than in the past because of environmental restrictions, local hostility toward Japanese nationals, and persistent rumblings of sociopolitical instability. Back in 1974, for instance, Japanese businessmen, uneasy about Thai politics, decided to shelve a $450 million petroleum project in Thailand. On the other hand, economic relations with China have undergone a transformation since diplomatic ties were established in the early 1970s.[67] Under terms of a long-range trade pact, Japan is committed to export plants and construction equipment to the PRC worth $10 billion. Nippon Steel is building a $3 billion steel mill in Shanghai, due for completion in 1980. In return the Japanese are hoping for substantial deliveries of oil and gas—in the case of oil, rising to 40 million metric tons or more by 1985. To develop Chinese petroleum reserves, Japanese banks have rushed forward to make financing proposals, encouraged by China's departure from a traditional policy of rejecting foreign loans as a means of capitalizing domestic projects. Peking is even ready to use funds from Japan's official aid-dispensing organization, the Overseas Economic Cooperation Fund. Oil refiners in Japan are not entirely happy about the Chinese connection. Taching oil (all of Japan's imports come from the Taching fields) is high in paraffin content, which means that something like 20 heavy oil crackers will have to be built at a cost of $11 billion to process the 40 million metric tons Japan is committed to take in 1985.

China wants to export surplus LNG to Japan in return for liquefaction technology. The motive behind such an arrangement is not purely economic, for the Chinese are anxious to lure the Japanese away from Soviet inducements to participate in the development of Siberian resources. Japan wants to increase natural gas imports beyond the 6 million metric tons consumed in 1975-76 to meet a demand estimated at 20.6 million tons in 1980 and 42 million tons in 1985, or 8 percent of all energy needs. Most of the 1975 supply of

LNG came from a Shell-Mitsubishi project in Brunei. Projected supplies are expected to come from long-term contracts with Malaysia, Indonesia, various Middle Eastern countries, and China. Imports from the PRC are based on confirmed natural gas deposits in the range of 700,000 million cubic meters, with total deposits that may equal those of the Middle East (1.9 million cubic meters). In return for Japan's help in the construction of a big LNG plant (300,000-ton capacity per year) near Tientsin, the Chinese are amenable to a 20-year contract for the sale of LNG at international prices.[68]

The other side of Japan's shift to the Asian energy scene concerns the Soviet Union. A once-promising liaison for energy development in Siberia and Sakhalin has foundered in uncertainty because of a Russo-Japanese dispute over the possession of four islands in waters north of Japan and uneasiness in Moscow about the Japan-PRC economic alliance, which is certain to strengthen Chinese military potential. These rough diplomatic edges notwithstanding, Japan has been counting on reliable Soviet deliveries of coal, gas, and oil from Siberian fields,* developed in part with Japanese capital. The outcome of this collaboration is doubtful for several strong reasons.[69] Some American industry people wanted to get credit financing to cooperate with the Japanese in the Siberian venture. The snag was congressional opposition to a subsidy used for the development of Soviet natural resources. With tight money, recession, and politics eliminating the United States and contracting capital in Japan, the Soviet adventure has seemed less attractive. Moreover, the Soviets do not appear to have ready capital for a solo performance. While Japan is hardly capital-poor, despite cutbacks since 1973, Japanese investment has been sliding steadily into the Chinese camp rather than trying to milk two cows at once. The market for Japan's technology in China is immense, and the two countries have historic cultural bonds. In the face of all complications and ambivalence, Japanese foreign policy prefers harmonious relations with both resource-rich communist countries, an attitude consistent with Japan's traditional policy since 1952 of economic self-interest before ideological commitment. The East Asia triangle, into which U.S. concerns are regularly insinuated, will be worth observing in the next

---

*Specifically the Tyumen oil fields and Yakut natural gas fields. Soviet coal exports (chiefly South Yakutian coking coal) were to have gone to Japan, starting in 1983, at the rate of 7 million tons a year for 20 years.

146 / NATIONAL ENERGY PROFILES

decade as Tokyo tries to please everyone while pressing for maximum leverage in the natural resource markets of the world.

The second investment-cooperation trend is toward a closer relationship with the Arab oil states, the "very happy marriage which will have no divorce" toasted in January 1974 by Foreign Minister Ohira of Japan and Sheik Yamani of Saudi Arabia. The reciprocal basis for the union is oil for Japan and capital investment spiced with technical assistance for Saudi Arabia. A "clarification"* of Japan's position on the Arab-Israeli conflict was followed by a brisk revival of business between Japan and the Arab states, thus illustrating the ease with which oil can sway politics in energy-dependent states. In the past five years the climate of amity has been sweetened by loans and projects. Both Iraq and Iran have received billion-dollar loans in return for an assured flow of oil. The sudden collapse of Iranian stability has been disconcerting, especially for Mitsui industries, whose joint petrochemical project with Iran at Bandar Shahpur was left in abeyance some 80 percent complete. The Iraqi government has promised to deliver 1.1 billion barrels of oil over a ten-year period; the loan is to be used for construction of a refinery and a gas liquefaction plant. Saudi Arabia, Kuwait, Iraq, and Iran, until recently, have been keen on Japanese assistance to engineer and build their petrochemical industries. An offshore oil exploration agreement has been negotiated with North Yemen, and in early 1974 about $280 million in "reconstruction" money went to Egypt as an incentive to open the Suez Canal. In Algeria a $276 million telecommunications system has been engineered by Fujitsu. An oil refinery in Syria has been financed with $100 million of Japanese money. These and other projects are part of an overall strategy that includes exporting more goods to the Near and Middle East, a flow that has been rising at some 50 percent a year since the mid-1970s, promotion of Arab investment in Japanese overseas projects, and the encouragement of substantial Arab deposits in Japanese banks. So far the strategy has been vindicated by results.

In the meantime Japan has to face the problem of maintaining good relations with familiar, older trading partners in North America and Europe, the former being an important source of food and raw

---

*Diplomatic relations with Israel were not broken, to keep the United States happy, though the government said, "Japan may have to reconsider its policy toward Israel." The rights of the Palestinian people were recognized and affirmed. This was in November 1973.

materials. A sore point on the international scene is Japan's ability to pile up enormous trade surpluses. At the close of fiscal year 1978 (March 31, 1979), the trade surplus was $249 billion, some 23 percent above the previous fiscal year. Japan's trade and payments surplus is a tribute to the energy and resourcefulness of Japanese business, but reactions have tended to be negative, especially in the United States where protectionist sentiment has been stimulated. The trade picture is an uncomfortable triumph for Tokyo, because a friendly export market in both Europe and North America is indispensable to the country's economic stability.[70] Good bilateral relations underlie the satisfaction of import needs as well, and there are various cooperative enterprises requiring an atmosphere of mutual respect. For example, most of Japan's coking coal has been shipped from the United States, Canada, and Australia.* There is a contract with the United States for 10 million tons of LNG and a $1 billion agreement (worked out in 1978) for energy research in nuclear fusion and coal liquefaction. For Mexican oil $1 billion in loans is promised.†

Japan's business success has been admired and envied. The dynamics and mechanisms of this success, such as consensus decision making, industrial planning, worldwide marketing, and "people-centered" business, are well understood and may in time come to be widely imitated.[71] The great issue for Japan is how long this success can be sustained when the country depends almost wholly on strategic imports through the medium of a fragile international order. It should be obvious that a lot can happen outside the perimeter of Japan's cognizance or control: wars, revolutions, unpredictable enmities flashing to the surface, resource shutouts for the sake of politics, ideology, or sheer pique, global recessions, or perhaps a supertanker breaking up in a narrow strait and snapping the petroleum lifeline. Should the United States and Europe conspire to humble OPEC in the event of further embargoes or ruinous price hikes, the Japanese would be placed in a delicate position. Competition for energy is certain to become more fierce and ungentlemanly as the world's nations attempt to absorb another 2 billion people in the course of a mere generation. An economy the size of Japan's, with its exceptional vulnerabilities, will have to rely for survival on luck, wise diplomacy, and a willingness to harmonize self-interest with

---

*Japan is prospecting heavily for uranium ore in Canada and Australia.

†The Mexican president said in 1978 that 20 percent of Mexico's output could be going to Japan by 1980.

all the subtle imperatives of international stability. In the past Japan has cut a poor, awkward figure on the international stage, seeming too often crude, opportunistic, narrowly economic in purpose, and indifferent to the legitimate interests of other nations, rich and poor alike. All of that will have to change.

ENERGY AND THE ENVIRONMENT

There is reason to believe that energy conversion in Japan will be checked in due time by physical limits that technology is quite powerless to conjure away. With about half the GNP of the United States, the Japanese homeland is already a celebrated environmental mess. The effects of an economy twice as large may be unendurable, both for nature and the people of Japan.[72] Only recently has there been a growing appreciation of the destructive environmental impact entailed by GNP, satirically rendered by some as "gross national pollution." A formerly exclusive focus on production has shifted noticeably to the "diseconomies" (costs not included in the calculus of productivity) of environmental disruption, known in Japan as kogai, or "public nuisance."*

The pollution level in Japan is somewhere between 10 to 40 times greater than it is in other industrial states. One price for economic renown has been the proliferation of bizarre pollution-related diseases—Minamata disease (mercury poisoning), Itai-Itai (cadmium poisoning), PCB (polychlorinated biphenyl) poisoning, and Yokkaichi asthma—which have killed hundreds and disabled thousands[†] of citizens.[73] These and other ailments are direct consequences of economic activity producing enormous quantities of industrial waste in the range of many hundreds of millions of tons a year.[74] The sewage and domestic waste of urban households add to the burden. Production, consumption, energy use, and pollution are inseparably bound together in the same growth phenomenon. The Central Council for Environmental Pollution has warned that "even if the growth rate turns sluggish in the future, the total release of environmental pollutants will increase drastically because of the expansion in the scale of economic activities."[75] In the case of air pollution, for example,

---

*The phrase kogai depato (pollution department store) has also been coined.
[†]In 1976 the "officially recognized" pollution victims came to 33,466.

it was estimated in the early 1970s that major contaminants—sulfur dioxide, nitrogen dioxide, carbon monoxide, suspended particulate matter, and dust fall—would swell 3.6 times between 1970 and 1985 on the assumption of an 8.4 percent growth rate for the economy with no change in industrial structure. Sulfur content of the air was about 200,000 tons in 1955, 2 million tons in 1970, and would reach some 7.25 million tons in 1985, the latter figure assuming the implementation of air pollution laws already on the books.[76] One can infer from all of this a doubling of air pollution at a 6 to 7 percent economic growth rate, with some modification of industrial structure. The problem of air quality in Japan is less severe than it might be because of recent laws, but the absolute quantity of pollutants is sure to increase as the number of vehicles, power plants, and industrial concerns multiply.[77] It is no accident that virulent air pollution is linked to the spread of energy-intensive producers of iron and steel, ceramics, paper-pulp, chemicals, petrochemicals, and power. As the economy expands there will be predictable effects on human health, local climate patterns, vegetation, and water systems, all of which receive a steady rain of contaminants from the sky.

The deterioration of fresh water supplies is more serious to industrial Japan than air pollution. Without large amounts of usable water the country's entire economic life would falter. At a time when the demand for water is soaring, the supply is shrinking in both quantity and quality.[78] The consumption of industrial water, which must be fresh, rose 33 percent from 1966 to 1972. Public consumption in urban areas nearly tripled from 1969 to 1976. At pre-1973 growth rates the need for water would approach 116 billion tons a year by 1985, compared with 69.5 billion tons consumed in 1965. Among other things that have to be done to raise the utilization of water, five times the number of earthquake-proof dams already in existence (about 1,100) must be constructed, at a huge investment of capital and energy resources.[79] Water shortages are in the offing because of unrealistic economic practices that simultaneously soak it up and pollute it.[80] Coastal water resources, which supply a large portion of Japan's nourishment in the form of plant and animal life, are threatened nearly everywhere by effluent, sewage, and oil contamination. The latter should come as no surprise in a land besieged with oil tankers, oil storage depots, and oil-dependent industries. Out of more than 2,000 coastal pollution incidents in 1975, nearly 80 percent were due to oil. The year before a refinery tank on the Seto Inland Sea burst and spilled 43,000 kiloliters, which spread over some of Japan's best domestic fishing grounds.[81] Coastal water pollution drives Japanese fishing boats far away from home in search of protein and necessitates heavier food imports. The thermal effects on water from power plants have been extensive and are expected to worsen if the nuclear industry achieves its future goals. Through the 1980s there is likely to be considerable tension between the greedy water needs of industry and those of an expanding urban population.[82]

Also entangled with energy conversion is the land crisis of the past decade. Only 30 percent of Japan's 370,000 square kilometers is habitable. The remainder is swallowed up by mountains and forests. If one deducts land under cultivation, roads, parks, golf courses, and the like, the actual "working and living space" comes to 46,000 square kilometers, a minuscule 12.5 percent of the habitable portion. Most of Japan's population, the world's eighth largest, is packed into an area the size of Switzerland and more than 60 percent live in a narrow Pacific corridor between Tokyo and Kita-Kyushu; known as the "Tokaido Megalopolis," the strip between Tokyo and Kyoto is especially congested. Usable land has become fantastically expensive as demand has swamped supply in the wake of intense economic, urban, and demographic growth. Land prices in 1974 were up 8,300 times over those of 1936 and are now 20 times or more those in West Germany. Values in the early 1970s were escalating at 30 percent a year. Current inflationary pressures are adding 15-18 percent annually to the cost of land, a variable more crucial to the price of power generation in Japan than in larger, more sparsely inhabited lands.

A second unique feature aggravating the land issue is the fact that ownership throughout the country is confined largely to small parcels such as farm plots. In 1972 land was owned by 29.5 million people and 640,000 companies, while in residential areas the respective numbers were 12.1 million and 490,000.[83] The consolidation of land units suitable for the construction of energy-generating facilities, for which water must also be available, is a major obstacle to rational energy planning. In 1970 a significant round of purchases either by government or industry would have cost about $500 billion; by now, of course, the figure is much higher. Land needs for industry and energy facilities are growing at the rate of 10,000 hectares a year. It is not very clear how electric power demand can nearly double between 1975 and 1985 when appropriate sites for power plants are needed desperately for urban, residential, transport, oil storage, and other purposes.[84] Where will hundreds of new power plants be sited? How will dense populations make room for a Gorgon's head of power lines? Where will the water for power generation come from? How can monstrous quantities of fuel be stored and moved around the country without transcending physical limits? So far Japan's planners have confined their discussion to the means of procuring more oil and tend to ignore the issue of how to accommodate it physically. All of these questions are mostly unresolved. A meaningful land use policy consonant with the carrying capacity of

Japan is still in the fledgling stage,* and industrial development proceeds wherever it can with only minor government interference.[85] The sole directions left for development are straight up (high rises) or into the sea by means of wholesale reclamation projects, which are achieved by lopping off entire mountains and dumping them into the sea or by piling up refuse as in the case of "Dream Island" in Tokyo Bay.

After 20 years of memorable economic growth, it is conceded widely in Japan that public welfare has lagged behind.[86] The rapid accumulation of wealth and material power has not closed the welfare gap (fukushi gappu) and benefited the Japanese people as one might casually expect. To be sure there are many quantitative indexes—per capita disposable income, possession of consumer durables like vacuum cleaners and television sets, the prevalence of formal education, longevity—that say the Japanese have not done badly, that the second industrial revolution has paid off. Judged by other criteria, however, industrial civilization in Japan has not fulfilled its promise and seems harried by constitutional defects. Frequent complaints are sounded about inadequate housing, endemic overcrowding in working and living space areas, perpetually intolerable commuting conditions (requiring professional "pushers" to jam the last possible body onto the trains), a shoddy social and natural environment, and the rising expense of food and other necessities. Since large-scale energy conversion is popularly associated with "progress," a true perspective on Japan's prosperity asks one to consider these deprivations of the affluent.

The land crisis and inflation have precluded ownership of homes on a plot of soil for all but the well-to-do. Millions have been shunted into cramped, poorly constructed tenements, and the rental market has eclipsed the buyer's market. In 1970 some 70 percent of employee households had television sets and 98 percent had washing machines, but at the beginning of 1976 only 0.6 percent had central heating and 75 percent of the entire population was without sewer service.[87] The spread of industry and its concentration in urban areas has meant blight, ugliness, and pollution. The city of Yokkaichi, famous chiefly for its respiratory ailments, displays an utterly irrational juxtaposition of heavy industry and urban life. Tokyo is the home of no less than 83,000 factories of varied size.

---

*A Land Utilization Bill was passed in 1974, whose effectiveness still remains to be seen.

Together they blanket the city with noise, vibration (two of the most common citizen complaints), and a wide spectrum of pollutants. The industrialization of Japan, or its advent as a "high entropy" culture, has damaged three-fourths of the country's vegetation and keenly affected the ecology of all but four of 51 major rivers.[88]

One might suppose that a country with the world's third largest GNP would be nutritionally advanced, yet the Japanese worker "is marginally nourished as compared with the working population of other industrial nations."[89] Farmers have been siphoned off the fields to work in factories, which have carried their own activities, along with urban sprawl, into pockets of irreplaceable arable land. The result has been a decline of food self-sufficiency, exacerbated by population growth far in excess of the land's carrying capacity. Fifty percent of Japan's food must be imported, a form of dependence and insecurity as basic as petroleum dependence.

More free time and income have produced the paradox of "leisure pollution."[90] Japan's traditional beauties have been undermined or destroyed by industry, urbanism, and domestic tourism. Eager to escape cramped homes and cities denuded of vegetation, millions of citizens take to the roads and rails in search of quiet diversion and legendary charms of nature, only to find parks and scenic places inundated by cars* and surging crowds. Automobile exhaust has destroyed tree cover in historic spots (such as the Saburu road to the fifth stage of Mount Fuji); wildlife habitats have been fragmented by roads, parking lots, and tourist accommodations; and panoramic views are frequently a jigsaw of power lines. It is ironic that harried travelers find away from home, more often than not, the unpleasantness they sought to escape. No wonder that polls have tended in recent years to show growing unhappiness, pessimism, insecurity, cynicism, and doubts about the future of the "economic miracle." As one writer put it: "The problem is the Japanese are being glutted with a surfeit of material possessions. Yet, despite these possessions and despite the rising GNP 'scores,' many Japanese do not see any improvement in the quality of their lives."[91]

One may surmise that the good life requires something more than an impressive productive system and lots of oil to burn. This is no gratuitous judgment by an outsider but the attitude of many thoughtful Japanese who have begun to resent the social price of a

---

*In a country poorly suited to automobile transportation, cars are seven times more numerous by unit of land area than in the United States.

production-first policy during the era of high growth. Others have come to realize that Japan's utilization of energy has already over-taxed physical and social systems of the nation, opening the disagree-able vista of a future disrupted by crisis storms.

FUTURE PROSPECTS

The argument of Japan's political and economic elite is that social and environmental needs cannot be satisfied without more growth in the vicinity of 4 to 7 percent of GNP annually; the former is thought to be a minimum rate, while the latter is not considered a ceiling by any means. There has been animated debate in recent years about the nation's domestic priorities and world aims, but actual policy and economic behavior have changed little since the promulgation of the New Economic and Social Development Plan, 1970-1975, which exalts the growth potential of the economy and con-siders a 10.6 percent growth "appropriate."[92] What has changed is the rhetoric, not goals or aims, about which there has been no fresh consensus: "Private quality of life aspirations still outstrip any gen-eral will for ecological rehabilitation."[93] Herman Kahn, the most optimistic champion of Japan's economic future, wrote before 1971 that "this very confusion about ultimate goals makes even more likely a heavy concentration in the immediate future on the simple—and intrinsically attractive—goal of catching up with or surpassing the West economically and technologically," a goal Kahn is reluctant to have the Japanese abandon in 1979, arguing that more growth can solve the problems of growth.[94] Although written in the early 1970s, these assessments still ring true as Japan enters the 1980s.

In spite of protestations by press and citizenry that the public interest has been sacrificed to big business, affirmations of contri-tion from the business moguls themselves, and routine promises from LDP politicians that reform and saner policies are imminent, the ship of state is carrying the economic machine along much the same course as before.[95] The Japanese are at odds with themselves. Not being able as yet to decide the extent to which material aspira-tions should be pared down or modified, the choice by default is to muddle along with self-serving gestures of restraint while absorbing as much energy as the global economy and international tolerance will permit. Flourishes of economic discipline in the past few years have been directed to the conventional, familiar purpose of controlling inflation by temporarily reducing consumer demand, not to the more radical purpose of deflecting the economy in a new direction. The government seems reluctant to show initiative, resting content with frantic activity when the heat is on. There is little inclination to seek bold, imaginative strategies in the arena of energy. The Japanese style hinges on expedients and ad hoc measures. The flexibility of the immediate postwar years has atrophied as the economy has grown in size and complexity. Japan's managers are committed, it seems, to wrestling with crises as they arise. Changes in the top leadership positions of the LDP have very little substantive meaning, the differences among Sato, Tanaka, Miki, Fukuda, and Ohira, to mention five recent prime ministers, being quite marginal when one cuts through the rhetorical dust. Masayoshi Ohira's* recently stated belief that "a strong leader is not needed for the Japanese people because they themselves are full of vitality" is wholly at odds with the grim reality of Japan's vulnerability in a dangerous age of ecological scarcity.[96] His up-to-date vision of a "cultural age" displacing the earlier "age of politics" (initiated by Prime Minister Yoshida in the early 1950s) and the "economic age" (initiated by Prime Minister Ikeda in the early 1960s) has all the marks of being a pleasant scrim behind which the main actors, in business, will get on with the task of making Japan as rich as possible.

Neither moderate planners nor ebullient "sweet singing" economists like Hsiao Kanamori have understood that "relations between the great systems on which society depends are upside down."[97] In a rationally organized industrial society one would expect the economic system, which distributes wealth, to be subordinate to the productive system, which yields wealth, and both of the foregoing to be subordinate to the ecological system that provides the basic resources. Like virtually all modern industrial societies, Japan has the triad reversed. The economic and productive systems take precedence over the environmental resource base; "the faulty design of the production system has been imposed by the economic system, which invests in factories that promise increased profits rather than environmental compatibility and efficient use of resources."[98] From this perspective Japan's environmental and conservation programs can be seen for what they are—cosmetic adjustments within a fundamentally unhealthy, topsy-turvy system of economic, productive, and ecological relations. The Japanese have before them two broad options. The first is to perpetuate the existing system of relations as long as possible by tinkering here and there with its flaws, the most evident being a nearly total reliance on nonrenewable sources of energy brought in from the outside. The second is to transform the system

---

*Ohira died June 11, 1980, amidst a crisis of confidence in the LDP.

itself in the direction of <u>sustainability</u>, to subordinate economic and productive activity to environmental limits and move decisively to use energy resources more efficiently. How might such a transformation take place?

The first priority is a strong push to supply the nation with forms of energy other than fossil fuels. It is unfortunate that nuclear power is an alternative form already leaned on, but it is probably unrealistic to expect nuclear power to play a lesser role for at least a generation. In the meantime there are many alternatives that Japanese capital and diligence could bring to fruition, including the use of alcohol for the transportation sector, solar energy for water and space heating, and geothermal wherever feasible. As other renewable alternatives appear on the horizon, or when there are technological improvements in existing ones, Japan should have an institutional and financial structure ready to exploit the new possibilities without delay. As was pointed out earlier, an immense reservoir of untapped energy in Japan is the amount wasted through carelessness and inefficiency. Through productive conservation Japan can measureably reduce imports, and the price of conservation is much less than that of developing or buying virgin resources. Energy systems generally, throughout the country, must be tailored to standards of thermodynamic efficiency: how much work is needed to perform a given task, and how much work is actually obtained from the work potential of a given energy source. While the steel industry has set high standards of energy efficiency, many other industries fall short. Compared to other industrial nations, Japan has an excellent public transportation system, but it must be supported and expanded to reduce the impact of more than 20 million vehicles. No one knows exactly how Japan should maneuver through this transition to greater energy self-sufficiency, nor can one even guess at the capital investment needed. The prime minister would do well to provide some of the leadership that he thought Japan could do without.

Japan's access to large amounts of fossil fuel and lesser amounts of uranium must be viewed as a short-term subsidy for partial investment in renewable energy, for unless fixed stocks of energy are invested prudently in the next 20 years to the end of creating a self-regulating industrial culture not dependent on fossil fuels, there will be no energy base for an orderly transition to renewable energy once the subsidy has been depleted. One can dispute the timetable for oil reserves to trickle away or become too expensive to extract, but few students of the subject expect them to support a 5 percent rate of global industrialization for more than a few decades.[99] Heavy oil users like the Japanese must bear in mind also the many applications of petroleum other than being consumed as fuel. Oil is a valuable petrochemical resource. Arab planners are aware that oil in the ground is likely to grow in value as additional uses are discovered for it. Why should good businessmen surrender a versatile nonrenewable resource to improvident nations wishing to burn it up? Petrochemical enterprises can be just as profitable and last much longer.

One can expect nuclear power to be a problem on several counts. As uranium is depleted in the world market, reactors of the light or boiling water types will have to be replaced with breeders capable of running on plutonium, which allows the nuclear fuel to be stretched out. Breeder reactors cannot operate without plutonium reprocessing plants, thus necessitating a "plutonium economy." Waste material from breeders is remarkably toxic and durable. Moreover, plutonium is the stuff bombs are made of. Given the social problems in Japan of siting nuclear power plants now, and all the genuine problems of safety and efficient operation stumbled over thus far, one wonders how much better the Japanese would fare with a breeder technology, which is far more complex and dangerous. The escalating expense of nuclear power plant construction in the United States is due partly to safety demands and tighter regulation of nuclear technology, all of which can be expected to close a more determined grip around the atom in the wake of Three Mile Island. The Japanese case is analogous. Public resistance to nuclear power is ardent if not entirely effective. Even though the construction of plants goes on despite protest, costs are driven up as government and industry try to mollify public fears. As more capital is invested to make light water reactors acceptable to the public, an intrinsically more expensive breeder technology is likely to be priced out of the market, leaving Japan with a nuclear white elephant, mounds of waste, serious thermal pollution, and a substantial loss of capital investment for more sensible forms of renewable energy. To invest in fusion without a viable breeder program to bridge the interim is to gamble away valuable capital resources. While drifting with the current toward an imagined nuclear sunrise, Japanese planners in business and government would do well to pause and rethink the nuclear strategy as a solution to greater energy independence. It has been said that Japan's nuclear power industry is the essential foundation upon which military applications of the atom can be built, the implication being that Japan may need nuclear weapons at some future time to defend and guarantee imports of energy and other resources. Be that as it may, such a road would lead the nation into even more dangerous terrain than commercial nuclear power. There are other ways of shoring up "national security" in the sense of protecting economic potential. The shortest route to economic security is reduction of energy demand supplemented by conservation and development of renewable energy alternatives.

Japan's impact on world resources and the environment is out of proportion to its physical scale:

> Were Japan merely a small and insignificant island na-
> tion in the far reaches of the Pacific, their growth-
> related attitudes and policies would not overly concern
> anyone else. But, on the contrary, the Japanese occupy
> a major portion of the world economy. What the Japanese
> think and do about economic growth and about prospects
> for the emergence of a world of growing resource scarci-
> ties and their economic consequences are of great im-
> portance to the entire world.[100]

It is fortunate that all things are no longer considered possible in Japan, but it is still expected that the economy will grow briskly in the next decade. Since it is reasonable to assume that means will be found to accomplish the growth, the most relevant question is not whether Japan will reform and erect a new industrial order, but whether the existing system can organize enough rational change to cut oil imports and soften pressure on the environment. When one discounts rhetoric and ritual gestures, behavior indicates only slight movement toward a conservation-resource recycling, knowledge-intensive economy less dependent on fossil fuels.[101]

Part of the reluctance for fundamental change lies with the conservative LDP, Japan's party of growth and the dominant political force since the mid-1950s. It was Shigeru Yoshida (prime minister from 1949 to 1955) who formulated the LDP's basic policy: a single-minded commitment to economic development under the protective shield of a mutual security treaty with the United States as the best strategy for restoring Japanese prestige and global influence.[102] One might say that Yoshida set his country the task of becoming a world leader in the conversion of energy. All subsequent heads of state have accepted and promoted the goal of economic preeminence, with some variations in political style and fortune.

Nobusuke Kishi, a Yoshida protégé and head of state from 1957 to 1960, lost his ministry over a revision and extension of the mutual security treaty. His successor, another Yoshida favorite, Hayato Ikeda, played with great astuteness on the related themes of "income doubling" and a "utility theory" of the mutual security agreement that justified the pact as a benign weapon of national defense requiring no defense expenditures, thus storing up scarce capital for investment in the economy. When Eisaku Sato became head of state in 1964, he defused criticism of the LDP over the defense treaty even further by adroitly linking it to the reversion of Okinawa, while at the same time running a government highly cooperative with domestic business interests. When Kakuei Tanaka came to power in 1972, he played a new variation on the economic theme with his proposal to remodel the Japanese islands, which was described by one critic as "a concept of maximizing the economic growth of Japan and turning it into one giant factory."[103] Tanaka's inflated scheme assumed, with wretched timing (just before the 1973 oil crisis), unlimited access to petroleum and other natural resources. Takeo Miki's ministry, beginning in late 1974, tied the postembargo slogan of "stable growth" to a "lifecycle" concept of welfare and security benefits. The next man, Takeo Fukuda, offered to fight inflation, improve international relations, and scrub more corruption out of the LDP-business network. We have already commented briefly on Ohira's "age of culture" notion, which implies quite falsely an end to zealous preoccupation with economic growth and material standards of life.

The central problem of the LDP is how to retain decisive power by legitimizing "stable growth" in a framework less naked than that of the "all-out" years of growth. Somehow the traditional party of growth and economic success must prove itself capable of flexible, creative leadership in coping with inflation, environmental disruption, social welfare expectations, and energy-resource vulnerability without abandoning the venerable course charted by Yoshida and his successors, lest the LDP find itself stripped of an identity. In past years the opposition parties of Japan have been sharply critical of growthism and improvisation as the value stance and procedure of government, but the LDP usually had a swelling GNP as its reply and trump card. The growth mystique appears to be wearing thin, and the heart of the party has been tarnished by a series of scandals involving high party members.[104] Some students of the Japanese political scene predict a disintegration of the LDP from within, because of factionalism and corruption, and from without, because of poor showings in elections,* opening the way for some kind of coalition among the opposition parties (which are Socialist, Komeito, Democratic Socialist, Independent, and New Liberal, in approximate order of strength in the Diet over the past few years).[105]

On the other hand, the LDP has some advantages to help it weather the heavy seas of retrenchment and falling expectations. The chief of these is simply being in power and having been in power for more than two decades. The party is structured on two levels: the central party headquarters and prefectural federations of chapters

---

*LDP slippage in elections is a long-term trend since the late 1960s.

(the surface structure), and the local support associations for individual Diet members (the deep structure).[106] The influence of the LDP lies in the ability of Diet members to channel benefits to their support associations. Opposition parties have failed to create stable support associations and, not being in power, have no benefits to distribute. Factionalism within the opposition parties* and strife between them inadvertently strengthen the LDP. So far an opposition coalition has not emerged, and none of the parties has come up with a cogently structured platform for domestic and foreign policy clearly distinguishable from that of the LDP. They merely snipe petulantly at the LDP's failures and discomfiture, none of which have obscured its success in leading Japan to the status of world economic power.

On balance the prospects of the LDP in the next decade seem good. Scandals, factionalism, and declining strength in national elections are compensated for by firm support from the bureaucracy, without which no party can rule,† tight collusion with the business world, and ambivalent public attitudes concerning growth and further industrialization. In the process of trying to have their cake and eat it too, the Japanese are likely to experience an edgy domestic tension as the social and natural environments continue to deteriorate under the pressure of additional physical growth.[107]

Barring regional wars in East Asia or the Middle East, a more cataclysmic world war, a world depression (or recession, as the new economic mystics prefer), or other such upheavals, Japan probably will go on walking the tightrope across the 1980s. It is likely that economics will continue to overshadow ideology and international politics, in spite of the expanding connection with China and much rhetoric about a new era for Japan as a mature international actor. The core of the old Yoshida policy will, in effect, persevere behind new trimmings and embellishments. Opportunistic nations may be tempted to use Japan as a political tool and take advantage of its extreme dependence on imports.[108] The most explosive issue for Japan and other affluent industrial societies is the desire of poorer countries to improve their material circumstances amidst the plenitude of a few economic giants. The issue is sure to become more insistent as global population edges toward 6 billion by the close of

---

*The opposition could not work together even while the LDP was blushing over the Lockheed scandal during the 1976 House elections.

†The bureaucracy controls information and commonly drafts policy papers.

the century, and as rich countries absorb even more fixed energy stocks to bolster corpulent GNPs and affluent styles of life. By 1980 Japan may be taking nearly 8 percent of the world's available energy, compared to 5.5 percent for the rest of Asia, 2 percent for Africa, and 4.5 percent for Latin America (see Table 3.5).

Sometimes Japan is referred to as a "model" for poor, small countries with few resources. What the down and out Japanese did, so the argument goes, others can do. Actually no worse model could be found (except perhaps the United States and the Soviet Union) in a tense, crowded world faced with shrinking resources and environmental decline. It would be unwise for developing countries to neglect agriculture—a self-regulating type of economic activity based on the photosynthesis of green plants—for the sake of building an industrial economy reliant on fossil fuels, but many are, of course, doing just that. If Japan is to be a model of some kind, let it be one of environmental prudence, energy conservation and efficiency, restraint in the use of nonrenewable resources, and self-conscious regard for the needs of unborn generations, Japanese or otherwise. The world is in need of a big industrial economy capable of setting an example of good sense in the utilization of energy, able to try novel social and economic patterns adaptive to ecological scarcity, willing to behave as though sustainability is the challenge facing industrial civilization at this historic moment. The Japanese have the makings of an appropriate response in their traditional values, hitherto diluted and blunted by sudden immersion in consumerism and heroic materialism. Those values include discipline, group solidarity (out of which a sense of the public interest might be forged), self-denial, frugality, and a religious veneration of nature. Of course the agricultural society out of which those values came has receded before a tidal wave of urbanization, but most Japanese are not so far removed from them as one might suppose; they were very much intact during World War II. The strains, uncertainties, and ambiguous fruits of industrialization might catalyze a resurgence of traditional values capable of neutralizing the worst manifestations of Japan's adventure with economic success.[109] In any event, by trying to join the exclusive club of superpowers at a time of ecological scarcity—a phrase suggestive of all resources essential to human welfare in a state of contraction—Japan would be following a few dubious examples in the West rather than shaping a creative one in the East.[110]

It is time to come full circle in this discussion. The central facts of life for Japan are petroleum dependence and vulnerability. Not a great deal can be done about dependence, but action can be taken to mitigate somewhat the nation's high degree of vulnerability. Some three-fourths of Japan's oil has been travelling from the Near and Middle East through strategic channels—the Strait of Hormuz at

TABLE 3.5

Projected World Demand for Primary Energy (10^{13} kcal)

	Energy Requirements			Percent Proportion			Percent Growth	
	1960	1970	1980	1960	1970	1980	1960–70	1970–80
Free world								
North America								
United States	1,129.4	1,709.2	2,585.0	33.8	31.9	28.5	4.2	4.2
Canada	114.4	186.3	297.7	3.4	3.5	3.3	5.0	4.8
Japan	84.4	284.1	717.0	2.5	5.3	7.9	12.9	9.7
OECD, Europe	673.2	1,145.1	1,932.7	20.1	21.4	21.3	5.5	5.4
Oceania	35.2	59.4	114.4	1.1	1.1	1.3	5.4	6.3
Asia	99.0	206.8	480.7	3.0	3.9	5.3	7.6	8.8
Africa	52.8	80.3	173.8	1.6	1.5	1.9	4.2	8.1
Latin America	106.7	196.9	413.6	3.2	3.7	4.6	6.3	7.7
Total	2,295.1	3,868.1	6,714.9	68.7	72.3	74.1	5.3	5.7
Communist bloc	1,050.5	1,490.5	2,357.3	31.4	27.8	26.0	3.6	4.7
World, total	3,345.6	5,358.6	9,072.2	100.0	100.0	100.0	4.8	5.4

Note: Figures are not precise and do not total 100 percent.
Source: A Long–Term Outlook of Japanese and U.S. Economies—1980 (Tokyo: Japan Economic Research Center, 1973), p. 112.

the mouth of the Persian Gulf and the Malacca Strait—whose accessibility might be blocked by premeditated design or by unexpected catastrophes involving the congested traffic of large tanker vessels. Japan's policy of "diversification" obviously needs implementation on two levels. The first entails less dependence on a few suppliers of petroleum like Saudi Arabia, Kuwait, and Indonesia. The second has to do more specifically with geography, the limitation of dependence west of the Malacca Straits.[111] In addition to diversifying specific types of energy used there is an urgent need to multiply the number of suppliers and to concentrate them east and south of Japan. Among the countries that might be wooed in such a diversification strategy are Australia, Thailand, Malaysia, Singapore, and the Philippines for onshore and offshore energy, and Taiwan and Korea for offshore fuel. Nearer to home are the PRC and the USSR, whose auspicious potential as suppliers is still being sorted out and tested. Across the Pacific lies the petroleum bonanza of Mexico. The Mexican government is sanguine about exchanging generous amounts of oil and natural gas for Japanese technology and economic assistance. Even farther to the west are Africa and the North Sea area. The delicate problem in all of this is how to steer a resolute course between suppliers too small or too big and how to fend off new relationships of one-sided dependence. While the PRC and the USSR are promising as suppliers of energy, they might also use their resource-supplying powers to exert arbitrary and unseemly pressure on Japan to choose sides in their quarrels.

From every viable perspective one can muster there is no painless relief from the dilemmas holding Japan captive. There can be little freedom of movement or choice in the shackles of radical energy dependence. One might attempt to soften the harshness of Japan's position by arguing that we live in an interdependent world that needs Japanese technology and productivity, and that other countries, like Italy and West Germany, are vulnerable as well. There is also the anomalous fact that Japan seems to go on accumulating riches despite the realities of energy dependence. On the other hand, complex systems of interdependence are extremely prone to dislocation and breakdown; the political and social management of the present world economy is markedly primitive and hindered at every turn by greed, unenlightened self-interest, ideological polarization, and the desperate logistics of keeping pace with growing population and rising economic expectations in poorer nations of the world. Should the existing system of interdependence come unraveled because of wars, depressions, ecological disasters, or other forms of unpleasantness, Japan will fall faster and harder than most industrial states. There will not be much comfort in the knowledge that a few countries like Italy will plummet less tumultuously.

NOTES

1. Kazushi Ohkawa and Henry Rosovsky, Japanese Economic Growth (Stanford, Calif.: Stanford University Press, 1973), pp. 8-21. See also Edward Denison and William Chung, How Japan's Economy Grew So Fast (Washington, D.C.: Brookings Institution, 1976), pp. 10-25.

2. White Papers of Japan, 1974-1975 (Tokyo: Japanese Institute of International Affairs, 1976), p. 48.

3. Hugh Patrick and Henry Rosovsky, eds., Asia's New Giant: How the Japanese Economy Works (Washington, D.C.: Brookings Institution, 1976), p. 386.

4. White Papers of Japan, p. 49.

5. Economic Survey of Japan, 1974-1975 (Tokyo: Economic Planning Agency, 1974), pp. 77-78, 81-85. Other sources of statistical information are OECD Economic Surveys 1977: Japan (Paris: Organization for Economic Cooperation and Development, 1977) and Japan Economic Yearbook, 1978/79 (Tokyo: Oriental Economist, 1978).

6. Hsiao Kanamori, "Slow Growth Proponents Again Are Mistaken: Growth Potential of the Japanese Economy and Policy Choices," Oriental Economist, July 1975, p. 27.

7. Japan Economic Yearbook, 1975-1976 (Tokyo: Oriental Economist, 1976), p. 80.

8. Japan's Economy in 1980 in the Global Context (Tokyo: Japan Economic Research Center, 1976), p. 53.

9. Jun Ui, ed., Polluted Japan (Tokyo: Jishu Koza, 1972), pp. 64-67.

10. Japan Economic Yearbook, 1972 (Tokyo: Oriental Economist, 1972), pp. 211-12.

11. Kakui Tanaka, Building a New Japan: A Plan for Remodeling the Japanese Archipelago, Trans. Simul International (Tokyo: Simul International Press, 1972), p. 102.

12. White Papers of Japan, p. 244, Table II-3-2.

13. Far Eastern Economic Review, August 25, 1978, p. 43.

14. Ibid., p. 47. See also Japan's Industrial Structure (Tokyo: Ministry of International Trade and Industry, 1978).

15. Japan's Sunshine Project (Tokyo: Ministry of International Trade and Industry, 1974), p. 2n.

16. Far Eastern Economic Review, December 15, 1978, p. 51.

17. Susuma Awanohara, "Rethinking Japan's Prospects," Far Eastern Economic Review, June 11, 1976, p. 98.

18. Kanamori, "Slow Growth Proponents," pp. 26-27.

19. "Outlook for Japan's Growth: Japan's Economy in Five Years, As Seen by Five Research Organizations," Oriental Economist, January 1975, p. 13.

20. Kanamori, "Slow Growth Proponents," p. 27.

21. Oriental Economist, July 1978, p. 4.

22. A Long Term Outlook on Japanese and U.S. Economies (Tokyo: Japan Economic Research Center, 1973), p. 34.

23. "Economic White Paper: In Pursuit of a New Stable Course," Oriental Economist, March 1975, p. 11.

24. Japan Environment Summary (Tokyo: Environment Agency, November 10, 1975), p. 4.

25. White Papers of Japan, pp. 90-92. Compare Oriental Economist, March 1975, p. 67, where the number cited is 126 million.

26. White Papers of Japan, p. 240.

27. On all of this, see Far Eastern Economic Review, April 13, 1979, p. 49; and February 6, 1976, pp. 38-39.

28. Eiji Ozaki, "Economic Policy in Japan," Oriental Economist, September 1975, p. 19.

29. Japan's New Energy Policy (Tokyo: Ministry of International Trade and Industry, 1976), pp. 24-26. See also John Surrey,

"Japan's Uncertain Energy Prospects: The Problem of Import Dependence," Energy Policy, September 1974, p. 224.

30. Japan's New Energy Policy, pp. 52, 55.

31. Ibid., p. 53.

32. Ibid., p. 59.

33. A quote from an American Physical Society study in Barry Commoner, The Poverty of Power: Energy and the Economic Crisis (New York: Alfred A. Knopf, 1976), pp. 38-41.

34. Wilson Clark, Energy for Survival: The Alternatives to Extinction (New York: Doubleday, 1974), pp. 234-51.

35. Ozaki, "Economic Policy in Japan," p. 22.

36. Ibid., p. 24.

37. Quality of the Environment in Japan, 1973 (Tokyo: Environment Agency, 1973), pp. 135-37.

38. "Japan Aims for a 90-Day Stockpile Target," Petroleum Review, January 1976, pp. 25-27. Useful publications for monitoring Japan's petroleum situation are Japan Petroleum Weekly (Tokyo: Japan Petroleum Consultants, 1975-79), Petroleum Intelligence Weekly, and the monthly Energy in Japan (Tokyo: Institute of Energy Economics, 1975-79).

39. Henri Hymans, "Japan's Search for Local Power," Far Eastern Economic Review, July 25, 1975, p. 42.

40. Japan's New Energy Policy, pp. 20, 31.

41. "Energy Situation in Japan and New Applications of Coal," Look Japan, October 10, 1974, p. 20. Japan Economic Yearbook, 1975-1976, p. 80. Japan's New Energy Policy, pp. 20, 32.

42. "Coal Mining Technology of Japan," Look Japan, November 10, 1974, p. 23.

43. Outline of Long Range Program on Development and Utilization of Atomic Energy (Tokyo: Atomic Energy Commission, 1972), p. 2.

44. Compare Far Eastern Economic Review, December 16, 1977, pp. 48-49, Japan's New Energy Policy, p. 87, and White Papers of Japan, pp. 242, 244.

45. "Nuclear Power in Japan: The Key in an Energy Dependent Economy," OECD Observer, April 1974, p. 39.

46. Susuma Nagai, "Going Ahead with Atomic Power—A Dangerous Choice," Japan Quarterly, January-March 1978. For a good review of the history and current status of nuclear power in Japan, see Masao Sakisaka, "Japan's Energy Policy Under Review," Energy Policy, December 1974, 346-49, and Henri Hymans, "The Japanese March to Nuclear Power," Far Eastern Economic Review, October 10, 1974, p. 39.

47. Outline . . . on Development and Utilization of Atomic Energy, p. 9.

48. Susuma Awanohara, "Japan Closes the Gap in Enrichment," Far Eastern Economic Review, July 16, 1976, p. 56.

49. Diana Lumsden, "Nuclear Fuel Enrichment: Nobody Wants to Do It," Electric Power and Light, December 1974.

50. New York Times, May 8, 1979, p. A9.

51. Takashi Mukaibo, Safety Research on Nuclear Facilities in Japan, presented at the 9th World Energy Conference, Detroit, September 22-27, 1974. See also "Reports on Nuclear Programs Around the World," Nuclear News, February 1975, and The Development of Atomic Energy in Japan, a bulletin of the Japan Atomic Energy Commission, 1973.

52. Ui, Polluted Japan, pp. 73-74.

53. Japan Times, November 23, 1975; October 25, 1975.

54. Outline . . . on Development and Utilization of Atomic Energy, pp. 13-14. For a more recent and no more satisfactory statement on waste disposal, see Japan Environment Summary, November 10, 1976, p. 4.

55. See Denis Hayes, Energy: The Solar Prospect (Washington, D.C.: Worldwatch Institute Paper No. 11, 1977). Also Clark, Energy for Survival, p. 511.

56. Tsutomu Inoue, Present Status and Future Prospects of Geothermal Development in Japan, presented at the 9th World Energy Conference, Detroit, September 22-27, 1974. A good recent summary of the geothermal situation is "Harnessing Geothermal Energy," Business Community Quarterly of Japan, Winter 1977.

57. See J. Robinson, "Is the Answer to the Energy Crisis Blowin' in the Wind?" Ocean Industry, January 1975.

58. Japan's Sunshine Project, Chapter 3.

59. Jon Halliday, A Political History of Japanese Capitalism (New York: Pantheon Books, 1975), pp. 285-86. See also "The Day Japan Dies For Want of Oil," Japan Echo 4:3 (1977).

60. Japan Times, August 16, 1975.

61. Compare Far Eastern Economic Review, December 5, 1975, p. 4; Japan Economic Yearbook, 1975-1976, p. 79; and Economic Information File—Japan, 1974-1975 (Tokyo: World Economic Services, 1974), p. 129.

62. Surrey, "Japan's Uncertain Energy Prospects," pp. 212-13, 225-26.

63. See Yuan-li Wu, Japan's Search for Oil (Stanford, Calif.: Hoover Institution Press, 1977), pp. 62-72.

64. Far Eastern Economic Review, February 2, 1979, p. 42.

65. Financial Post, September 21, 1974.

66. Far Eastern Economic Review, November 3, 1978, p. 40.

67. Shin'ichiro Shiranishi, "The Potential for Economic Co-operation," Japan Quarterly, January-March 1979.

68. Henri Hymans, "Japan Eyes Peking's Liquid Gas," Far Eastern Economic Review, February 20, 1976, pp. 41-42. On various aspects of energy politics between China and Japan, see Selig Harrison, China, Oil, and Asia (New York: Columbia University Press, 1977), Chapter 7.

69. Koji Nakamura, "A Rethink in Japan," Far Eastern Economic Review, January 31, 1975, pp. 62-63.

70. On the trade issue generally, see Oriental Economist, April 1979, p. 14, and Far Eastern Economic Review, April 6, 1979, pp. 124, 125.

71. See Ezra Vogel's Japan as Number One: Lessons for America (Cambridge, Mass.: Harvard University Press, 1979), and Rodney Clark, The Japanese Company (New Haven, Conn.: Yale University Press, 1979).

72. There is a detailed account of Japan's environmental problems in a broad context in Donald Kelley, Richard Wescott, and Kenneth Stunkel, The Economic Super Powers and the Environment: The United States, the Soviet Union, and Japan (San Francisco: W.H. Freeman, 1976).

73. Pollution Related Diseases and Relief Measures in Japan (Tokyo: Environment Agency, 1972).

74. Japan Environment Summary, July 10, 1976, p. 3.

75. Long Term Prospectus for Preservation of the Environment (Tokyo: Environment Agency, 1972), p. 7.

76. See Quality of the Environment in Japan, 1976 (Tokyo: Environment Agency, 1976), pp. 114-32.

77. Ibid., p. 106. Japan Environment Summary, May 10, 1975, pp. 4-5.

78. Kelley, Wescott, and Stunkel, The Economic Super Powers, pp. 85-91.

79. Tanaka, Building a New Japan, pp. 74-77.

80. Ui, Polluted Japan, pp. 74-77.

81. Japan Environment Summary, April 10, 1975, pp. 2-3. A follow-up study of that spill is reviewed in the August 10, 1975 issue, pp. 1-2.

82. Economic Survey of Japan, 1973, p. 34.

83. "Japan's Peculiar Land Problem," Japan Quarterly, July-September 1976, pp. 217-19.

84. Japan Environment Summary, July 7, 1975, pp. 1-3.

85. Japan Environment Survey, August 10, 1975; February 10, 1976. See also Environmental Policies in Japan (Paris: OECD, 1977).

86. John Bennett and Solomon Levine, "Industrialization and Social Deprivation: Welfare, Environment, and Post-industrial Society in Japan," in Hugh Patrick, ed., Japanese Industrialization and Its Social Consequences (Berkeley: University of California Press, 1976).

87. Quality of the Environment in Japan, 1976, p. 154.

88. Japan Environment Summary, February 10, 1975, pp. 2-4.

89. Bennett and Levine, "Industrialization and Social Deprivation," p. 452.

90. White Paper on National Life, 1973: Life and Its Quality in Japan (Tokyo: Economic Planning Agency, 1973), p. 120.

91. Edward Olsen, "Japan: Economic Growth and Cultural Values," Asian Forum, Spring 1976, p. 7.

92. New Economic and Social Development Plan, 1970-1975 (Tokyo: Economic Planning Agency, 1970), pp. 99-101.

93. Richard Ellingworth, Japanese Economic Policies and Security, Adelphi Papers, No. 90 (London: International Institute of Strategic Studies, 1972), p. 9.

94. Quoted in ibid., p. 10.

95. Koji Nakamura, "An Old Order Does Not Change," Far Eastern Economic Review, May 9, 1975.

96. Far Eastern Economic Review, April 6, 1979, p. 28.

97. Barry Commoner, "Energy," New Yorker, February 2, 1976, p. 38.

98. Ibid.

99. See Albert Parker, "World Energy Resources," Energy Policy, March 1975, pp. 61-62, for an optimistic picture, and Desmond

Dewhurst, "World Energy Resources—The Looming Gap," Petroleum Review, May 1976, p. 277, for the pessimistic side of the issue.

100. Olsen, "Japan: Economic Growth," p. 3.

101. "Is Japan Too Sanguine on Growth?" New York Times, February 4, 1979, p. 47. See also Shigeto Tsuru, "In Place of Gross National Product," Area Development in Japan 3 (1970).

102. Eiji Tominomori, "Stability of the Conservative Regime," Japan Quarterly, January-March 1976, pp. 53-54.

103. Ibid., p. 56.

104. See Michael Blaker, "Japan 1976: The Year of Lockheed," Asian Survey, January 1977, pp. 53-54.

105. For example, Takeshi Hatakayama, "One-Party Rule in Japan Nears Its End," Japan Quarterly, October-December 1976, pp. 382-83. See also Hong M. Kim, "The Crisis of Japan's Liberal Democratic Party," Current History, April 1975.

106. Tominomori, "Stability of the Conservative Regime," pp. 58-59. For a brief but good analysis of the basis of LDP power, see also Hiroshi Imazu, "The Political Structure of LDP Rule," Japan Quarterly, April-June 1979, and Sadayuki Sato, "The Economic Foundations of the Conservative Resurgence," in ibid.

107. See Kenneth R. Stunkel, "The Growth Issue in Japanese Public Opinion and Politics," Asian Profile, October 1975, and the perceptive disillusionment of Pauline Bush, "To Japan—With Love and Concern," Japan Quarterly, April-June 1979.

108. Susan Rolez, "Japan's Foreign Policy," Asian Survey, November 1976, p. 1041.

109. On some of these strains, see Solomon Levine, "Japan's Economy: End of the Miracle?" Current History, April 1975, and Norie Huddle and Michael Reich, Island of Dreams: Environmental Crisis in Japan (New York: Autumn Press, 1975).

110. See John Copper and Kenneth Stunkel, "Super Power Status: The Elusive Goal," Japan Interpreter, Winter 1975.

111. See Wu, Japan's Search for Oil, pp. 73-75.

4

THE PEOPLE'S REPUBLIC OF CHINA
Samuel S. Kim

How does a nation possessing only 7 percent of the world's arable land manage to feed nearly 20 percent of the world's population in an age of scarcity? To put the problem in more measurable comparative terms, China's total land area is about the same as that of the United States (3.7 and 3.6 million square miles, respectively), both occupying similar latitudes. Yet Chinese agriculture with only 10 percent of its land under cultivation, as compared to about 22 percent in the United States, has to support four and a half times as many people. This adverse relationship between population pressure and arable land resources is perhaps the most pressing reality conditioning the nature and development of Chinese ecopolitics.

This is not a technical study of Chinese energetics per se, concerned with attempts to capture all scientific and statistical data in full detail.[1] Instead, the principal purpose of this study is to use China's energy development as an empirical basis for disciplined macro-inquiry into its authoritative allocation of resources and values. In pursuit of this objective, several lines of inquiry will be made based on the methodological premise that a nation's energy profile is rich in heuristic power and normative implications. Only a few broad questions and their underlying assumptions need to be specified at the outset.

First, the pattern of Chinese energy use may be accepted as a barometer showing the extent to which Chinese society has been

----

Much of the research on which this study is based was supported by a faculty creativity grant from Monmouth College.

modernized. Second, domestic energy policy may be accepted as an indicator of China's overall growth policy since there is a generalized ratio between growth in GNP and energy requirements (so-called energy-GNP elasticity). Likewise, there is an empirical relationship between increase in per capita GNP and increase in per capita energy consumption; hence, per capita energy consumption may be accepted as an index of lifestyle in Chinese society. The energy policy process has always been permeated by dialectics of Chinese politics. Chinese energy development thus provides a measure of performance of Chinese communism and the strengths and weaknesses of its political and economic systems.

Third, given the universality of energy use, China's energy policy lends itself to comparative study. Part of the answer to the continuing debate among economists about the relevance of China's development experience for other developing countries lies in Chinese energy policy. Fourth, the quality of life in Chinese society can be delineated by examining the relationship between energy and the environment in the context of the Chinese politics of scarcity. Finally, China's energy trade and energy-related transnational activities shed much light on the change and continuity in the Maoist principle of self-reliance, on the one hand, and on China's role in the elusive struggle of the global underdogs to establish a New International Economic Order (NIEO), on the other. In sum, this study attempts to use China's energy development as a frame of reference to see where China came from, where it is now, and where it is heading, in its long march toward modernization.

THE PATTERN OF ENERGY USE

For millennia the Chinese economy precariously depended on subsistence agriculture, drawing its kinetic energy requirements from human and animal muscle power and its thermal energy requirements from forest fuel. The most salient feature of the traditional Chinese energy profile has been the heavy reliance on solar energy converted by green plants (photosynthesis) to produce not only human food and animal feed but also fuels and raw materials. This traditional energetics made extensive and reckless use of wood for household, commercial, and industrial uses, creating dangerous deforestation over vast areas throughout China.[2]

The coming of Western and Japanese imperialism discredited the traditional Sinocentric world order beyond redemption, stimulating the birth of modern Chinese nationalism. The presence of foreign imperialism hardly transformed the traditional agriculture, although it established a small modern industrial base in Manchuria

and turned some coastal cities into processing centers and entrepôts
for trade. Still, China's vast mineral and energy resources that
could have been used for domestic industrialization or foreign exploitation remained largely unknown, unexplored, and untapped by the
time the unequal treaty system came to an end in 1943. As a result,
the highest pre-1949 output (1936-43) in primary energy production
stood as follows: 6.0 billion kilowatt hours (kWh) in electric power;
62 million metric tons (mmt) in coal; and 0.3 mmt in crude oil.[3]

Of the primary fuels during the pre-Communist period, then,
only coal had been produced in large quantities. Natural gas was
negligible. Hydroelectric power was developed in only Manchuria
to any significant degree as part of the economic development plans
of Japanese colonialism. The petroleum industry was in its embry-
onic stage. The total exploratory drilling for oil between 1907, when
the first drilling started at Yanchang in northern Shaanxi Province,
and 1948 was no more than 34,000 meters, with the cumulative total
output of only 2.78 mmt.[4] In the pre-1949 period, about 85 to 90
percent of the country's petroleum needs had to be met through foreign
sources.

With the establishment of the People's Republic of China (PRC)
on October 1, 1949, the leadership of Chinese communism inherited
a shattered economy with a weak, primitive, and lopsided energetics.
The effect of the Sino-Japanese War and the ensuing civil war was to
cause further damage and dislocation in the already deformed industrial infrastructure as shown in the disparities between the highest
pre-1949 output figures in primary energy production and those of
1949 in Table 4.1. It should also be noted here that the Soviets,
taking advantage of their occupation of Manchuria to disarm Japanese
troops, dismantled heavy industrial facilities and shipped to the Soviet
Union about 50 percent of the Manchurian industrial capacity including
1,000 megawatts (MW) of turbogenerating capability of both thermal
and hydro units.[5]

The PRC's energy development since 1949 is summed up in
Table 4.1. The table shows the extent to which the Chinese energy
profile has been transformed in each of the four primary energy
categories. Coal was the king of the Chinese energy system in the
1950s but the unparalleled exploratory drilling in search of oil deposits ushered in a transition to liquid and gaseous fuels in the early
1960s. The discovery and development of Daqing—literally meaning
"great celebration" but the name given to this famous oil field in
Heilongjiang Province—represent a landmark in the history of Chinese
energy development. Thanks to Daqing, China achieved self-sufficiency
in oil by 1964 and began to export "black gold" by 1973. China has
already risen to the third place in primary energy production in the
world, following behind the United States and the Soviet Union but
equalling Saudi Arabia.

TABLE 4.1

China's Primary Energy Production, 1949-79

	Coal (mmt)	Crude Oil (mmt)	Electric Power (in billion kWh)	Natural Gas (billion cubic meters)
1949	32.43	0.12	4.30	0
1952	66.49	0.44	7.30	0
1976	483.00	87.00	203.00	45.00
1977	550.00	93.64	223.40	55.00
1978	618.00	104.05	256.55	60.50
1979	628.20	106.10	278.87	61.53

Sources: For Coal: Figures for 1949, 1952, and 1978 are based on Xinhua release on September 22, 1979, in Foreign Broadcast Information Service-PRC, September 24, 1979, p. L2; figure for 1976 is taken from Premier Hua Guofeng's report on the work of the government, in FBIS-PRC, June 18, 1979, p. L5; and figure for 1979 is based on Xinhua release on March 19, 1980, in FBIS-PRC, March 20, 1980, p. L5.

For Crude Oil: Figures for 1949, 1952, and 1978 are based on Xinhua release on September 22, 1979, in FBIS-PRC, September 24, 1979, p. L2; figure for 1976 is taken from Premier Hua Guofeng's report on the work of the government, in FBIS-PRC, June 18, 1979, in Renmin Ribao, January 1, 1980, p. 1.

For Electric Power: Figures for 1949, 1952, and 1978 are based on Xinhua release on September 22, 1979, in FBIS-PRC, September 24, 1979, p. L2; figure for 1976 is taken from Premier Hua Guofeng's report on the work of the government, in FBIS-PRC, June 18, 1979, p. L5; and figure for 1979 is based on Xinhua release of February 27, 1980, reporting 1979 power output increase of 8.7 percent over the 1978 figure, in FBIS-PRC, February 27, 1980, p. L3.

For Natural Gas: Figures for 1976-77 are taken from Vaclav Smil, "China's Energetics: A System Analysis," in U.S. Congress, Joint Economic Committee, Chinese Economy Post-Mao: A Compendium of Papers, 95th Cong., 2d sess. (Washington, D.C.: U.S. Government Printing Office, 1978), p. 367; the 1978 figure comes from FBIS-PRC, January 5, 1979, p. E28, reporting the total output increase in natural gas in 1978 by 10 percent over that of 1977; and the 1979 figure is computed based on Xinhua release of December 31, 1979, reporting 1979 natural gas output increase of 1.7 percent over that of 1978, in Renmin Ribao, January 1, 1980, p. 1.

In absolute terms, Chinese primary energy consumption too has increased dramatically in the last 30 years from over 20 mmtce (million metric tons coal equivalent) in 1949 to nearly tenfold in a decade to 300 mmtce in 1972 to over 600 mmtce in 1980.[6] China has already surged ahead of Japan as the world's third largest energy consumer behind the two superpowers. Yet China still remains a poor developing country in terms of per capita energy consumption. In 1979, for example, China had approximately 690 kg of coal equivalent per capita, as compared to 1,100 kg in South Korea, 1,300 kg in Mexico, and 1,600 kg in Taiwan.[7] It is to change this reality and to make China a powerful and advanced industrialized country by the end of the century that serves as the central raison d'être of the ambitious modernization drive that is currently being carried on by the post-Mao leadership.

In the Chinese energy profile coal is still the most dominant component, representing more than two-thirds of the country's primary energy use. Although the coal output averaged an annual growth rate of 9 percent between 1953 and 1978, its share of primary energy use has steadily declined from 96 percent in 1949 to 70.91 percent in 1978.[8] It should be noted that one-third of China's raw coal is produced by some 100,000 small mines with a typical annual output of about 1,000 tons.[9]

The declining share of coal in the changing pattern of Chinese energy use is largely accounted for by the entry of oil into the Chinese energy profile. The share of primary energy provided by crude oil increased from 2 percent in 1952 to 4 percent in 1957, 8 percent in 1965, 14 percent in 1970, 20 percent in 1974, 23 percent in 1978. The faltering growth rate (5.6 percent) in coal production coupled with the extremely high growth rate (23 percent) in oil production in the period 1971-74 led a large number of industrial facilities to switch from coal burning to oil burning, a development that will restrain export potential of Chinese oil.

Due to the infant infrastructure of Chinese energetics, the use of natural gas is largely confined to Sichuan Province. This province alone accounts for over 90 percent of total production of natural gas. Still, natural gas has steadily improved its share of primary energy use in China, moving up from 0.8 percent in 1957 to 7 percent in 1965, and 10 percent in 1974. Electric power is a weak link in the Chinese energy profile. The share of primary energy use provided by hydroelectric power, according to outside estimates, remained in the range of 0.5-0.8 percent in the period 1952-74. However, China's release of statistics on primary energy outputs in mid-1979 showed that U.S. (CIA) estimates, while remarkably accurate on oil and coal outputs, have grossly underestimated China's electric power output. China's power output in 1978 reached 256.55 billion kWh instead of 160 billion kWh projected by U.S. estimates. In 1979 China's output in electric power reached 278.87 billion kWh.[10]

While primary or commercial energy sources such as coal, oil, natural gas, and hydroelectric power are indispensable in modern commercial and industrial activities, the Chinese energy profile will be greatly distorted if we exclude traditional noncommercial and unconventional energy sources that dominate China's rural energy flows. That is, rural China still lives in the solar-dominated ecosystem with only marginal inputs of commercial energy sources. "Even for the nation as a whole," observed one recent study commissioned by the Joint Economic Committee of the U.S. Congress, "solar energy recently transformed by green plants still predominates: approximately 4.1×10^{15} kcal of phytomass energy—as food, feed, fuel, and raw materials—were used to support China's people and animals in 1974, while the total flow of fossil fuels and primary electricity amounted to less than 2.65×10^{15} kcal."[11] This study also estimated that approximately 0.14 percent of solar energy reaching the surface is being converted by autotrophs into new plant mass, resulting in a net production of 15×10^{15} kcal.[12] China also uses such an unconventional energy source as biogas but this will be discussed later in the chapter. In sum, then, China uses three types of energy sources: primary or commercial, traditional noncommercial, and unconventional.

The most dramatic development in the distribution of primary energy use is the industrial consumption. Even when power generation requirements are subsumed under a separate category as in Table 4.2, the industrial consumption increased from 13 percent in 1950 to 51 percent in 1976. This highlights the overall trend (despite politically induced fluctuations) of rapid industrial expansion at an average annual growth rate of 13 percent during the first 26 years of the PRC. The industrial consumption is as prominent in each of the four primary energy sources. In 1974, for example, industry consumed 56 percent of coal, 73 percent of crude oil, 60 percent of natural gas, and 76 percent of hydroelectricity.[13]

The reverse trend is at work for residential and commercial use as its share of primary energy plummeted from 64 percent in 1950 down to 26 percent in 1976. Clearly, China is not a consumer society. In absolute terms, however, energy consumption by residential and commercial sectors increased at an annual rate of 4.2 percent in 1958-74. What is worth noting here in terms of Chinese energy balance and efficiency is that residential and commercial sectors make little use of crude oil (1 percent of the total) and hydroelectricity (6 percent of the total) while consuming 34 percent of coal and 40 percent of natural gas.

TABLE 4.2

Sectoral Consumption of Primary Energy in China, 1950–76
(all values in million metric tons of coal equivalent
and, in parentheses, in percent)

	Total	Electricity Generation	Industry	Transportation	Agriculture	Residential and Commercial
1950	30.4 (100)	3.6 (12)	3.9 (13)	3.3 (11)	Negligible	19.6 (64)
1952	47.5 (100)	4.4 (9)	12.6 (27)	5.0 (11)	0.1 (1)	25.4 (53)
1957	96.5 (100)	8.8 (9)	28.9 (30)	9.3 (10)	.6 (1)	48.9 (51)
1960	198.3 (100)	20.0 (10)	105.3 (53)	15.5 (8)	5.5 (3)	52.0 (26)
1965	178.4 (100)	16.7 (9)	75.2 (43)	14.3 (8)	6.1 (3)	66.1 (37)
1970	251.4 (100)	21.6 (9)	115.4 (46)	19.9 (8)	13.0 (5)	81.5 (32)
1974	377.0 (100)	33.6 (9)	193.5 (51)	26.4 (7)	18.2 (5)	105.3 (28)
1976	445.0 (100)	42.0 (9)	228.4 (51)	31.5 (7)	27.6 (6)	115.5 (26)

Source: Vaclav Smil, "China's Energetics: A System Analysis," in U.S. Congress, Joint Economic Committee, Chinese Economy Post-Mao: A Compendium of Papers, 95th Cong., 2d sess. (Washington, D.C.: U.S. Government Printing Office, 1978), p. 354.

The transportation sector has maintained its share of primary energy use in the 7-8 percent range in the period 1960-76, although its consumption in absolute terms increased from 15.5 mmtce in 1960 to 31.5 mmtce in 1976, growing at an average annual rate of about 4.7 percent. This sector relies exclusively on coal (84.5 percent) and oil (15.5 percent). As of the mid-1970s about 80 percent of oil consumed in the transportation sector was for highway and inland waterway transport with the remaining 20 percent for railroads. However, dieselization of the railroad system, which started in the late 1960s, has steadily increased petroleum requirements for the transportation sector as a whole.

Although agriculture consumed only about 6 percent of primary energy in 1976, the growth rate of this sector has been more rapid than any other sector in the Chinese energy profile. This growth trend is accounted for by a steady growth in Chinese agricultural mechanization. During the period 1958-74, for example, energy use in agriculture increased at an average annual rate of 26 percent, compared to 11.6 percent in industry.[14] An increasing use of diesel and electric motors for a variety of farm tasks—tilling, sowing, irrigating, pumping, harvesting, processing, and transporting—has generated heavy petroleum requirements for agriculture. More than 83 percent of the primary energy used in agriculture comes from petroleum.

The Chinese industrial structure is marked by a separate if not always equal development of both large- and small-scale enterprises. An outgrowth of the "walking on two legs" policy, small-scale industries occupy a prominent part of the Chinese industrial structure, producing (in 1973) 54 percent of nitrogen fertilizer, 50 percent of cement, 5 percent of hydroelectric power, 30 percent of coal, 21 percent of pig iron, and 15 percent of crude steel.[15] The development of small-scale industries is a rational application of "soft technology" to enhance local self-sufficiency, to overcome regional imbalances in resource endowment, and to minimize the time, capital, and infrastructure costs inherent in large urban industrial enterprises.

In spite of these advantages, however, small-scale industries are more energy-intensive (or less energy-efficient) than are larger plants, which benefit from economies of scale and modern technology. Energy efficiency differences between small-scale and large industrial plants are demonstrated in the Chinese fertilizer industry. Small-scale fertilizer plants had to use 23,000-31,000 kcal of energy inputs to produce 1 kg of nitrogen, compared to imported modern ammonia and urea plants that required only 11,300-17,700 kcal to produce the same amount of nitrogen.[16]

Even larger Chinese industrial plants do not enjoy high energy efficiencies compared to their counterparts in the West and Japan. The iron and steel industry stands out as an energy-inefficient example. According to an investigation on thermal utilization in 27 key steel mills conducted by the Ministry of Metallurgy in 1978, for example, their coal consumption per ton of steel was over 600 kg higher than Japanese counterparts of similar caliber. That is, the thermal energy consumed in China in producing 30 mmt of steel could produce 50 mmt in Japan.[17] Likewise, typical Chinese turbogenerators are too small to reach the thermal efficiency levels of large Western or Soviet units.[18]

While serving a variety of social, economic, and political needs, China's industrial structure makes an inordinate and inefficient use of its primary energy. Still, we should not overstate the case here since the impact of industrial structure on China's overall energy consumption is not adversely reflected in the energy-GNP elasticity coefficient (that is, the ratio between a percentage growth in energy consumption for each percentage growth in GNP). The Chinese energy-GNP elasticity coefficient has been leveling off from 2.57 in 1953-57 to 1.87 in 1958-65, 1.42 in 1966-70, and 1.42 in 1971-74.[19] This is a good ratio when compared with other countries whose energy-GNP elasticity coefficient is between 1 and 2.

DOMESTIC ENERGY POLICY

Like its great rivers, China's growth or development policy also has demonstrated its unpredictable quality: rapidly surging forward, suddenly halting or even receding, regaining momentum, and then shifting its course under the impact of changing political and environmental pressure. The volatility of China's growth policy is shown in the fate of the five five-year plans. The First Five-Year Plan (1953-57), formulated in the context of a close Sino-Soviet alliance, was largely modeled after a Stalinist development strategy of concentrating on the heavy-industry-oriented pattern of economic growth. Mao gradually realized that this model was not too congenial to his values or to China's resource endowments and realities and it has not been revised. The Second Five-Year Plan (1958-62) was miscarried by the Great Leap Forward disaster, and the Third Five-Year Plan (1966-70) was interrupted by the Cultural Revolution. Finally, the Fifth Five-Year Plan (1976-80) was drastically redrawn and proclaimed as the Ten-Year Plan (1976-85) at the Fifth National People's Congress in February 1978. At the second session of the Fifth National People's Congress held in June 1979, however, the

more ambitious targets of the Ten-Year Plan were abandoned as the
session adopted a policy of a three-year readjustment.

The record of economic performance, though uneven in sectoral development and erratic in its pattern of growth, has nonetheless been quite impressive in its long-term and comparative perspective. Between 1952, when the immediate postcivil war reconstruction
of the country was largely completed, and 1978 the average annual
rate of overall economic growth was about 6 percent or about 4 percent in per capita terms—an economy doubling every 12 years. An
overwhelming share of this growth is attributable to industry, whose
annual growth rates averaged around 11 percent. Agricultural growth
has managed to keep pace with the population growth of about 2 percent per year, a remarkable feat given the adverse relationship
between population pressure and arable land resources in China. In
contrast, India, like most other developing countries, achieved the
average annual rate of per capita economic growth of only 1.9 percent
between 1950 and the early 1970s, approximately one-third that of
China's.[20]

Before mentioning the growth policy of the post-Mao leadership embodied in the Ten-Year Plan, it is necessary to point out
some heroic assumptions and dominant principles in the Chinese
image of development during Mao's lifetime. Clearly, the principle
of independent and self-reliant development of national economy
stands out as one dominant component in the Chinese image of development. Self-reliance is conceptualized as the only way that developing countries can keep the initiative in their own hands, preserve
their resource sovereignty, prevent the structural penetration of
their economies by imperialist predators, and liberate themselves
from the vicious process of exchange of unequal values. Interdependence in the contemporary world economic system, according to
the Chinese, could easily turn into an interdependence "between a
horseman and his mount."[21]

Self-reliance never implied autarky. It was never a rigid
dogma but an operational principle of maximal self-realization. The
Chinese term zili gengsheng literally means "regeneration through
one's own efforts," connoting a means, not an end. It means that
each country should rely primarily on its own efforts and resources,
using foreign aid as a supplementary means for developing its own
national economy. Guided by this principle, China's growth was
achieved independently and self-reliantly with only marginal external
help. Even during the 1950s, total Soviet loans amounted to no more
than about $1.5-$2 billion over a period of ten years (which China
paid off from 1955 to 1965). China's foreign debt service ratio—the
ratio of payments on foreign loans to GNP—was minimal. Since the

Soviet withdrawal of industrial technicians, managers, and blueprints from China in mid-1960, the PRC allowed no foreign equity participation in its economy.

China's growth policy has often exhibited the "breakthrough complex." In part this was a conceptual legacy of the Long March and the final victory of Chinese communists over innumerable odds and obstacles—or the strategic triumph of "millet plus rifles over Chiang Kai-shek's aeroplanes plus tanks." In greater part this reflected Mao's messianic exaltation of the omnipotence of human willpower. In a country with an agricultural economy that is unevenly endowed with the factors of production, it perhaps made practical sense to tap the power of the human will as the motive force of societal development; hence the stress on "men over machines" and "men over weapons." If the Great Leap Forward was Mao's attempt to substitute human willpower for technology, the two developmental models in the 1960s and the first half of the 1970s—Daqing (or "In Industry Learn from Daqing") and Dazhai (or "In Agriculture Learn from Dazhai")—represent Maoist examples of a conquering human spirit over extremely adverse material and technological bottlenecks. Such breakthrough mentality is also embodied in the current Ten-Year Plan.

The Chinese dialectic on economic growth has little to do with the Western polemics of the growth versus no-growth debate. In fact, the Chinese reject the notion of "energy crisis" or the kind of growth crisis projected by the Club of Rome's Limits to Growth.[22] Rather the debate has been over the priority of two competing goals: equity (redistributive justice) versus efficiency (vertical economic growth). What is unique about the Maoist image of development in this connection is the set of normative assumptions it embodied: that development is both desirable and feasible only within a system of populist feedbacks, mobilizing the creative energies of the people to serve the people; that the horizontal distribution should not be separated from the vertical growth of the national pie; that equity is as important—sometimes more important and desirable—as efficiency; that an egalitarian social and economic order cannot be maintained too long without a continuing struggle to keep the superstructure pure and proletarian. In short, economic growth was conceptualized as a value-enhancing and value-realizing process.

The sharp decline of economic performance in the last two years of Mao's lifetime (1974-76) provided a necessary excuse or justification for the post-Mao leadership to quickly purge the "Gang of Four" and to initiate a series of new plans and programs for economic growth. The year 1977 witnessed a stream of national conferences dealing with all aspects of the four modernizations (industry, agriculture, science and technology, and national defense), culminating in the Ten-Year Plan approved by the Fifth National People's Congress in February 1978. Table 4.3 outlines key components in the plan. Within a year it became clear that the industrial targets in the plan were too high in light of available and projected energy supply.

It also became clear to the post-Mao leadership that the ambitious goals of the Ten-Year Plan could easily be derailed by an unchecked demographic pressure. Conscious of this race between population growth and economic growth, the post-Mao leadership has already shown a renewed vigor in the birth control program, or what the Chinese call "childbirth planning" (jihua shengyu). The 1978 Constitution even codified in Article 53 the state's advocacy and encouragement of family planning. The rate of natural population growth has actually gone down from 2.34 percent in 1971 to 1.2 percent in 1978. The current policy on population control pivots around the concept of an only child accompanied by a strong sanctions system (negative and positive incentives) and is to be carried out in two stages: in the first stage, the present growth rate of 1.2 percent is to be reduced to 0.5 percent by 1985; in the second stage, the projected growth rate of 0.5 percent is to be reduced to a zero growth rate by the year 2000.[23]

What is most striking about the post-Mao leadership is not only its monumentally ambitious plan for modernization but also the manner in which economic growth is being pursued as an almost intrinsic value in and of itself. Deng Xiaoping's oft-quoted statement (made in the early 1960s) about cats being good, whether black or white, so long as they caught mice, seems to have become the operational code in the implementation of the Ten-Year Plan. The Third Plenary Session of the 11th Central Committee of the Communist Party of China adopted in December 1978 a historical decision to shift "the emphasis of our Party's work and the attention of the people of the whole country to socialist modernization" as of 1979. In a word, China shifted from the "politics take command" approach to the "economics take command" approach as of January 1, 1979. A frustrated Chinese posted a challenging normative question to the post-Mao leadership on Peking's "Democracy Wall": "What kind of modernization does China hope to realize? The Soviet type, the American type, the Japanese type, the Yugoslav type? . . . on these issues the masses know nothing."[24]

What impact, if any, did the international energy crisis have on the development of China's domestic energy policy? Coming in the wake of the gathering storms of the global monetary crisis, food crisis, recession coupled with inflation (stagflation), and another round of internecine war in the Middle East, the 1973 oil embargo marked a turning point in the transformation of world politics from a geopolitical to an ecopolitical axis. The price of crude oil, which in real terms had actually declined in the period 1950-70, was quadrupled

TABLE 4.3

Profile of China's Economic Growth as Reflected in the Ten-Year Plan, 1975–84

Sector	Project Category	Growth Rate Projection	Output Projection
Agriculture	over 85 percent mechanization	4–5 percent per year	400 billion kg of grain by 1985
Industry	120 large projects	over 10 percent per year	
	10 iron and steel complexes		60 mmt by 1985
	8 coal mines		1,000 mmt by 1984
	9 nonferrous metal complexes		
	10 oil and gas fields		
	30 power stations		
	6 new trunk railways		
	5 key harbors		

from $2.60 on January 1, 1973, to $11.66 on January 1, 1974. The impact on most industrialized nations as well as on oil-poor Third World countries (who were about 75 percent energy dependent on imported oil) was shattering. [25]

Against this backdrop, the Sixth Special Session of the UN General Assembly was convened in April-May 1974 to review the basic structural problems in the world economic system that confronted the world community with a crisis of global dimensions. The United Nations inaugurated global ecopolitics when the special session adopted a Declaration on the Establishment of a New International Economic Order and a Programme of Action on the Establishment of a New International Economic Order. [26]

The oil embargo was most congenial to the Chinese image and strategy of world order, serving at once both Chinese principles and interests. The Chinese greeted the embargo as a historic milestone in system transformation, a breakthrough in the protracted struggle of the global underdogs to destroy the old economic order characterized by exchange of unequal values (that is, imperialists buying cheap and selling dear) and to establish a new order. The embargo was not seen as causing but merely reflecting a deeper crisis embedded in the internal and external contradictions of the capitalist system. The so-called energy crisis has nothing to do with the exhaustion of the world's fossil fuel resources. This line of reasoning is consonant with the Maoist image of international crises as recurrent and protracted phenomena, generated by economic factors and related to the domestic crises of political actors. [27]

The Chinese position on the energy crisis in 1973-76 needs to be only briefly summarized here. Capitalism is incapable of rational exploitation and use of energy resources. Driven by the desire to seek fabulous profits, the monopoly capitalist class in the United States shifted from coal mining to indiscriminate and reckless drilling, recovering only 35 percent of its oil from underground. Much worse, monopoly capitalists for strategic and economic reasons leave domestic oil unexploited and go abroad to plunder the oil of the Third World. "The temporary and false prosperity of the imperialist countries in postwar years is built on the natural resources and the blood and sweat of the people of the Third World."[28]

To complete the cycle of contradictions, what monopoly capitalists plundered abroad is then drained off in massive arms expansion and war preparations as well as in the frightful waste endemic in the bourgeois lifestyle. In short, the energy crisis is a function of built-in contradictions of the capitalist system. There is no such thing as a global energy crisis, as revealed in the following statement:

As for talk in Western countries that the world's energy

resources are becoming exhausted, it is nothing but
pessimistic groaning by the decadent class. It always
describes its own crisis and doom as the approaching
doomsday of the whole world. Under the law of conservation of matter, matter does not die. Nature provides
mankind with unlimited energy resources and mankind's
ability to understand and conquer nature is unlimited and
will not remain at a particular level permanently. [29]

The most dominant value in Chinese global policy is oriented
toward a protracted struggle to weaken the strong and the rich, and
to strengthen the weak and the poor in the global community. How
does China reconcile this value with the harsh impact of higher oil
prices on the oil-poor Third World countries? This dilemma is explained away by a long-term trend analysis: the great historic significance of the OPEC action should not be negated by "some temporary
difficulties" for nonoil-producing countries of the Third World. [30]

Even in the short term, the Chinese encouraged and applauded
OPEC's foreign aid for the oil-poor Third World countries (which increased from $1 billion in 1973 to $5. 6 billion in 1975) as a significant contribution to help the poor and strengthen the unity of the Third
World. This argument is widely echoed in Third World politics, too.
For the Third World, the OPEC action, in spite of its deleterious effect on the economies of Third World oil-importing countries, represented a strategic opportunity for self-assertion. Greater economic
disparities among Third World countries themselves strengthened
their collective solidarity against the global topdogs for a fundamental structural reform in the old economic order. This is the essence
of NIEO politics. [31]

From the standpoint of economic interests, the OPEC action
came precisely at a time when China was entering the world petroleum market as an exporter. This created an ideal situation for China's
foreign exchange earning. The price of Chinese crude oil soared from
the 1973 preembargo level of $3. 94 per barrel to an early 1974 figure
of $14. 80 per barrel before settling back to $12. 85 per barrel during
the course of the year. [32] Nonetheless, China's 1974 balance-of-payments deficit of $1. 3 billion could have been met in 1975 by the sale
of only 15 mmt of Chinese crude oil. Higher oil prices also broadened
the resource base of Chinese diplomacy as Peking began to charge
"friendship prices" to certain preferred customers.

Just as China has declined to join the Group of 77, the most
powerful and dominant pressure group of the Third World in global
ecopolitics, so has China shown no interest in OPEC membership.
Chinese membership in OPEC would present a variety of economic
and political constraints; hence, China is most likely to remain an

outside spectator in global oil politics, giving moral and symbolic support to the oil struggle while keeping the initiative in its own hands as far as its own domestic and foreign energy policy is concerned. This is a vivid example of how the Chinese principle of maximum flexibility and self-realization (self-reliance) works.

China's development of domestic energy sources can begin only from its resource base. It is therefore necessary to outline briefly the scope and extent of this energy resource base. Although China is already the third largest producer of primary energy in the world, there is a wide gap between its energy potential and its current output. In absolute terms, China's fossil fuel and hydropower resources rank with, or perhaps even above, those of the two superpowers. China's hydroelectric potential of 580 million kilowatts (compared to 390 million kilowatts for the United States) is the greatest in the world. Proven reserves of Chinese coal, which are claimed to be about 600 bmt, are third in the world after the United States and the Soviet Union. The "oil-poor" China has only recently awakened to the potential of its petroleum resources. Most foreign analysts place China's recoverable onshore and offshore petroleum resources at between 10 and 20 billion metric tons (70–140 billion barrels). China has yet to publish its own estimates of proven oil reserves, although it claimed in late 1979 that the country ranked 13th in the world "in workable oil reserves."[33] China's natural gas resources are estimated by outside experts at 25 trillion cubic feet.[34] In short, China possesses a necessary energy resource base to achieve a superpower status.

Coal is still China's principal fossil fuel, accounting for about 71 percent of the country's primary energy use. In spite of marked fluctuations in the pattern of growth, the general and long-term trend has been exponential with 9 percent annual growth rates. China claimed in early 1978 that its coal output during the first 28 years "has gone up 16-fold as compared with the early post-liberation years, ranking third in the world as against tenth in 1949."[35] As already noted, the growth rate in Chinese coal output began to falter in the early 1970s but has regained its momentum in recent years. In 1979 China surged ahead of the United States to become the second largest coal producer in the world (after the Soviet Union).

In relative terms, the coal industry received a low priority in Chinese economic planning until recently. As a result, a number of technological problems and bottlenecks have remained unresolved: low investment in the development of new coal mines and intensive exploitation of old mines; low mechanization; shortage of coal preparation capacity; low share of large-scale surface mining, the most efficient method of extraction; and weak infrastructure (unserviceable freight cars that hinder coal transportation).[36] God's act did not help China's faltering coal industry either, as the July 1976 earthquake totally demolished the Kailuan coal mines in Tangshan, China's oldest and largest commercial mine concentration.

In the mid-1970s the Chinese leadership began to realize the ominous implications in the gradual shift from a coal-based to a petroleum-based economy. In January 1975 China reestablished a separate Ministry of Coal Industry and soon launched a program of general mechanization of large coal mines. This program was further refined and redrawn under the Ten-Year Plan. The plan recognized the coal industry as an important but weak link, calling for the construction of eight major coal centers, each with a designed capacity of over 10 mmt in ten years. The eight coal fields have a verified aggregate reserve of 12,000 mmt. The goal is to double coal output during the plan period (that is, from about 500 mmt to 1,000 mmt) and to reach 2,000 mmt by the end of the century.

In pursuit of this ambitious goal, the post-Mao leadership instituted three key measures in 1978: tapping the potential of old mines by renovating their old equipment and technological processes; developing and transforming the country's 20,000 small coal mines in 1,100 counties so as to enhance their promotion of the economic development of the rural areas; and improving distribution of the coal industry so as to make northeast, north, east, central-south, northwest, and southwest China basically self-sufficient in about ten years. [37]

Under the readjustment policy of the Ten-Year Plan launched in mid-1979, coal has gained an even more prominent place in the modernization drive. The current policy is to rely on coal as the most important energy source for a considerable period of time and to encourage as many industries as possible to shift from oil burning to coal burning. In 1979, 22 new coal mines with a total capacity of 9.5 mmt were newly opened and put into operation while 10 old mines were enlarged, resulting in the 1979 coal output of 628.8 mmt, 13.82 mmt above the projected target. [38] The major plans for 1980, as revealed by Gao Yangwen, minister of the coal industry, are to develop coal gasification and liquifaction plants, to expand coal dressing facilities, and to cooperate with foreign countries in coal mining and increased exports of coal. [39]

Although Western experts estimate that a developing country such as China must increase its electric power at a rate of about 1.3 times that of industry generally, China's own past experience suggests that a 1 percent growth in the national industrial output required 1.2 times the electric power output. [40] China's electric power industry generated 278.87 kWh in 1979. According to Renmin Ribao, this has elevated China's world ranking in the electric power output from 25th in 1949 to 7th in 1979. [41] As Table 4.1 shows, this amounts to an impressive 38-fold increase in the period 1952-79. In the period 1953-78, the average annual increase of electricity was 14.2 percent.

However, China today suffers from a widespread shortage of electric power, judging from repeated calls for conservation, fuller utilization of existing generating capacity, more efficient operation, and staggered hours of operation in some industrial plants. The post-Mao leadership has admitted that electric power is a "vanguard industry" but a very weak link in the current modernization drive. Technologically, China's power industry is believed to be about 15-20 years behind the advanced industrialized countries. Although China has an estimated reserve of 580 million kilowatts from its hydropower resources (the largest in the world), only about 2.5-2.7 percent is being utilized today to generate electricity. It is estimated that the development of one extra percent of this theoretical potential would bring an additional 4.5 million kw of electricity or 10 mmtce. [42]

As of the end of 1977, there were 61 large thermal and hydroelectric power stations in China with a power-generating capacity of more than 250,000 kilowatts each. What is rather unique in the Chinese electric power profile is the rapid development and proliferation of small hydroplants with the average capacity of only 50 kilowatts. In the period 1971-78, small hydroplants increased from 15,000 to about 65,000. [43] In 1979, 7,214 new small hydroelectric power stations were built, adding 1.07 million kilowatts of generating capacity. By the end of 1979, the number of small hydropower stations in China exceeded 90,000 with a total generating capacity of 6.33 million kilowatts. This represents 40 percent of the power used for agricultural production in China in 1979 or 3.5 times the country's combined generating capacity of the hydroelectric and thermal power stations during "the early period" of the founding of the PRC. [44]

While inefficient in cost per kilowatt of installed capacity and in kilowatt hours of power generated, these small hydroplants continue to provide several important economic services: they bring power to rural communes and industries (providing about 40 percent of the electricity needed in the countryside); they reduce demands on the overworked central power stations and the underdeveloped power transmission systems; and they perform such additional functions as flood control, irrigation, timber sawing, and fish breeding. Today 1,500 counties throughout China have their own small hydroelectric power stations performing the above-mentioned services.

In order to lead the 10 percent industrial growth projected in the Ten-Year Plan, the electric power industry has to grow at the rate of about 12-14 percent per year. It was in this context that Premier Hua Guofeng issued directives in November 1977, setting forth the dual theme of more conservation and more electric power. [45] The Ten-Year Plan followed up on this by calling for the completion of 30 power stations. The development strategy for the power industry, as revealed in early 1978 by Qian Zhengying, minister of water conservancy and power, followed the long-standing Chinese practice of "walking on two legs" by developing simultaneously large, medium-sized, and small plants as well as thermal and hydropower stations. [46]

However, the post-Mao leadership is attempting to overcome the backwardness of China's power industry by placing the major emphasis on the construction of large hydroelectric power stations. New hydro stations are to be built, according to Minister Qian, based on the following guidelines: they do not pollute the environment; they are easy to operate; and they serve all-around purposes of flood-control, irrigation, navigation, and raising of aquatic products. Thermal stations are to be built in those places where water resources are unavailable but coal abundant, but only low-quality coal is to be used for thermal stations so as to strike a proper balance between the power industry and other industrial sectors.

Clearly, the inability to sustain the 12-14 percent annual growth rate in electric power needed to fuel the 10 percent industrial growth rate was one of the major factors contributing to the downward readjustment of the more ambitious targets in the Ten-Year Plan. As Table 4.1 shows, the growth rate in electric power dropped from 14.9 percent in 1978 to 8.7 percent in 1979. At the National Work Conference on Electric Power held in early 1980, it was noted that the government could not provide more funds for the electric power industry even though it was lagging behind the growth rate of the national economy as a whole. In order to resolve this "contradiction," the conference mapped out a strategy of translating weakness into strength by stressing the necessity of maximal and efficient utilization of existing resources and facilities. Specifically, the electric power industry was to use less capital and fuel to produce more electricity by strengthening central and unified management over power grids and by making every effort to increase power generating by practicing economy (conservation). [47]

Today China is carrying out a dual strategy of vigorously pushing the rapid completion of projected hydropower stations while at the same time calling for an intensified conservation drive. China is now constructing a huge multipurpose hydraulic project, consisting of a 2.56-kilometer dam and two hydroelectric power stations with a combined capacity of 2.7 million kilowatts, located in Yichang, Hebei Province, around the middle reaches of the Yangtze River (Chang Jiang). When completed, this Gezhouba multipurpose hydro project, the biggest ever attempted in China, would provide 13,800 million kilowatt hours of electricity a year. It is also revealing that nearly half of 44 large- and medium-sized thermal and hydroelectric power generation projects completed in 1979 (with a total generating capacity of 4 million kilowatts) were built in the vicinity of coal mines in various parts of the country, suggesting that a reconversion from an oil-burning to a coal-burning economy has already taken place. [48]

The breakthrough mentality is most dramatically reflected in the saga of China's oil discovery and development: Daqing ("Great Celebration") in 1959; Shengli ("Victory") in 1962; and Dagang ("Great Harbor") in 1964. As of 1975, these three major oilfields, all concentrated in northeast China, produced 80 percent—54 percent in Daqing, 20 percent in Shengli, and 6 percent in Dagang—of China's total petroleum output of over 74 mmt. By the mid-1970s Western and Japanese circles were already talking about China emerging as the Saudi Arabia of the Far East. Such "great leap forward" speculation was fueled and excited by the following developments: the 1968 geophysical survey by a UN agency (CCOP) that gave a high probability that the continental shelf between Taiwan and Japan "may be one of the most prolific oil and gas reservoirs in the world";[49] the phenomenal growth rate of 23 percent per year in the Chinese petroleum sector in 1971-74; and China's recent move toward offshore exploration.

However, China's oil production curve peaked in 1974 at 20 percent annual growth over 1973 and fell off to about 13 percent in 1975 and 1976 and further down to 8 percent (less than the growth rate of China's industry as a whole) in 1977. The 1978 oil production registered only 11.1 percent increase over that of 1977. The imperative of the readjustment policy of 1979 is made clear in the 1979 oil output of only 1.97 percent increase over that of 1978, the smallest increase in the decade of the 1970s. In a more euphoric atmosphere of the mid-1970s, different Western and Japanese observers projected China's 1980 oil production in the range of 130-400 mmt.[50] It has now become obvious that even the lowest of the projected range is too high a goal for China to achieve in 1980.

Several explanations may be offered here for this turn of development in the volatile Chinese energy profile. First, the 20-23 percent growth rate is unsustainable as the statistical base expands. A visiting American petroleum equipment mission to China in September 1978 noticed a chart at the Underground Palace in Daqing that "showed the crude production curve [of Daqing] at almost level for 1975-78," confirming recent Western estimates that Daqing's production has been falling off since 1972.[51] Second, the heavy concentration of Chinese oil production in the major oilfields in the Northeast has already begun to press on their proven reserves, setting in motion a declining reserve/output ratio. One recent study suggested that "China's eastern fields could be exhausted by 1990 unless new sources are developed and that nearly all post-1980 production over 4 million bpd must come from development of either offshore or interior fields."[52] Third, there have been no more Daqing-type breakthroughs in the discovery of new oil fields since the mid-1960s. Finally, given the long lead times involved in offshore oil development, China's offshore oil production is not likely to alter the present energy profile until at least the mid-1980s.

The Chinese petroleum industry now stands at the threshold of a difficult transition in the face of cross pressures from both demand and supply sides. On the demand side, there are sharply rising claims from industry, agriculture, transportation, and the military, all vying to meet their targets under the Ten-Year Plan. The rapid development of oil and gas fields in the 1960s led the government to encourage coal-burning industrial enterprises to convert to oil. As a result, the oil consumption by industrial and transportation systems increased by more than 16 times in the 11-year period 1966-77. The distribution of oil consumption is also revealing. Over 65 percent of the total oil consumption in 1979 was used to heat boilers, with the remaining 35 percent for other industrial purposes. [53]

On the supply side, however, the petroleum industry is hampered by a multitude of technological and infrastructural problems: shortage of sophisticated geophysical equipment; rudimentary offshore drilling and production technology; a thin pipeline network of only 7,500 km; shortage of refining capacity; and underdevelopment of oil ports to accommodate modern supertankers. [54] These problems are accentuated as China's exploration and production shift to deeper strata, less accessible fields, and offshore waters. In short, Chinese hydrocarbon production, in order to keep up with rising demands, requires expensive technological modernization. It will take $20 billion per year to reach 3-5 million bpd by 1980 from onshore sources alone, and $7-$13 billion per year to attain that production by 1990. [55]

The Chinese leadership realizes that industrial modernization would require "extensive use of new materials and new sources of energy." [56] Yet the target for the petroleum industry is identified as the construction or completion of ten oil and gas fields subsumed under the 120 large projects of the Ten-Year Plan. The July 6, 1978, editorial of Renmin Ribao, entitled "Strive To Build Ten More 'Daqing Oilfields,'" uncouples oil and gas fields and describes the target for the petroleum industry in two ways: to reach "advanced world levels" in oil output by 1985; and to build ten more "Daqing oilfields" before the end of the century. The editorial also calls for a thorough survey of the country's petroleum resources onshore and offshore so as to "swiftly increase our oil reserves and production." It then summoned the breakthrough spirit: "In order to develop rapidly the petroleum industry, we must achieve a major breakthrough (yige dade tupo) in the speed of drilling oil wells." [57] In the wake of the Third Plenary Session of the 11th Chinese Communist Party's Central Committee in December 1978, the Ministry of Petroleum Industry adopted the decision to concentrate all efforts on the triple goal of production, construction, and technological revolution for the development of the petroleum industry.

However, as part of a general retrenchment in the Ten-Year

Plan—or what Peking calls "the eight-character principle of readjusting, restructuring, consolidating, and improving the national economy"—many of the 120 industrial projects were delayed or canceled in 1979. The current policy on oil places greater emphasis on efficient utilization and conservation than on accelerated exploration and production. The key to the oil problem is seen in a rapid reconversion from an oil-burning to a coal-burning economy. All oil-burning boilers are prodded to switch back to coal-burning ones. Beginning 1980, industrial units that were planning to install oil-burning equipment should change to coal-burning equipment. Even if the petroleum industry expands in the future, it is now argued, the Chinese should not return to the practice of "gorging from the big pot" (chi daguofan), treating oil as China's strategic resource. Because excessive oil consumption results in less foreign exchange and slower economic growth, the government is now moving in the direction of enforcing a system of rewards and punishments in fuel consumption. [58]

Unfortunately, China has published virtually nothing about its natural gas industry. Table 4.1, which should be read as rough outside estimates, suggests that this sector too has shown a fast but volatile growth in recent years. The annual growth rate for the period 1971-78 has fluctuated between 25 percent and 10 percent, yielding the average annual rate of 18.2 percent for the period. It is significant to note, however, that the growth rate has declined rather substantially in the last six years except for 1977. The 10 percent increase in gas output for 1978 over that of 1977 is the lowest annual growth rate registered up to that point. The 1979 gas output increased only 1.7 percent over that of 1978, in line with the coal output (1.6 percent increase) and crude oil output (1.97 percent increase).

China's natural gas industry is predominantly concentrated in Sichuan Province, which is believed to account for over 90 percent of China's natural gas production. It has more than 200 gas-bearing structures located and more than 1,000 kilometers of pipelines laid. Moreover, gas is used extensively in the industrial activities of the province: steel and iron plants, salt facilities; fertilizer plants; cement factories; power generation; and local manufacturing. In Chengdu, Sichuan Province, a visiting delegation from the National Committee on U.S.—China Relations noticed the many public buses with rubber bags for natural gas on top. In addition to the Sichuan Basin, the Bohai fields and the newly opened Daqing zone could increase the total gas output to 100 bm^3 by 1980, according to one estimate, and possibly lead to exporting liquefied natural gas from Dagang to Japan. [59]

The reticence of the Chinese press on the supply, demand, and developmental strategy of the natural gas industry suggests that this sector is not regarded as an important national link to the Ten-Year Plan. Alternatively, it may be argued that the development of natural gas is being absorbed in the development of the petroleum industry (as implied in the projected target of ten more oil and gas fields) since natural gas is often associated with various oilfields, particularly with the Daqing and Dagang oilfields as well as with the seabed of the continental shelf between Taiwan and Japan.

The policy of "walking on two legs" has led China to experiment and develop in varying degrees such novel or unconventional energy sources as tidal power, geothermal power, solar energy, nuclear power, and biogas generation. With the exception of biogas, however, these novel energy sources—or what the Chinese refer to as "new energy sources"—remain in their experimental stage.

China's tidal energy potential is relatively small and insignificant, since China has shown no interest in this energy source. There is just one tidal power station in Guangdong Province. Its capacity, depending on flow, is between 200 and 250 kw. On the other hand, China's disarranged geologic structures, diverse land formations, and severe earthquakes all lend themselves to one of the world's greatest geothermal potentials. China began to make use of geothermal power in 1970. As of September 1979, there were seven small experimental geothermal power stations with a total capacity of 2,200 kw in the provinces of Guangdong, Jiangxi, Shandong, Liaoning, and Xizang, with their capacity ranging from 50 kw to 1,000 kw. [60] The rationale for developing geothermal stations is that they will "enable China to make full use of 'low-level energy sources' such as the numerous hot springs and underground hot water found in many parts of the country and industrial waste heat." [61] China has so far discovered 2,300 geothermal sites where underground hot water comes to the surface.

The use of solar energy began as early as 1966 in Xizang (Tibet) Autonomous Region, where the most favorable conditions are found. Two-thirds of China has sunshine for over 2,000 hours per year. Xizang receives around 3,000 hours of sunshine annually. The climatic conditions in Xinjiang Uygur (Sinkiang Uighur) Autonomous Region are equally good as the region receives anywhere from 2,500 to 3,200 hours of sunshine annually and the intensity of solar radiation is as high as 130–159 kcal per square centimeter. Today the use of solar energy in the form of simple solar stoves and solar water heaters is extended to some large cities as well as to remote areas of China. The total nationwide area of solar energy collectors for water heating in 1979 reached 70,000 square meters, with more than 2,000 solar energy stoves. [62]

China's interest in the development of solar energy has continued unabated under the retrenchment policy. Solar energy research is listed as one of the key state projects in the 1978–85 National Science-Technology Outline Plan. In May–June 1979 China's New Energy

Sources Delegation, headed by Lin Hanxiong, deputy director of the Second Bureau of the State Scientific and Technological Commission, toured U. S. installations under sponsorship of the China Committee of the National Academy of Sciences. In September 1979 the Chinese Solar Energy Society was founded. Today there is widespread use of solar ovens, solar boilers, and solar public baths in Xizang; Xinjiang boasts that it has a "solar bath train" that moves up and down the Xinjiang-Lanzhou line providing its workers with refreshing baths. Solar water heaters have been installed in many hotels, barber shops, and bath houses in Peking. In February 1980 the lake resort city of Hangzhou started operating a sightseeing boat powered by solar energy. It is now argued that conventional energy sources would certainly be exhausted at some point in the future, hence "efforts should be made to develop new energy supplies and to facilitate replacement of conventional fuel with inexhaustible natural energy supplies. "[63]

China's interest in the development of nuclear power plants did not get underway until 1978 when it became part of the modernization drive. In his address to the Fifth National People's Congress on February 26, 1978, Premier Hua Guofeng defined modern science and technology as being "characterized mainly by the use of atomic energy and the development of electronic computers" and then called for the development of nuclear power stations. It was against this backdrop that a seven-year Sino-French trade agreement was signed on December 4, 1978. The agreement calls for almost $14 billion in trade between the two countries. Included in the agreement is a sale of two French-built (but U. S. -designed) nuclear power plants costing more than $1 billion each. France was supposed to supply the enriched uranium for the two nuclear plants, each of which could generate 900, 000 kw of electricity.[64] However, this preliminary agreement was cancelled in mid-May 1979 as part of the retrenchment policy.

Still, China's current thinking on nuclear power development reflects the "technology-decides-everything" approach. Repeating all the standard arguments of nuclear power advocates in the West, a Chinese publicist has recently argued that China "must go ahead with such projects so that nuclear energy will take up a certain portion in the total amount of power produced in our country." "In the wonderful and varied world of energy," this publicist concluded his plea, "Nuclear energy is a rising star."[65] Likewise, prominent Chinese nuclear physicists, obviously reflecting the government's thinking and approval, gave their support to the proposal to build nuclear power stations in China at the founding congress of the Chinese Nuclear Society in February 1980. [66] The delay in the development of Chinese nuclear power plants caused by the cancellation of the two French nuclear power plants cannot be accepted as a basic policy or conceptual change.

China's most remarkable contribution to the development of novel energy sources is biogas generation. Indeed, this is an application of "soft technology" par excellence, worthy of emulation in many developing countries. Essentially, it is anaerobic fermentation of organic wastes (human, animal, and crop wastes, garbage, vegetation, waste water, and so on) to produce a gas that is a mixture of methane (60–70 percent) and carbon dioxide (30–35 percent). Common combinations of raw materials for anaerobic fermentation in digesters are: a mixture of 20 percent urine, 30 percent human excreta, and 50 percent water; 10 percent human excreta, 30 percent animal dung, 10 percent straw and grass, and 50 percent water; 20 percent human excreta, 30 percent pig manure and urine, and 50 percent water; 10 percent each of human and animal wastes, 30 percent marsh grass, and 50 percent water. [67]

The major advantages and benefits of biogas are economical, ecological, and energy-saving. It is a cheap source of energy for household cooking and lighting as the cost of installing a typical 10 m^3 digester is about 40 yuan ($23.50), primarily for the cement. Such a digester, when properly maintained, can provide an average Southern Chinese family of five with sufficient fuel for cooking and lighting. This in turn leads to the conservation of human (cooking time) and natural (firewood) resources for other uses. Biogas is ecologically beneficial, too, as the manure obtained from biogas digesters is richer in nitrogen content and as the biogas process greatly improves rural environmental sanitation. Biogas is also energy-saving as it is being used for such nonresidential purposes as a cheap and renewable source of energy for diesel engines, generators, crushing machines, and pumps and machines for processing agricultural products. Its chief disadvantage is the impossibility of using this process efficiently in colder regions of China due to the thermal requirements of the fermentation process.

The first Chinese attempts to convert organic wastes into biogas date from 1958 during the Great Leap Forward, but it was not until the early 1970s that a well-organized mass campaign to popularize the biogas technology and use spread across most of southern China. In Hebei Province, for example, 40,000 out of 56,000 households have biogas digesters. By the end of 1975, Sichuan Province had close to 500,000 biogas digesters. The current modernization drive has not ignored this novel source of energy, judging by the convocation of the Second National Marsh Gas Conference in mid-1978, which called for greater popularization of biogas throughout rural China. Plans from all provinces indicate, according to a Xinhua announcement of August 22, 1978, "that the country will have 20 million marsh gas [biogas] generating pits by 1980 and 70 million by 1985, which means that by 1985, 70 percent of peasant households will use this form of energy for cooking."[68] By 1979 there were over 8 million biogas digesters in the Chinese countryside, providing some 30 million people with methane gas for cooking and lighting. [69]

INTERNATIONAL ENERGY POLICY

During the three decades of its existence, the PRC has managed to transform its international energy position from a dependent consumer to an independent exporter. At the peak of its dependency on the Soviet Union in the 1950s and early 1960s, China imported about half of its total domestic consumption of petroleum products (though this constituted less than 3 percent of total energy consumption); in addition, China's petroleum and refinery equipment was imported almost exclusively from the Soviet Union. Thanks to Daqing, Shengli, and Dagang, China made a successful transition from dependency to self-sufficiency. The period of self-sufficiency (1964-72) witnessed China's gradual undoing of the Soviet technological base for the petroleum industry. By 1973, however, the Chinese petroleum industry had come of age, setting the stage for its debut in the world market as a modest but expanding exporter of crude oil. In 1978 China made yet another transition into a complex web of "interdependent" relationships with foreign countries by signing a variety of contracts for the accelerated development of primary energy sources as well as for the renovation of infrastructural requirements as an integral part of the modernization drive. [70]

Even during the period of self-sufficiency, China engaged in small-scale energy trade with some 45 countries on every continent. However, Japan occupied the most prominent position in China's energy trade. Hence China's energy trade with Japan serves as a good indicator of Chinese energy trade policy. The commodity mix of Chinese energy trade during this period consisted of solid fuel and petroleum by-products exports and liquid fuel imports. Coal was the principal energy commodity. China began to export coal to Japan in 1954 and its volume greatly fluctuated, reaching as high as 1 mmt in 1967 and then falling to 0.25 mmt per year in the following years under the impact of political and economic constraints. [71]

Japan also played a key role in China's dramatic entry into the world petroleum market in 1973. China initially offered 0.2 mmt of crude oil for export to Japan but Zhou Enlai soon raised this to 1 mmt. The first contract was signed in April 1973 at the then current price of $3.80/bbl, the price that was pegged to Indonesian Minas crude, which in turn was pegged to Arabian light. Since then the price of Chinese crude has closely followed world market prices. Chinese crude exports to Japan rose to 4 mmt in 1974, 8.1 mmt in 1975, dropped to 6.15 mmt in 1976, rose slightly to 6.53 mmt in 1977, 7 mmt in 1978, and 7.6 mmt in 1979.

   The implementation process of the Ten-Year Plan has brought
into sharper focus international dimensions of Chinese energy policy.
The principle of self-reliance as the Maoist model of development
was subjected to so many quantitative and qualitative changes in 1978
and 1979 as to call into question its viability as the normative guide
in China's current international energy policy. The tensions, contradictions, and discrepancies between the functional requirements of
energy development and the normative requirements of self-reliance
have been explained away by a "have cake and eat it too" double talk.
   A major international link to China's modernization drive lies
in the vigorous pursuit of barter trade: balancing Chinese exports of
raw materials and crude oil and Chinese imports of advanced foreign
technology and complete plants. The China-Japan Long-Term Agreement signed on February 16, 1978, is an apt illustration of commodity
complementarity between the two countries, as it specifically stipulates in the preamble, "Japan's exporting technology, plants, and construction materials and machines to China, and for China's exporting
crude oil and coal to Japan, with the support of the respective Governments." While the types of technology and equipment to be imported
from Japan are left unspecified, the agreement mentions the commodity items and even their amounts to be exported to Japan from the first
year to the fifth year as follows:

	Crude Oil (mmt)	Coking Coal (mmt)	Coal for General Use (mmt)
1978	7.0	0.15-0.3	0.15-0.2
1979	7.6	0.5	0.15-0.2
1980	8.0	1.0	0.5 -0.6
1981	9.5	1.5	1.0 -1.2
1982	15.0	2.0	1.5 -1.7

   Besides Japan, the list of customers for China's crude oil includes the Philippines, Thailand, North Korea, Brazil, France, and
Italy. On November 17, 1978, the United States was added to the growing list of Chinese customers when the Coastal States Gas Corporation
signed a contract for 3.6 million barrels of China's low-sulfur, light-
gravity crude to fuel power stations of the Los Angeles Water and Power Company and San Francisco's Pacific Gas and Electric Company.
In 1979 China exported $96.5 million worth of crude oil and gasoline
to the United States. On January 3, 1980, China and Thailand signed
an oil-supplying contract, under which the former will sell to the latter 950,000 tons of oil (700,000 tons of crude oil and 250,000 tons of
light diesel) in 1980.[72] Although small in volume, China's energy commodity trade also involves exporting petroleum products to Hong Kong.

Peking's oil diplomacy has now been extended to Taiwan. "Only by conducting economic interchanges with the mainland," says a recent Radio Peking Mandarin broadcast to Taiwan, "supplying each other's needs, learning from the other's strong points to offset weaknesses and establishing joint efforts to exploit and enjoy the inexhaustible, precious deposits in the motherland, can Taiwan extricate itself from the economic predicament caused by external factors and embark again on the broad road of economic development. This is the only way of solving its energy problem."[73]

In spite of the growing list of Chinese customers cultivated by active oil diplomacy, China's crude oil exports cannot generate sufficient earnings to pay for the massive importation of foreign technology and plants. Rekindled by the prospects of "fabulous profits" in China's modernization program, Japanese bankers came up with forecasts in late 1978 that showed that "China will need at least $600 billion in investment funds over the seven years up to 1985, of which $200 billion will have to come from overseas earnings, foreign loans or equity participation."[74] On the other hand, a 1978 Japan External Trade Organization (JETRO) study showed that China's earnings from oil and coal exports to Japan over the first eight years of the China-Japan Long-Term Trade Agreement could generate only $13.7 billion, based on a 5 percent annual price increase.[75]

The gaps between projected oil (and other export) earnings and projected costs of technology and plant imports under the Ten-Year Plan raise a number of key policy and normative questions. Since the Sino-Soviet split in the early 1960s, China has followed a long-term policy of developing its national economy independently and self-reliantly. This policy has used foreign trade as a balancing factor in the planning process designed largely to eliminate some weak spots in the economy and to avoid long-term dependence. Foreign trade as a percentage of GNP reached the peak 3.11 percent in 1959 and gradually dropped to the low 1.87 percent in 1970, then rising up again to slightly over 4 percent in 1974-76 and reaching its all-time high of 5 percent in early 1980.[76] In addition, China managed to balance its exports and imports, avoiding foreign aid and debts. Technological dependence was conceptualized as an easy path of structural penetration and plundering of the national economy by foreign imperialists.

In the drastic redefinition of the principle of self-reliance by the post-Mao leadership, technology is now claimed as having no class character, and it is claimed that the Gang of Four had distorted the principle into self-sufficient backwardness and isolationism. Be that as it may, the deviations of the post-Mao leadership from the established patterns and practices of the past are unavoidable corollaries of unprecedented importation of foreign technology. What used to be called "exchange of unequal values" during the self-sufficiency period has now become "international trade practices" just as the word "loan"—which used to be <u>verboten</u> in Chinese trade practices—has now become respectable, as shown in the new policy statement of Li Qiang, China's minister of foreign trade, in December 1978: "China respects international trade practices. As long as the conditions are appropriate, we can consider accepting government-to-government loans. Both government-to-government and nongovernment loans are acceptable."[77]

China's bargaining for both bilateral and multilateral aid from capitalist countries, multinational corporations, and international bureaucrats and bankers since late 1978 marks a significant turning point in Chinese developmental politics. Since its entry into the United Nations in October 1971, China has represented the most powerful symbol of self-reliance in the global community, a poor and developing country that admirably meets the basic human needs at home and gives both bilateral and multilateral aid at one and the same time, yet absolutely declines foreign aid in any form whatsoever. (Note the decline of all external offers of assistance in the wake of the Tangshan earthquake—perhaps the most devastating earthquake of this century in terms of human and industrial destruction—in July 1976.)

In an unprecedented and surprise move, however, China broke away from the tradition by asking the United Nations Development Program (UNDP), the largest source of UN multilateral aid for technical assistance, for $100 million in aid in late 1978. In February 1979, UNDP approved a $15 million aid program for China.[78] China has also turned to other agencies of the UN system for a variety of technical assistance. The United Nations Fund for Population Activities will spend $50 million over the next four years (1980-83) to help China with its census in 1981 as well as with other technical problems related to China's drive toward a zero growth rate by the end of the century. On April 17, 1980, China was admitted into (and Taiwan expelled from) the International Monetary Fund, opening up the door to China's quota in the Fund of 550 million Special Drawing Rights, the equivalent of about $693 million.[79] China's admission into the World Bank now seems assured.

The roughly $3 billion worth of capital equipment China purchased during the period 1972-77—including some $1.3 billion in complete energy plant purchases from Japan and Western Europe—was paid off through export earnings, although China had to resort to a combination of deferred payment and progress payment schemes. China managed to adhere to the principle of self-reliance by avoiding foreign borrowing in its determination not to repeat the bitter historical lesson of being deeply in debt to foreign interests. In order to avoid foreign borrowing, which is the most direct and simple, but most feared, means of financing balance-of-trade deficits, China in-

troduced in 1978 a variety of innovations, including joint ventures
with foreign firms in Hong Kong, introduction of compensation and
by-product schemes and counterpurchase arrangements, establishment of export-oriented areas and factories, and plans to establish
a franchise system with temporary concessions involving natural re-
source-related projects. [80] All of these had been justified by an expansive interpretation of the principle of self-reliance.

In December 1978, however, China crossed the Rubicon on the
loan policy by going directly into foreign borrowing. The Sino-French
Trade Agreement signed on December 4, 1978—another long-term
trade accord—commits France to provide $6.8 billion in export cred-
its over the next ten years at the interest rate of 6.5 percent. Two
days later, a consortium of seven British banks signed with China a
$1.2 billion loan agreement at the interest of 7.25 percent. This is
the first time China has accepted government-backed credits from a
Western country. On March 8, 1979, China and Japan reached a "basic agreement" on a $2 billion loan at the interest rate of 6.25 per-
cent. In August 1979 U.S. Vice-President Walter Mondale announced
in China that the Export-Import Bank of the United States will make
available $2 billion in export credits over a two- to five-year period.
The total sum of government-guaranteed export credit offered by the
United Kingdom, France, Italy, Australia, Sweden, Japan, Canada,
and the United States between December 1978 and August 1979 amounted to $15.1 billion. [81]

The year 1978 also marked the beginning of foreign participation
in the development of Chinese energy projects. In mid-1978 China invited five U.S. oil companies (Pennzoil, Exxon, Union Oil, Phillips
Petroleum, and Mobil) to discuss the possibility of offshore oil
development on a service contract basis. In late 1978, China agreed
to give Japan exclusive rights for exploration, development, and production of oil resources in a 20,000 square km section of Bohai Bay.
The Japanese will finance the costs of development and will be
compensated in crude oil shipments when and if the fields begin producing. [82] As of the end of 1979, joint offshore development activity
seems to be moving at a measured pace, involving a dozen U.S.,
French, British, and Japanese companies in seismic surveys of designated zones stretching from the Bohai Gulf to Shanghai and from an
area southwest of Taiwan to the Gulf of Tonkin. [83]

China has also been seeking foreign help and advice in carrying
out a $30 billion hydro project. U.S. assistance is being sought in the
development of a high-voltage transmission system, while Japanese
experts are already in the field advising the Chinese on the construction of four large dams and hydroelectric power stations with an estimated capacity of 32,000 MW. In August 1979 Vice-President Mondale
signed in Peking a hydropower and water conservation agreement. On

March 15, 1980, China and the United States signed an agreement (an annex to the U. S. -China Hydroelectric Protocol the vice-president signed) that provides U. S. government help in the design of four hydroelectric power projects in China and that "could eventually lead to multimillion dollar business deals with American companies."[84]

For the development and renovation of the coal industry China signed its first service contracts with European countries since the departure of Soviet technicians in mid-1960. In late August 1978, ten European companies disclosed contracts for coal mining equipment and services that together promise to be one of the largest purchases of mining equipment by a single country ever made. In a joint venture arrangement West Germany's Krupp Industries and Stahlbau will join Demag-Lauchhamer to open up China's largest brown-coal deposit, a 55 sq km field in Jilin Province in northeastern China with potential reserves of some 2,000 mmt. On September 22, 1978, China signed yet another trade protocol with West Germany guaranteeing German coal-mining and processing equipment manufacturers sales of at least $4 billion. West Germany is the most dominant foreign participant in China's modernization of the coal industry. [85] In February 1980 China asked Japan to send a research team to develop the Jungar mines with a huge coal reserve in the Inner Mongolia Autonomous Region, whose planned production scale is estimated at 40 mmt per year. [86]

Foreign participation has also extended to the modernization of energy-related infrastructural requirements. A group of Dutch contractors and dredging companies signed "letters of intent" with China in mid-October 1978 to build a coal port on the Yangtze River and to dredge a channel in the mouth of the river. This project, valued at $2-3 billion, would take some six years to complete. In addition, a deep-water port would be built at the coastal city of Lianyunggang, Jiangsu Province, some 400 miles from Shanghai. In December 1978 Hitachi Shipbuilding of Japan announced that it was asked by China to construct a shipyard in China with the capacity to build 100,000 dwt cargo vessels. During his official visit to China in late 1979, Japan's Prime Minister Massayoshi Ohira signed Joint Projects, Loan Agreements to help China build six industrial projects through a government loan of 50 billion Japanese yen. Significantly, every project, with the exception of one project for a hydroelectric power plant in the lower reaches of the Yuanshui River in Hunan Province, is related to beefing up China's weak energy infrastructural systems in port and railway facilities. [87]

Under the mounting pressure of the modernization drive, China's foreign economic policy made some significant departures from the principles and practices of self-reliance. In order to attract direct foreign participation in the development of Chinese energy resources, China put into effect the Law on Joint Ventures on July 8, 1979. As if to further guarantee property rights to foreign companies, their right of remittance of profits to their home countries, and the protection of their technology and patents, China decided to join the World Intellectual Property Organization (WIPO) in March 1980, depositing the government's instrument of accession to the WIPO Convention. Gu Mu, vice premier and chairman of the newly established Foreign Investment Control Commission, disclosed in October 1979 that some 30 joint-venture projects were under consideration, adding: "China will insure more rights, give more convenience and be more generous" to joint ventures than is true in other countries. [88]

The retrenchment policy halted a blind rush into massive credit purchase of foreign technology and complete plants. The Chinese were able to suspend "already signed contracts" because of the insertion of a provision (ironically made at Japanese insistence) requiring approval of both governments. In late February 1979, the Chinese notified a number of Japanese trading firms that they had been denied approval to some 22 contracts for 29 plants worth over $2.6 billion signed since December 1978. However, by October 1979, only one contract out of 29 had been fatally wounded (a $76 billion deal for a 300,000-ton ethylene plant). [89] In short, the retrenchment policy slowed down the pace of credit purchase of foreign technology and complete plants, but it has not affected the basic policy choice on the question.

However, the Chinese political sensitivity to foreign dependence expressed itself in diversifying the sources of foreign technology and participation. A close examination also reveals Chinese attempts to complement ecopolitical interests with geopolitical interests in the conduct of international economic policy. The China-Japan Long-Term Trade Agreement has the obvious Soviet connection just as the Chinese choice of France as supplier of two nuclear reactors underlines the complementarity of the two countries' policies on arms control and disarmament issues in global politics. A noticeable slippage in Chinese activism in NIEO politics, which still lies in the realm of symbolic politics with little substantive impact on Chinese economic and strategic interests, highlights the transition of the post-Mao leadership from the value-oriented strategy to the power-oriented strategy of world order. [90]

ENERGY AND THE ENVIRONMENT

The human environment can easily be impaired by man's—and a nation's—maladaptive behavior. In modern times the common hazards to the environment come from unchecked population growth, undisciplined urbanization, unregulated growth and industrialization, and rampant consumerism, all of which make excessive claims on the carrying capacity of ecosystems. To be sure, the PRC as a late-comer in the industrial rat race has the hindsight advantage to avoid the ecological disaster hovering over advanced industrialized coun-tries. Still, China's approach to the environment has a number of novel features worthy of careful study and possible emulation by both developing and developed countries.

For analytical convenience the development of China's environ-mental policy can be divided into three periods. During the first pe-riod (1949-70), the main concern centered largely on the elimination of traditional hazards to the environment and the improvement of pub-lic health (environmental sanitation). In the second period (1971-77), China made frontal attacks on environmental pollution. In fact, China's debut in global conferences was made at the United Nations Conference on the Human Environment, held in Stockholm in 1972. This period also witnessed a steady stream of articles in the Chinese press on environmental problems as well as the establishment of an Office of Environmental Protection directly under the State Council, not as a part or appendage of a cabinet department. The third period (1978 on) marks the continuation and codification of the policy of the second pe-riod. The 1978 Constitution stipulates in Article 11: "The state pro-tects the environment and natural resources and prevents and elimi-nates pollution and other hazards to the public." At the same time, the four modernizations have been defined in complementary terms with environmental protection. In this spirit, China promulgated in September 1979 the Law of Environmental Protection of the People's Republic of China for trial use. [91]

Until about 1971 China had no environmental policy in the con-ventional sense of the term as the country's main efforts were con-centrated on the improvement of environmental sanitation. China had pursued the seemingly impossible goal of exterminating all flies, mo-squitoes, rats, and sparrows. The success of this "four evils" ex-termination campaign had brought about an unforeseen disruption of the natural equilibrium of the environment with the elimination of the sparrow population. Belatedly, China had to "rehabilitate" sparrows, deleting them from the "four evils" list. [92] Mass sanitation campaigns in the elimination of such primary causes of death in rural China as schistosomiasis (snail fever), hookworm, and other parasitic dis-eases—all related to peasant ignorance and the low level of communal hygiene coupled with the widespread use of animal and human wastes as fertilizer—greatly improved the rural environment.

Moreover, China made remarkable progress in minimizing such traditional hazards to the environment as floods, droughts, famine, deforestation, and breakdown of civil and social order. To cite one example, the menace of flood, which for centuries plagued the lower reaches of the Yellow River, has been brought under control and the danger of two dyke breaches every three years has been done away with for good. [93] While these efforts have undoubtedly improved the environment, they have also increased demographic pressure on the resources of the environment by drastically reducing death rates. For the crux of the Chinese population problem lies in a modest progress in reducing birth rates (until recently), coupled with a phenomenal progress in reducing death rates.

In the 1950s and 1960s China did not have an environmental policy specifically aimed at industrial pollution, as the leadership was too preoccupied first with survival and later with economic development. To improve the health and living conditions of the people and to improve the quality of the land became the twin goals of the "developmental-environmental" policy. In mid-1971, however, Peking ordered some sweeping new measures to combat major sources of industrial pollution. The new measures called for incorporation of means of eliminating the "three wastes"—gas, liquid, and slag—into new industrial projects. [94] This marks a turning point in the development of Chinese environmental politics. The Chinese leadership suddenly became aware of pollution that plagued industrialized societies in the West and showed determination to check its spread to Chinese society.

The assumptions, norms, and values underlying Chinese ecologism deserve a brief mention. First, in contrast to Western debates in which ecologism and developmentalism are often juxtaposed in competing or conflicting terms, the Chinese have conceptualized the two in mutually complementary terms. In the tug of war between extreme environmentalists and extreme developmentalists at the 1972 UN Conference on the Human Environment, for example, China took a middle-of-the-road position, stressing mutual harmony between environment and development. Every call for environmental protection and improvement in the 1970s has consistently stressed the theme that pollution will harm the protection of the people's health, the promotion of the well-being of future generations, and the development of socialist production. [95] We may characterize this as a systems approach in which the environment is defined as an integral part of the Chinese development system.

Second, the Chinese conceptualize environmental protection as a social responsibility. In this view, pollution in capitalist countries is an inevitable corollary of the laissez-faire approach, a manisfestation of the sharpening contradiction between the private ownership of the means of production and the social character of production. Put simply, profiteering is the prime cause of pollution. The Chinese carried this concept of social responsibility to the Stockholm Conference and achieved a measure of success in having it incorporated in the final Declaration. [96] As already noted, the 1978 Constitution firmly codified this principle in Article 11. The newly promulgated environmental law takes a step further by stipulating that citizens have the right to keep watch on the environment, report to the authorities, and file charges against organizations or individuals whom they deem to be causing harm to the environment.

Third, Chinese environmental politics is firmly linked to the supreme value in Mao's worldview, namely, his populism as epitomized in the slogan "Serve the people" (wei renmin fuwu). This populist ethos has been embodied in the Chinese "basic human needs" approach as another expression of environmental policy, that is, to enhance such biophysical requirements of basic human needs as the amount of food, clean water, adequate shelter, health services, and basic educational opportunity. Western environmentalists are notoriously prone to abstract ecologism in a manner that makes it almost irrelevant to the basic human needs agenda for the Third World. The Chinese integration of populism and ecologism also serves as a powerful implementation weapon by making environmental protection everybody's business.

Fourth, Chinese environmental politics stresses the "prevention first" approach. China has advocated the policy of putting antipollution measures into practice simultaneously with the design, construction, and commission of the industrial projects. It is admitted that this preventive method is more costly in the short run but "the cost will be much smaller than the price which has to be paid for keeping the pollution under control after it has occurred, and the results have proved much better."[97] In a similar vein, industrial hygiene and environmental protection are an important part of China's preventive medicine.[98] As noted earlier, one of the three operational guidelines for the construction of new hydro power stations as part of the modernization program is that they do not pollute the environment.

Fifth, China applies Maoist dialectics to the environmental problem by posing problem and progress in mutually contradictory/complementary terms. Development and environment are conceptualized as a unity of opposites, in which the two are mutually contradictory and mutually complementary at the same time. The logic of this dialectical approach becomes a bit more clear in the "turning the harmful into the beneficial" (the waste into the useful) campaigns. Specifically, the three evils or wastes (waste gas, liquids, and solids) are supposed to be transformed into treasures of renewable resources. Actually, this is the Chinese approach to conservation and recycling. To cite just one recent example, the rising city of Changzhou in Jiangsu Province is claimed to have generated electric power from exhausted heat, utilizing the steam produced by medium- or low-pressure boilers operating on "exhausted heat" to generate power.[99] According to incomplete statistics obtained for 1970-78, China recovered 115 mmt of

scrap iron and steel, of which 80 mmt were used in steel refining.[100]
The Chinese are seasoned practitioners of the "making-virtue-out-of-
the-necessity" approach. In Chinese ecological thinking, the pollution
problem will not be solved until all wastes are productively utilized.

Finally, China's "walking on two legs" strategy, the xiafang
(sending down or deurbanization) movement, and local self-reliance
all contribute to environmental protection by diffusion and dispersal.
China takes special pride in its overall planning, which works in the
rational and wide distribution of new industries in the various provinces. The basic policy of China's approach to the human environment was summed up by the head of the Chinese delegation to the 1972
UN Conference on the Human Environment as follows: "Our Government is now beginning to work in a planned way to prevent and eliminate industrial pollution of the environment by waste gas, liquid and
residue in accordance with the principles of overall planning, rational
distribution, multiple utilization, turning the harmful into the beneficial, relying on the masses, everybody taking a part, protecting the
environment and benefiting the people."[101]

How well, effectively, and widely is China's environmental policy being implemented? This question runs into the data problem.
Still, based on recent (and more candid) Chinese press accounts, we
may conclude that there are some major gaps between policy pronouncements and policy performance. In the absence of any systematic environmental impact statements from the Office of Environmental Protection, we can rely only on Chinese press accounts, visitors'
impressionistic observations, secondary works by foreign specialists,
and circumstantial evidence.

It may be safely argued that there is no air pollution for about
85 percent of the population who live in the countryside. Even in the
cities, where vehicular traffic is insignificant and population growth
diffused through the xiafang movement, air pollution may not be as
threatening as that of their counterparts in the West and Japan. Still,
some Chinese cities suffer from air pollution that is believed to be
as bad as—or even worse than—anywhere in the world due mainly to
the extensive use of coal for power and heat in both industrial and
residential uses.[102]

In spite of some valiant efforts since the early 1970s to improve
air quality, many comrades apparently failed to grasp the dialectical
relationship between environment and development. In certain locali-
ties and enterprises comrades treat as unimportant "trash collecting,"
while another common complaint or excuse from Chinese plant managers is that they are too busy in production to pay much attention to
environmental protection. This clearly shows that the production quotas in the planned economy pose as great a threat to the environment
as profiteering in capitalism. In the face of this problem, however, a

new warning was issued in mid-1978 that those industrial enterprises that cause serious pollution must stop production until they have solved the pollution problem within a stipulated period of time. [103] In early 1980 the government took another step by ordering 167 chronic polluters, including the Daqing oil field, to clean up by 1982 or face fines, criminal charges, or closure. [104] It seems highly unlikely that the government would really go as far as closing down the Daqing oil field. The inherent contradiction between environmental protection and high growth scenarios of the Ten-Year Plan may resist resolution by hortatory dialectics of Chinese ecopolitics.

In Chinese ecosystems, water pollution creates a more direct, widespread, and immediate impact than does air pollution on the health of the people and the productivity of the local and national economy. China's main concern about water pollution in the 1950s and 1960s was largely sanitational; but with the industrial growth and expansion came industrial waste discharged into rivers and streams of the country. Although great progress has been made in eliminating the traditional water-borne diseases and controlling urban sewage, making drinking water available to the great majority of the populace, industrial waste emerged in the 1970s as China's prime bête noire of water pollution.

A few examples of water pollution show the challenge and how the Chinese are going about dealing with this threat to the environment. The river (Nen Jiang) running through the industrial city of Qiqihar, Heilongjiang Province, received so much industrial waste from numerous industries in the city as to reduce the catch of fish in 1969 to about one-fifth of the 1960 output. Between June and November 1970, a massive campaign was undertaken with results that not only cleaned up the river, raised the oxygen content of the water, and returned the fish, but also recovered cadmium, oils, acids, alkali, paper pulp, and silver from the waste waters. Fields irrigated by waste water were also claimed to have produced greater yields than ever before. [105]

Another example relates to the pollution of the river (Li Jiang) that runs through Guilin, Guangxi Zhuangzu Autonomous Region, the southern resort city renowned for its dramatic craggy landscape, clear river, and magnificent caves. On February 7, 1979, three factories in Guilin were ordered to stop operation and another closed permanently. Industrial wastes discharged by these plants have so badly polluted the river that some of the rocks have whitened and many trees withered. Quantities of such harmful chemicals as phenol, arsenic, chloride, and cyanide in the river have risen beyond the safety margin, causing three out of the four fishing teams in a small town close to Guilin to turn to other occupations. [106]

Still, Li Chaobo, director of the Environmental Protection Office of the State Council, conceded in a recent interview with Xinhua that many of China's 400,000 enterprises do not have pollution-control facilities and that "the amount of soot released annually by the country's industry and the domestic burning of fuel tops 10 million tons, while sulphur dioxide released nationally is estimated at 15 million tons." He further conceded that a huge amount of industrial liquid waste is discharged every day to rivers, lakes, and seas and that about 90 percent is not treated. The annual volume of solid waste now amounts to 200 million tons. [107]

While the main emphasis on the first two of the three wastes—gas and water—is on prevention, China's approach to the third—slag and other solid waste materials—is on multipurpose utilization through recovery and recycling. This is conservation and recycling à la China worthy of emulation throughout the world. The primary responsibility for the "turning waste into treasure" process—that is, to recover and recycle solid waste—is being placed on the enterprises who originate the waste in the first place. In order to generate maximum participation in this recovery-recycling process, however, the commercial departments throughout the country are actively encouraged to purchase waste materials for multipurpose utilization.

Thus the recycling process has a built-in economic incentive for capital accumulation for industrial enterprises as well as for residential neighborhoods. The Chinese politics of mass mobilization is also at work here, as factories, enterprises, government organs, schools, and communes are encouraged to be involved in the process as part of their daily routine chores or through special recovery drives. The "turning waste into treasure" campaigns are claimed to have produced in 1974 alone some 6.25 million tons valued at nearly 1,000 million yuan, including 2.87 million tons of scrap iron and steel, 50,000 tons of nonferrous scrap metals, nearly 1 million tons of paper-making materials. [108]

As part of the three-year readjustment of the national economy, the current environment policy is also linked to the energy policy. The principal aims of the environmental protection law are specified as having four interrelated components: strengthening the socialist legal system; eradicating environmental pollution; safeguarding people's health; and promoting socialist modernization. [109] At the operational level, energy efficiency, energy conservation, and energy recycling are stressed as the most effective means of solving both energy and environmental problems. In this spirit, one of the major tasks specified for all trade and industrial enterprises for 1980 is to conserve 10 percent of oil, 5 percent of coal, and 3 percent of electricity.

China's traditional energetics, as earlier noted, caused widespread deforestation of the country with adverse effects on farming ecosystems. The Chinese leadership recognized a vital role of afforestation in maintaining ecological equilibrium. The First Five-Year Plan, it was reported, resulted in planting 169. 3 million mu of trees throughout China. [110] Although the mass planting campaigns have continued ever since with varying intensity, many of the early, labor-intensive efforts were not too successful, increasing the proportion of the country covered by forest only from 8 percent to 12. 7 percent. Today China ranks 120th in the world in the ratio of forest cover.

It was against this backdrop that another massive afforestation campaign has been launched as part of the new Long March toward modernization. In March 1979 the Standing Committee of the National People's Congress adopted on a trial basis a Forestry Act of the People's Republic of China, establishing March 12 as National Arbor Day. In doing this, China again established the dialectical relationship between forestry and agriculture, drawing an analogy between fish and water. However, afforestation is publicized as playing "the vital role of conditioning the climate, nurturing water sources, conserving soil and water, preventing windstorms and solidifying sands, preventing and eliminating the pollution of our environment and changing our natural conditioning. "[111]

More recently, the government has proclaimed an ambitious new target of afforesting 231, 000 square miles, or more than double the total planted with trees since 1949, with the aim of covering about 20 percent of China with trees by the end of this century. By sometime in the twenty-first century China hopes to cover 30 percent of the country with forest, a proportion about equal to that in Europe and the United States. [112] To the extent that this new campaign succeeds in what it set out to do, it will enhance the environmental integrity of Chinese agricultural ecosystems.

By way of conclusion we may say that the most distinctive feature of the Chinese dialectics between energy and the environment is the systems approach of environmental protection as an integral part of the national development process. China may be able to avoid the kind of environmental degradation that now plagues most of the world's advanced industrialized societies. This is not because China has singled out ecologism as top-priority project. In spite of demographic pressure on the environment—and we should keep in mind the 25:1 consumption ratio between a newborn child in the United States and its counterpart in Asia, Africa, or Latin America[113]—the integrity of the environment in China may survive through the cumulative impact of a variety of behavioral norms and enforcement measures advocated by the leadership. Of course, this kind of assessment is based on the cautiously optimistic assumption that China's environmentalists would be allowed to play their countervailing role and that China's leadership would be sensitive to bridging the gaps between behavioral norms and policy measures in the course of the modernization drive.

FUTURE PROSPECTS

What are the future prospects and directions of the Chinese energy policy? How realistic are the ambitious targets projected in the Ten-Year Plan? Specifically, can China discover ten more Daqing oilfields? Can China double its coal output to 1,000 mmt by 1985? Can China increase its electric power output by about 12-13 percent per year, which is needed to sustain a 10 percent rate of industrial growth projected in the Ten-Year Plan? Above all, can China increase its agricultural production from 285 million tons in 1977 to the projected 400 million tons by 1985?

Of course, energy plays a crucial role in acting out such a high growth scenario embodied in the original Ten-Year Plan. The post-Mao leadership seems to realize fully the supply-demand linkage of energy to the requirements of the modernization drive. Within less than a year, the Chinese leadership reached a sobering conclusion that the ambitious targets in the Ten-Year Plan were unattainable due largely to the difficulty of sustaining high growth rates in primary energy production. The retrenchment policy initiated in mid-1979 was designed to reestablish more realistic targets in line with declining growth rates in energy output. In implementing the readjustment policy in 1979, for example, the government decided to cancel or postpone some 330 projects out of the 561 large and medium-sized projects that had been targeted for reassessment. [114]

China has launched a two-pronged attack on the energy question. On the demand side, the strategy is energy conservation through multipurpose utilization of what can be recovered and recycled, especially "waste heat." As an incentive for such recycling approach, the government would now allow Chinese factories that make profits from products reclaimed from industrial wastes to keep those profits for themselves rather than submitting them to the state.

On the supply side, China is broadening and embracing all sources of energy—traditional, primary, and unconventional—as revealed in the following public appeal: "We should tap all energy sources, including geothermal energy, solar energy, atomic energy, wind power and tidal energy, while stepping up the production of coal, petroleum and natural gas, expanding hydroelectric power generating, developing marsh gas [biogas] and making multipurpose use of such low-calorie fuels as gangue, bone coal and oil shale."[115] The most notable aspect of this latest campaign to deal with the supply problem is an accelerated switch from an oil-burning to a coal economy. In 1979 the power industry is reported to have saved a total of 3.63 million tons of coal by improving the efficiency of generating plants and converting some industrial units from oil-burning to coal-burning. [116]

Although the Chinese government released a substantial body of statistical data in 1979 for the first time since 1959, predicting China's future energy balance is still an extremely hazardous exercise, for it requires acceptance of too many nonquantifiable assumptions. Given the volatility of Chinese ecopolitics in the past (which inevitably affected economic performance) and the pronounced tendency in Chinese superstructural politics to turn today's heroes into tomorrow's villains (or vice versa)—note the fate of the Gang of Four and of Deng Xiaoping—we should be wary of any heroic methodology in predicting the future of tomorrow's China, even such less rigorous species of forecasting as extrapolation. Moreover our assessment of future prospects of Chinese energy development must take into account the role of imported foreign technology. Yet the full scope of China's credit purchase of foreign technology was considerably muddled in the latter half of 1979 as part of the retrenchment policy, and the full extent to which China can effectively absorb and utilize the already imported plants to meet the changing targets in energy development is still not too clear.

With the above caveats in mind, it is possible to argue that the targets in Chinese energy development are unattainable during the original plan period. The readjustment policy is the most eloquent testimony to the unattainability of the original targets in the Ten-Year Plan. The 1979 growth rates of 1.6 percent for coal, 1.97 percent for crude oil, 1.7 percent for natural gas, and 8.7 percent for electric power stand in sharp contrast with the overall industrial output growth rate of 8 percent for the same year. Even if all of the optimistic assumptions of developmentalists at home and abroad prove to be correct, China cannot be expected to emerge as another Saudi Arabia because of rising requirements of domestic consumption. [117]

The development of oil may serve as a key indicator of China's energy future. However, the future prospects of Chinese oil development are beclouded. Unless China makes Daqing-like breakthroughs onshore or accelerates offshore production in a significant way soon, the growth rate in the petroleum industry will decline even further at the very time when it is subjected to the dual claims from increasing domestic consumption (especially from the mechanization of agriculture) and increasing pressure to generate more oil export revenues to pay for imported technology and plants. Recently China had to inform Japan that it could not deliver all the oil it had pledged for 1980.

The difficult task assigned to China's "black gold" is further complicated by two additional problems. First, the accelerated development of oil, especially offshore development, is extremely capital-intensive. It takes money to make money out of China's oil development. Second, Japan as China's most important energy trade partner has proved to be somewhat unreliable and unpredictable. In September 1978, nine electric power companies in Japan formally advised their government that they would not want to buy more Chinese crude oil because of its quality (high wax content). This may very well be a bargaining ploy to cut the prices down since the same companies are using Indonesian crude oil with similar quality problems. At any rate, China, apparently trying to strengthen its relations with Japan and also to make up the shortfall from 1979, decided to decrease its price of crude oil from Daqing by 7.5 cents per barrel (from $33.20 down to $33.12 per barrel) effective as of April 1, 1980. [118]

Given the difficulties and dilemmas inherent in the ambitious goals and targets projected in the original Ten-Year Plan, it is not surprising that the post-Mao leadership has reformulated the principle of self-reliance in such a drastic manner as to widen the discrepancy between rhetoric and reality in China's developmental politics. The unavoidable value tradeoff in retailoring self-reliance to meet the operational requirements of the modernization drive calls into question the durability of the present coalition politics. Legitimization of the Ten-Year Plan, especially the means and manner with which it is now being pursued, is problematic. At the beginning of 1979 the government decided not only to restore old Chinese capitalists (with all their expropriated properties returned including fixed interest owed to them prior to September 1966) but also to allow them to live off their profits. Once the camel's nose of an embourgeoisement process is allowed to enter the tent of China's modernization drive, its corrosive influence on Maoist values cannot be easily checked.

In the course of implementing the Ten-Year Plan China has also opened itself wide to (foreign) capitalist ideas, institutions, values, and lifestyle as well as to advanced technology and complete plants. What would be the medium- and long-term demonstration (osmosis) effects of this opening? For one thing, it certainly provides a powerful ammunition for dormant Maoists to stage their comeback at an opportune moment in the future. Given the repeated failures of all previous reformers in the history of modern China to synthesize Western ideas successfully with Chinese values, the critical question remains as to whether Deng Xiaoping, an old man in a big hurry to reach the promised land, can accomplish this task of synthesis. Reflecting Deng's impatience, the Chinese press has been complaining lately that China has to overcome its vague concept of time, presumably referring to Mao's protracted sense of time in social and historical change.

Another demonstration effect of opening the Chinese window to the capitalist West and Japan would be to set in motion a "revolution of rising expectations" with all of its violent revolutionary potential for the body politic. The Maoist model of development seldom incurred this risk because it controlled man's expectations by defining the human ideal in terms of his capacity for self-denial in the service of a good society. Even the total and unqualified success of the modernization drive cannot possibly satisfy the rising materialistic appetite of the Chinese masses. Mao's populism and egalitarianism have already suffered symbolic slippage in the first phase of the modernization program. If the present trend continues unchecked—and there is no evidence of any normative revision in the readjustment policy—the center-periphery, urban-rural, and elite-mass contradictions may sharpen, challenging the legitimacy of Deng's leadership.

Deng Xiaoping's modernization program seems to reflect a nationalistic drive to rectify China's status inconsistency between achieved and ascribed rank dimensions in the international system. China is now one of the Big Five in the UN Security Council and has diplomatic relations with 120 nations and trade relations with more than 170 countries and regions (as of September 27, 1979)[119] and enjoys a high ascribed status in the international system. However, China as a poor, backward, and developing country does not enjoy a high achieved status. The repeated theme of making China a powerful socialist country by the end of this century is designed to redress this status inconsistency.

The pursuit of this status drive has allowed the resurgence of atavistic Han chauvinism with a discernible value shift in the Chinese image of world order. Mao's value-oriented image of world order carefully kept Han chauvinism in check. As a result, China during his reign was admirably faithful to the pledge never to be or act like a superpower. Under Deng's leadership, however, the Chinese image and strategy of world order shifted to the power-oriented direction, as evidenced in the lowering of Chinese activism in Third World politics, on the one hand, and in the growing ties with the United States, Japan, and Western Europe, on the other. Can Deng (or his successor) live up to this behavioral pledge of never acting like a superpower even if he succeeds in turning China into a powerful country by the end of the century? In the wake of China's invasion of Vietnam in early 1979, this has indeed become a troublesome question of global significance and deserves close attention as we study the future development of Chinese energy policy.

NOTES

1. It is no longer necessary to provide a long defensive dissertation on the data problem for any study of Chinese energy policy. For the first time since 1959, China has resumed the publication of its detailed statistical reports on its economy. On June 27, 1979, the State Statistical Bureau issued a detailed report entitled "Communique on Fulfillment of the National Economic Plan of 1978," providing production figures for primary energy outputs for 1977 and 1978. Curiously,

however, this report omits any figures for natural gas. See Renmin Ribao [People's Daily], June 28, 1979, p. 2, and Zhongguo duiwai maoyi [China's Foreign Trade], No. 6 (1979), pp. 5-9; for an English translation of this report, see Foreign Broadcast Information Service: Daily Report—People's Republic of China [hereafter cited as FBIS-PRC], June 27, 1979, pp. L11-L19.

It should be noted that all Chinese personal and place names with a few well-Anglicized exceptions—Peking, the Yellow River, the Yangtze River, Hong Kong, and Canton—are transliterated in accordance with the newly adopted Pinyin romanization system.

2. See Vaclav Smil, "China's Energetics: A System Analysis," in U. S. Congress, Joint Economic Committee, Chinese Economy Post-Mao: A Compendium of Papers, 95th Cong. , 2d sess. (Washington, D. C. : U. S. Government Printing Office, 1978), pp. 331-32.

3. CIA, People's Republic of China: Atlas (Washington, D. C. : U. S. Government Printing Office, 1971), p. 69.

4. See Yuan-li Wu, Economic Development and the Use of Energy Resources in Communist China (New York: Frederick A. Praeger, 1963), p. 1975; Vaclav Smil, China's Energy: Achievements, Problems, Prospects (New York: Praeger Publishers, 1976), p. 31.

5. William Clarke, "China's Electric Power Industry," in Chinese Economy Post-Mao, p. 406.

6. See Smil, "China's Energetics," p. 351, and FBIS-PRC, January 30, 1980, p. L1.

7. Vaclav Smil, "China Now Ranks No. 7 in World Power Generation," Far Eastern Economic Review [hereafter cited as FEER] October 5, 1979, p. 81.

8. See FBIS-PRC, November 19, 1979, p. L15.

9. Smil, "China's Energetics," p. 334.

10. For comparative study of U. S. estimates and actual Chinese output of electric power, see Clarke, "China's Electric Power Industry," and Renmin Ribao, June 28, 1979, p. 2.

11. Smil, "China's Energetics," pp. 332-33.

12. Ibid. , p. 328.

13. Ibid. , p. 354.

14. CIA, China: Energy Balance Projections, A(ER) 75-75 (November 1975), pp. 7-8.

15. Frederic M. Kaplan, Julian M. Sobin, and Stephen Andors, eds. , Encyclopedia of China Today (New York: Harper & Row, 1979), p. 179.

16. Smil, "China's Energetics," p. 357.

17. Renmin Ribao, editorial, November 2, 1979, p. 2.

18. Smil, "China's Energetics," p. 357.

19. China: Energy Balance Projections, p. 9.

20. See Nicholas R. Lardy, "Recent Chinese Economic Performance and Prospects for the Ten-Year Plan," in Chinese Economy Post-Mao, p. 49, and The Economist (London), October 6, 1979, p. 84.

21. See Huang Hua's speech before the plenary of the Sixth Special Session of the UN General Assembly in GAOR, 6th Special Sess., 2229th plenary meeting, May 1, 1974, para. 47.

22. Donella Meadows et al., The Limits To Growth (New York: Universe Books, 1972).

23. See Chen Muhua, "Controlling Population Growth in a Planned Way," Beijing Review [hereafter cited as BR], No. 46 (November 16, 1979), pp. 17-20, and Renmin Ribao, editorial, February 11, 1980, p. 1.

24. The Korea Herald [hereafter cited as KH] (Seoul), December 21, 1978, p. 4.

25. For a graphic illustration of varying degrees of dependence of selected developed countries on imported energy, see Committee for Economic Development, Research and Policy Committee, International Economic Consequences of High-Priced Energy (New York: CED, September 1975), p. 77.

26. For a detailed analysis, see Samuel S. Kim, China, the United Nations, and World Order (Princeton, N.J.: Princeton University Press, 1979), Chapter 5.

27. John A. Kringen and Steven Chan, "Chinese Crisis Perception and Behavior: A Summary of Findings," Paper delivered at the workshop on Chinese foreign policy sponsored by the Joint Committee on Contemporary China of Social Science Research Council, Ann Arbor, Michigan, August 12-14, 1976, p. 3.

28. Chang Chien, "Behind the So-Called 'Energy Crisis'" Peking Review [hereafter cited as PR], No. 11 (March 15, 1974), p. 6.

29. Ibid., p. 7.

30. "Oil Struggle Develops in Depth," PR, No. 4 (January 27, 1978), pp. 22-24.

31. Kim, China, the United Nations, and World Order, pp. 242-53.

32. Petroleum Intelligence Weekly 13, no. 11 (March 18, 1974): 10.

33. FBIS-PRC, November 19, 1979, p. L15.

34. Clarke, "China's Electric Power Industry," pp. 405-06.

35. Hsiao Han, "Developing Coal Industry at High Speed," PR, No. 8 (February 24, 1978), p. 5.

36. See Smil, "China's Energetics," p. 341; FBIS-PRC, January 7, 1980, p. L8; FBIS-PRC, March 5, 1980, p. L5.

37. Hsiao, "Developing Coal Industry at High Speed," p. 6.

38. FBIS-PRC, January 22, 1980, p. L16.

39. FBIS-PRC, March 5, 1980, p. L4.

40. See Clarke, "China's Electric Power Industry," p. 404, and FBIS-PRC, February 27, 1980, p. L3.

41. Renmin Ribao, September 28, 1979, p. 1.

42. FBIS-PRC, March 27, 1980, p. L16.

43. Clarke, "China's Electric Power Industry," p. 418.

44. FBIS-PRC, January 17, 1980, p. L4.

45. "Chairman Hua's Call for More Electricity," PR, No. 50 (December 9, 1977), pp. 3-4.

46. See Chien Cheng-ying [Qian Zhengying], "Power Industry—the Vanguard," PR, No. 3 (January 20, 1978), p. 16-20.

47. See FBIS-PRC, January 14, 1980, p. L21; FBIS-PRC, February 27, 1980, p. L2; and FBIS-PRC, February 28, 1980, p. L11.

48. FBIS-PRC, January 9, 1979, pp. E14-E15.

49. Cited in Selig S. Harrison, "China: The Next Oil Giant," Foreign Policy, No. 20 (Fall 1975), pp. 7-8.

50. See Choon-ho Park and Jerome Alan Cohen, "The Politics of China's Oil Weapon," Foreign Policy, No. 20 (Fall 1975), p. 33.

51. The China Business Review [hereafter cited as CBR] 5 (November-December 1978): 13.

52. Randall W. Hardy, China's Oil Future: A Case of Modest Expectations (Boulder, Colo.: Westview Press, 1978), p. 8.

53. Renmin Ribao, September 12, 1979, p. 1.

54. Smil, "China's Energetics," pp. 342-43.

55. A. A. Meyerhoff, "Chinese Petroleum Industry Potential," Working Group from California Arms Control and Foreign Policy Seminar on U.S. Relations in the Pacific, May 24, 1974, p. 5.

56. Chi Ti, "General Task for the New Period: Industrial Modernization," PR, No. 26 (June 30, 1978), p. 7; emphasis added.

57. Renmin Ribao, editorial, July 6, 1978, pp. 1-2.

58. Renmin Ribao, September 12, 1979, p. 1.

59. See Smil, China's Energy, pp. 44, 46; "China's Energetics," p. 344; "Szechuan's Gas Fields Making Headway," PR, No. 19 (May 9, 1975), p. 30; K. W. Wang, "China's Mineral Economy," in Chinese Economy Post-Mao, pp. 392-93.

60. FBIS-PRC, September 25, 1979, p. L7.

61. "Geo-Thermal Power Station," PR, No. 20 (May 17, 1974), p. 31.

62. FBIS-PRC, September 18, 1979, p. L15.

63. Ibid., p. L17.

64. CBR 5 (November-December 1978): 79.

65. FBIS-PRC, March 28, 1980, pp. L24-L25.

66. FBIS-PRC, February 27, 1980, p. L6, and FBIS-PRC, April 3, 1980, p. L7.

67. "China: Recycling of Organic Wastes in Agriculture," FAO Soils Bulletin, No. 40 (1977), p. 50.

68. FBIS-PRC, August 23, 1978, p. E11.

69. Kevin Fountain, "New Energy Sources in China," CBR 6 (September-October 1979): 30.

70. For a comprehensive analysis of China's international energy policy, see Kim Woodard, "China's Energy Development in Global Perspective," in James C. Hsiung and Samuel S. Kim, eds., China in the Global Community (New York: Praeger Publishers, 1980), Chapter 5.

71. Vaclav Smil and Kim Woodard, "Perspectives on Energy in the People's Republic of China," Annual Review of Energy 2 (1977): 320.

72. FBIS-PRC, January 4, 1980, p. E3.

73. FBIS-PRC, February 8, 1980, p. K1.

74. New York Times [hereafter cited as NYT], December 27, 1978, p. D8.

75. CBR 5 (September-October 1978): 67.

76. FBIS-PRC, March 6, 1980, p. L8.

77. PR, No. 52 (December 29, 1978), p. 4.

78. For background on this latest Chinese move see KH, November 21, 1978, p. 4; CBR 5 (November-December 1978): 72; The Interdependent, December 1978, p. 5; NYT, February 14, 1979, p. A11. For China's behavior in UNDP up to mid-1977, see Kim, China, the United Nations, and World Order, pp. 315-28.

79. NYT, April 18, 1980, p. D1.

80. Melinda Liu, "Keeping the Semantic Balance Regarding Loans," FEER, September 22, 1978, p. 37.

81. CBR 6 (September-October 1979): 21.

82. CBR 5 (November-December 1978): 76.

83. See Roger Gale, "Offshore China: Oil Surveys Nearly Complete," Energy Daily (Washington, D.C.), December 5, 1979, pp. 1-3.

84. See NYT, March 16, 1980, p. 12; FBIS-PRC, March 4, 1980, p. B1; FBIS-PRC, March 17, 1980, p. B1.

85. CBR 5 (September-October 1978): 52-53, 71-73.

86. FBIS-PRC, February 28, 1980, p. D2.

87. For details, see FBIS-PRC, December 11, 1979, pp. D2-D3.

88. NYT, October 2, 1979, p. D1.

89. CBR 6 (September-October 1979): 58-60.

90. For my further elaboration of this thesis, see Samuel S. Kim, "China and the Third World in NIEO Politics," Contemporary China 3 (Winter 1979).

91. For the full text of this 33-article law, see Renmin Ribao, September 17, 1979, p. 2.

92. Leo A. Orleans, "China's Environomics: Backing Into Ecological Leadership," in U.S. Congress, Joint Economic Committee,

China: A Reassessment of the Economy, 94th Cong., 1st sess. (Washington, D. C. : U. S. Government Printing Office, 1975), p. 120.

93. "Environmental Protection in China," NCNA-English, Peking, March 25, 1976, in Survey of People's Republic of China Press, No. 6067 (April 1, 1976), p. 156.

94. NYT, September 18, 1971.

95. See Commentator, "Great Importance Should Be Attached to Environmental Protection," Renmin Ribao, July 11, 1978, p. 2.

96. For detail, see Samuel S. Kim, "China and World Order," Alternatives: A Journal of World Policy 3 (May 1978): 569-73.

97. Chu Ko-ping, "Environment and Development," PR, No. 20 (May 14, 1976), p. 20.

98. Dr. Wu Chieh-ping, "Put Prevention First," PR, No. 9 (February 28, 1975), p. 31.

99. Commentator, "Make Good Use of Waste Heat," Renmin Ribao, July 14, 1978, p. 3.

100. FBIS-PRC, January 12, 1979, p. E15.

101. "China's Stand on the Question of Human Environment," PR, No. 24 (June 16, 1972), p. 8.

102. See Orleans, "China's Environomics," p. 134, and NYT, April 6, 1980, pp. 1, 14.

103. Commentator, "Great Importance Should Be Attached to Environmental Protection," p. 2.

104. NYT, April 6, 1980, p. 14.

105. Orleans, "China's Environomics," pp. 135-36.

106. FBIS-PRC, February 8, 1979, pp. H2-H3; NYT, February 8, 1979, p. A10.

107. FBIS-PRC, March 5, 1980, p. L7.

108. "China's 1974 Successes in Recovering Wastes," NCNA-English, Peking, February 22, 1975, in Survey of People's Republic of China Press, No. 5805 (March 6, 1975), p. 141.

109. See Commentator, "Seriously Enforce the Environmental Protection Law," Renmin Ribao, September 18, 1979, p. 1.

110. PR, No. 8 (April 22, 1958), p. 15.

111. FBIS-PRC, March 9, 1979, p. E2.

112. NYT, April 7, 1980, p. A12.

113. Orleans, "China's Environomics," p. 117.

114. FBIS-PRC, January 3, 1980, p. L18.

115. FBIS-PRC, July 20, 1978, p. E8.

116. FBIS-PRC, December 28, 1979, p. L20.

117. For computer-based projections of China's future energy balance, see Smil and Woodard, "Perspectives on Energy in the People's Republic of China," pp. 333-40.

118. KH, April 1, 1980, p. 3.

119. See Renmin Ribao, September 27, 1979, p. 5.

5

BRAZIL
Kenneth Paul Erickson

The energy crisis of the 1970s has posed serious problems and
challenges for all the countries considered in this book. The strains
that it imposes upon Brazil, however, are surely different and perhaps greater than those faced by the others. Brazil is caught between
First and Third Worlds, dependent upon the industrial capitalist powers for the capital and technology that underpin its program of industrialization, and dependent upon the OPEC nations for the energy to
fuel its modern sector. [1]

The oil crisis hit Brazil at a time when many Brazilians optimistically believed that their nation was about to escape its historical
dependence upon the world's dominant industrial nations, a dependence
that had shaped its underdevelopment. Brazil's military rulers, whose
aspirations do not suffer from an excess of modesty, had even proclaimed that their nation would attain world-power status by the year
2000. The claims, in fact, are not implausible. Brazil possesses
some of the natural attributes of major-power status, attributes that
only China and India among nonindustrial countries also possess. It
is the world's fifth largest nation in terms of land surface and sixth
largest in terms of population (120 million in 1979). What distinguishes
Brazil from China and India, however, is its sustained economic
growth. Its gross domestic product has grown by an average of nearly
7 percent per year since 1947. Indeed, Brazil recently passed India,
a country almost six times as populous, to become the world's tenth
nation in terms of economic output. [2] The key to the bid for major-
power status, therefore, is economic.

The successful growth policies of the postwar period, however,
contained a serious flaw that became apparent only after the oil embargo and price rise of 1973-74: the basic energy to fuel Brazil's dy-

namic and expanding industrial sector was oil, and increasingly this oil was imported. By 1974, when the price of crude oil quadrupled on the world market, Brazil was importing 85 percent of its petroleum, and these imports accounted for 39 percent of the total energy consumption of the nation. [3] Brazil is the largest oil importer in the Third World.

Brazil's problems, then, are in important ways different from those of the world's dominant industrial nations upon which the U. S. press has focused. Brazilian policy makers must of course find ways to use energy more efficiently, but, if they are to break out of their historic cycle of dependence, they must do this while sustaining an economic growth rate roughly twice that of the already industrialized nations.

THE PATTERN OF ENERGY USE

Brazil has gone almost overnight from a predominantly rural agrarian society to an urban industrial and commercial one. The nation's dramatic increase in petroleum use and the accompanying evolution—or, more accurately, revolution—in energy-consumption patterns reflect this socially wrenching transition. One very perceptive account highlights the speed of this process: "In the postwar period, Brazil leaped from a wood-burning to an oil-driven economy in the course of a generation, a transition that had taken Europe three centuries to achieve. As recently as 1946, 70 percent of Brazil's energy supplies came from firewood and charcoal. "[4] Even as late as 1966, these two sources supplied over 40 percent of Brazil's energy requirements. As Table 5. 1 makes clear, only in 1968 did petroleum, then providing 38 percent of total energy consumption, overtake firewood.

Brazil's third major source of energy, in addition to oil and firewood, is hydropower. Together, these three sources have supplied more than 85 percent of the nation's energy over the last ten years (see Table 5. 1), even though their relative shares have been changing. During the past decade, the energy supplied by petroleum more than doubled, and that supplied by hydropower nearly tripled; firewood, while supplying a slightly greater absolute amount of energy, slipped from first to third place as its share fell from 38 to 20 percent.

Table 5. 2 presents official projections of energy production, by source, from 1978 through 1987. Brazil's policy responses to the world oil price rises become apparent in these projections. The same three energy sources occupy the first three places, but petroleum's relative growth slows, while hydropower more than doubles its output and moves into first place in 1986. Firewood consumption shrinks a bit, and its share drops from 19 to 10 percent. Also significant in

TABLE 5.1

Primary Energy Consumption in Brazil, in Tons of Petroleum Equivalent, 1967–77

Year	Hydroelectricity		Petroleum		Natural Gas		Alcohol		Shale Oil	
	1,000 tons	Percent	1,000 tons	Percent	1,000 tons	Percent	1,000 tons	Percent	1,000 tons	Percent
1967	8,465	16.5	17,371	33.8	105	0.2	367	0.7	—	—
1968	8,860	16.6	20,279	37.9	93	0.2	160	0.3	—	—
1969	9,481	16.7	21,993	38.7	96	0.2	27	0.0	—	—
1970	11,560	18.9	23,311	38.1	104	0.2	155	0.2	—	—
1971	12,549	19.1	26,186	39.9	140	0.2	213	0.3	—	—
1972	14,918	21.3	28,740	41.0	166	0.2	328	0.4	—	—
1973	17,055	21.9	34,240	43.9	178	0.2	260	0.3	—	—
1974	19,011	22.5	36,947	43.8	339	0.4	160	0.2	—	—
1975	21,412	23.7	39,300	43.5	369	0.4	136	0.1	—	—
1976	23,626	23.8	42,894	43.3	367	0.4	144	0.1	—	—
1977	26,943	26.1	43,063	41.7	505	0.5	537	0.5	—	—

(continued)

Table 5.1, continued

Year	Firewood		Bagasse		Charcoal		Coal		Uranium		Total	
	1,000 tons	Percent	1,000 tons	Percent	1,000 tons	Percent	1,000 tons	Percent	1,000 tons	Percent	1,000 tons	Percent
1967	19,291	37.4	2,825	5.5	1,003	1.9	2,048	4.0	—	—	51,475	100.0
1968	18,048	33.8	2,564	4.8	1,094	2.1	2,317	4.3	—	—	53,415	100.0
1969	18,999	33.4	2,762	4.9	1,191	2.1	2,342	4.0	—	—	56,891	100.0
1970	18,809	30.8	3,356	5.5	1,484	2.4	2,391	3.9	—	—	61,170	100.0
1971	18,862	28.8	3,559	5.4	1,655	2.5	2,431	3.8	—	—	65,595	100.0
1972	17,661	25.2	3,990	5.7	1,822	2.6	2,491	3.6	—	—	70,116	100.0
1973	17,429	22.4	4,459	5.7	1,897	2.4	2,493	3.2	—	—	78,011	100.0
1974	18,541	22.0	4,361	5.2	2,536	3.0	2,469	2.9	—	—	84,364	100.0
1975	19,328	21.4	4,032	4.5	2,897	3.2	2,850	3.2	—	—	90,324	100.0
1976	21,294	21.5	2,166	4.2	3,154	3.2	3,435	3.5	—	—	99,080	100.0
1977	20,885	20.2	4,714	4.6	2,489	2.4	4,106	4.0	—	—	103,252	100.0

Source: Brazil, Ministério das Minas e Energia, National Energy Balance, 1978. (Rio de Janeiro, 1979), p. 12.

TABLE 5.2

Projected Primary Energy Consumption in Brazil, in Tons of Petroleum Equivalent, 1978-87

Year	Hydroelectricity 1,000 tons	Percent	Petroleum 1,000 tons	Percent	Natural Gas 1,000 tons	Percent	Alcohol 1,000 tons	Percent	Shale Oil 1,000 tons	Percent
1978	28,088	25.6	46,452	42.4	614	0.6	1,461	1.3	—	—
1979	30,934	26.5	49,297	42.5	659	0.6	1,967	1.6	—	—
1980	34,066	27.5	50,269	40.6	677	0.6	2,479	2.0	—	—
1981	39,886	30.3	51,180	39.0	736	0.6	2,521	1.9	—	—
1982	45,059	32.3	51,823	37.2	976	0.7	2,598	1.9	—	—
1983	49,410	33.3	54,481	36.8	1,109	0.8	3,001	2.0	—	—
1984	53,252	33.9	56,069	35.6	1,142	0.7	3,357	2.1	1,154	0.7
1985	57,816	34.6	58,478	35.0	1,172	0.7	3,541	2.1	1,154	0.7
1986	61,626	34.7	61,288	34.6	1,218	0.7	3,735	2.1	1,732	1.0
1987	65,516	34.8	64,477	34.2	1,268	0.7	3,941	2.1	2,310	1.2

(continued)

Table 5.2, continued

Year	Firewood 1,000 tons	Percent	Bagasse 1,000 tons	Percent	Charcoal 1,000 tons	Percent	Coal 1,000 tons	Percent	Uranium 1,000 tons	Percent	Total 1,000 tons	Percent
1978	20,676	18.8	5,058	4.6	2,554	2.3	4,830	4.4	—	—	109,733	100.0
1979	20,469	17.5	5,602	4.8	2,655	2.3	4,793	4.1	137	0.1	116,513	100.0
1980	20,265	16.4	6,168	5.0	2,939	2.4	5,736	4.6	1,114	0.9	123,713	100.0
1981	20,062	15.3	6,600	5.0	3,086	2.3	6,172	4.7	1,114	0.9	131,357	100.0
1982	19,861	14.2	7,013	5.0	3,152	2.3	7,878	5.6	1,114	0.8	139,474	100.0
1983	19,663	13.3	7,534	5.1	3,284	2.2	8,496	5.7	1,114	0.8	148,092	100.0
1984	19,466	12.4	8,023	5.1	3,481	2.2	8,886	5.7	2,412	1.6	157,242	100.0
1985	19,272	11.6	8,405	5.0	3,600	2.2	10,004	6.0	3,517	2.1	166,959	100.0
1986	19,079	10.8	8,805	5.0	3,600	2.0	10,647	6.0	5,545	3.1	177,275	100.0
1987	18,888	10.0	9,224	4.9	3,600	1.9	11,244	6.0	7,761	4.2	188,229	100.0

Source: Brazil, Ministério das Minas e Energia, National Energy Balance, 1978 (Rio de Janeiro, 1979), p. 14.

224

these projections, however, is the rise in hitherto unimportant domestic sources that public policy has recently supported, namely alcohol, shale oil, and nuclear power. Policy changes in 1979 further increased the share of alcohol, while decreasing that of nuclear.

The doubling of Brazil's overall energy consumption between 1967 and 1977, and the projected 72 percent increase from 1978 to 1987, reflect the effectiveness of the nation's postwar industrial policies. In Brazil a powerful technocratic elite has used its control of the state's economic policy-making machinery to initiate and expand modern industry, both directly, as entrepreneurs and managers, and indirectly, by catalyzing private-sector activity.

The 20-year history of the Brazilian automobile industry illustrates the numerous ways in which the technocrats and their industrial allies used public policy to foster a new industry,[5] one that surely increased the energy intensiveness of the transportation sector. By the late 1970s, with annual output around 1 million vehicles, the Brazilian auto industry had become the world's tenth largest. Ninety-nine percent of the components are manufactured locally.

In a country whose per capita income averaged $300 in 1960 and $400 in 1970, such a dramatic expansion in vehicle production did not happen by accident or because of "impersonal market forces." It was a result of public policy. In the years after the military coup of 1964, the government fostered a policy of income concentration, in effect taking money out of the pockets of the poor and transferring it to the rich.[6] In the 1950s and 1960s, gasoline prices were held down by government subsidy, and a massive highway program continually improved conditions for the motorist. The country counted only 3,100 kilometers of paved highway in 1955, but this figure steadily rose, to 11,500 in 1959, 26,500 in 1965, and 73,300 by 1974.[7] The railroads, most of which the state owned or acquired during this period, were allowed to deteriorate to the point that, even for haulage, they could not compete with highway transport. By the early 1970s trucks carried nearly 75 percent of Brazil's freight, while railroads carried only 17 percent.[8]

These policies, it should be clear, reflect an uncritical adoption of models from the industrialized, capitalist world, and particularly from the United States. That the United States should serve as a model is not surprising, for it, like Brazil, is a New World nation peopled mainly by immigrants, and it is, by economic-growth criteria, a success story. The implications of this model for Brazil's energy profile, however, are particularly unfortunate, because U.S. public policy has created the most conspicuously energy-inefficient economy and society in the world.[9] Brazilian policy makers, like their North American counterparts, paid little attention to the energy implications of their development plans prior to 1973. Their supply-oriented energy policies simply took for granted the availability of cheap imported petroleum and abundant domestic hydropower.

Brazil's surging industrial expansion has occurred quite predictably in the most energy-intensive sectors. In a country that measures its progress by the pace at which it paves over and builds up, it is not surprising that cement production has grown dramatically, from 4.4 million tons in 1960 to 20.5 million in 1977. By 1979 the cement industry consumed 20 percent of all the fuel oil used in Brazil, far surpassing its nearest competitors: petroleum refining (14 percent), steel (11 percent), and petrochemicals (11 percent). Brazilian steel production rose from 1.8 million tons in 1960 to 8.3 million tons in 1975, and planners projected 22 million tons in 1980 and 40 million in 1985. Brazil was twentieth among world steel producers in 1970, but it will move into the top ten in the 1980s if it attains its goals. [10] The worldwide slowdown of the late 1970s will undoubtedly set back the timetable, but the trends nevertheless will not change. The nation has been rapidly expanding its production of nonferrous metals, including the notably energy-intensive smelting and refining of aluminum. By the mid-1970s it had already become the world's tenth largest producer of petrochemicals, with major expansion plans nearing completion and others on the drawing boards. [11] Oil refining, itself an energy-intensive activity, rose from 54,800 barrels per day in 1954 to 412,000 in 1971. It then soared to 1.1 million in 1975 and to over 1.26 million in 1977. By then Brazil's refining capacity was sufficient to produce virtually its entire consumption of petroleum derivatives, so additional increases merely reflect annual increases in consumption. [12]

One rather curious element appears in Brazil's pattern of energy use. Although national energy consumption has never ceased its sharp rise, the ratio of energy consumption to gross domestic product has been declining over the last 10 years. In 1965 it took the equivalent of .94 tons of petroleum equivalent to generate $1 million of output, but in 1970 it took only .84 tons, and in 1976 only .80. [13] One would have expected the contrary, especially considering the dramatic growth of the most energy-intensive industrial sectors and the least efficient forms of transportation. Moreover, the ratio declined during the late 1960s when international petroleum prices were dropping and thus were unlikely to induce energy-efficient behavior. Since the energy/ GDP ratio was on the rise in most industrial countries during the 1960s, and since Brazil has imported its industrial processes from these industrial nations, it seems unlikely that the figures reflect a uniquely Brazilian bent for energy efficiency. A more likely explanation is that offered by an Indian author to explain a similarly puzzling trend in his own country. As commercial sources of power increase their share of total energy consumption at the expense of noncommercial fuels such as firewood and agricultural residues, there occurs a one-time rise in apparent energy efficiency. This occurs because modern equipment using commercial fuels or electricity captures more of the heat produced by the fuel than does traditional equipment—for example, open hearths in residences. Less waste heat escapes. [14]

Finally, in a nation of continental dimensions like Brazil, one must consider the way energy is distributed regionally and socially. In Brazil social and regional inequalities are pronounced, and energy-use patterns reflect this. The data in Table 5.3 show that the relatively wealthy and industrial Southeast consumed over four times as much energy per capita as any other region, and over six times as much electricity per capita as the North and Northeast. Only 54 and 44 percent, respectively, of the urban dwellings in the North and Northeast had electrical connections in 1970, compared with 80 percent in the Southeast. Brazil was still 44 percent rural in that year's census, and the vast majority of rural homes, especially those outside the Southeast, had no electrical service.

In terms of social class, of course, it is the 15 to 20 percent at the top—who live and work in space-conditioned homes and offices, drive the rapidly proliferating automobiles on the expanding highway network, and use consumer durables made of energy-intensive metals or plastics—who enjoy the creature comforts and satisfaction of caprice that modern energy can provide. The urban workers—who ride mass transit, live in cramped quarters, and consume only a limited range of consumer durables—surely use only a small fraction, per capita, of the energy consumed by the rich. The rural poor employ virtually no modern forms of energy. In effect, energy consumption patterns merely reflect the inequitable distribution of wealth in Brazil, where the 1970 census showed that the top 5 percent received 36 percent of the income, while the bottom 40 percent received only 9 percent, and the next-lowest 40 received 28 percent. [15]

DOMESTIC ENERGY POLICY

Ever since the Portuguese landed, writers have extolled Brazil's richness in mineral resources. A recent study, for example, expressed it thus: "In quantity, quality, and variety of mineral resources, Brazil stands among the world's leaders."[16] The nation's subsoil riches, unfortunately, do not include an abundance of commercially utilizable fossil fuels, a deficiency of relatively minor significance until OPEC quadrupled world petroleum prices in 1973-74.

In the present era of high oil prices, Brazilian policy makers have only three feasible energy strategies at their disposal, if they hope to maintain or increase their nation's level of economic output: they may increase production of domestic energy resources, for these do not cost scarce foreign currency; they may increase nonenergy exports significantly, in order to pay for rising levels of energy imports;

TABLE 5.3

Indicators of Energy Consumption in Brazil, by Region, 1971

Region	Population (1970)		Energy Consumption Per Capita		Electricity Consumption		Percent of Urban Dwellings with Electrical Connections
	Millions	Percent Regional Share	Tons	Percent Regional Share	kWh/ Inhabitant	Index (SE=100)	
North	3.6	3.9	1.0	1.6	114	14	54
Northeast	28.1	30.2	9.1	14.2	131	16	44
Southeast	39.9	42.8	41.4	64.9	820	100	80
South	16.5	17.7	10.0	15.7	260	32	66
Center–West	5.1	5.5	2.3	3.6	146	18	48
Brazil total	93.2	100.1	63.8	100.0	—	—	—
Brazil average	—	—	—	—	450	55	71

Note: In official Brazilian usage, the regions are comprised of the following states or territories: North: Rondônia, Acre, Amazonas, Roraima, Pará, Amapá. Northeast: Maranhão, Piauí, Ceará, Rio Grande do Norte, Paraíba, Pernambuco, Alagoas, Fernando de Noronha, Sergipe, Bahia. Southeast: Minas Gerais, Espírito Santo, Rio de Janeiro (which absorbed the former state of Guanabara in 1975), São Paulo. South: Paraná, Santa Catarina, Rio Grande do Sul. Center–West: Mato Grosso, Mato Grosso do Sul, Goiás, Federal District.

Sources: Population figures from 1970 census, in Fundação IBGE, Anuário Estatístico do Brasil (Rio de Ja-neiro, 1974), p. 43. Other data are derived from Rosa Maria Fucci, "Energia," in Fundação IBGE, Geografia do Brasil: Região Nordeste (Rio de Janeiro, 1977), Vol. 2, pp. 247, 268–69; and from the "Energia" chapters by José Cézar Magalhães in the other four volumes of the same source: Região Norte, Vol. 1, pp. 325, 338; Região Sudeste, Vol. 3, p. 377; Região Centro-Oeste, Vol. 4, pp. 231, 247; and Região Sul, Vol. 5, pp. 307, 333–35.

or they may decrease consumption through some form of conservation, thus reducing the growth of imports. This section and the next will discuss policies and practices designed to implement these strategies.

The very essence of Brazil's industrial, agricultural, population, and internal land-settlement patterns can be summed up in one word: growth. Sustained rapid growth dates back at least to Getúlio Vargas' authoritarian rule between 1930 and 1945. Vargas and a group of modernizing nation-builders built a strong centralized interventionist state that has systematically fostered industrialization ever since.

The resultant industrialization and urbanization dramatically increased the need for commercial fuels, and to meet this need, state policy makers and technocrats fashioned entrepreneurial roles for themselves. This is most clearly apparent in the energy sectors that surged forward after the war. Petrobrás, the state oil monopoly, was created in 1954; and in electricity, technocrats in 1962 created Eletrobrás, a holding company that coordinated the expanding state and federal hydroelectric generating agencies. Eletrobrás set national electricity goals and tied the existing small and medium-sized generating and distribution companies into large regional grids.

Brazilian nationalism played a major role in the creation and policies of both organizations. Indeed, nationalist sentiment was so strong that Brazil even nationalized its oil before any of it had been discovered. In the first three decades of this century, Brazilian administrative policies effectively discouraged foreign oil companies from exploring or refining in Brazil, although they were allowed to import and market their refined products. Decree laws issued by Vargas in the 1930s made official their exclusion from exploration and refining. Only afterward, in 1939, was oil discovered in Brazil. The monopoly role given to Petrobrás 20 years later was, of course, the logical conclusion to this process. Foreign firms dominated commercial electric power in Brazil from the turn of the century until the 1950s and 1960s, when public agencies developed a major role in hydro generation. These agencies soon overshadowed the foreign companies and then progressively bought them out. The government took over the final foreign-owned company in 1978. [17]

Technocratic policies have produced dramatic economic growth. Brazilian gross domestic product has grown by an average of 7 percent per year since 1947, setting one of the world's outstanding sustained economic growth records. During the most dramatic surge in growth, from 1968 to 1974, when output boomed ahead by 10 percent annually, Brazilians and many foreign observers spoke in glowing terms of an "economic miracle." Accompanying the "miracle," however, were some often-overlooked costs. The poor paid a very high price: real income declined, causing diet and, ultimately, health to deteriorate so that by the mid-1970s epidemic diseases ravaged many urban areas.

The model that produced the "miracle" also imposed a cumulative
cost upon the nation's economic autonomy, because it increased Brazil's dependence on foreign oil.

Only after 1974, when the cost of imported oil appeared likely
to put the brakes on Brazil's growth engine, did the nation's leaders
make energy a priority issue. President Ernesto Geisel's inauguration in March 1974 coincided with the first major oil shock. The Second National Development Plan (1974-79), elaborated under his supervision in 1974, declared that energy policy had become a "decisive
element" in national development and stressed that "Brazil must, in
the long run, supply internally its vital energy needs."[18] Basically,
the plan called for rapid increases in the production of domestic liquid
fuels and the substitution of petroleum by relatively cheap hydroelectricity in the transport sector. It earmarked large sums to explore
for domestic oil. For motor fuels, it funded pilot programs in shale
oil and alcohol to replace gasoline and diesel fuel. It promised major
new resources for upgrading and electrifying the decaying railroads,
particularly lines serving the nation's steel industry and handling the
bulk of its ore and agricultural exports. For the megalopolises of Rio
and São Paulo, it supported mass transit, particularly by expanding
their new subway systems, reequipping the old commuter rail lines,
and improving bus service.[19]

The Second Economic Development Plan served more to articu-
late a set of aspirations than to provide a blueprint for the next five
years, because Brazil failed to meet its ambitious 10 percent annual
growth goals (although its annual average GDP increase of 6.9 percent
from 1974 through 1978 far surpassed the performance of the world's
industrial leaders). Lower growth meant lower tax revenues, forcing
many of the investment projects to be scaled down, even though most
still received sizable sums. Despite these efforts, in 1979 when Geisel's term ended, Brazil had become even more dependent upon imported oil. Further complicating matters, the oil-induced balance-
of-payments deficit had made the country increasingly dependent upon
the bankers of the world's dominant capitalist nations.

In Brazil as well as the United States, therefore, exhortation
and mild conservation measures had proved ineffective policy tools
for stemming the growth of petroleum imports. Moreover, Brazil's
more concrete policy responses to the first oil shock—the long-term
infrastructural investments in petroleum and rail transport—had not
yet begun to show results and would not do so until the 1980s.

The second oil shock, following the Iranian revolution, thus hit
Brazil even harder than the first. Indeed, the impact was so severe
that one high Central Bank official lamented that it had "rendered our
economic model completely unviable," requiring "a redefinition of
governmental priorities and a restructuring of the economy."[20] In July

1979, President João Baptista Figueiredo addressed the nation with his energy program, warning that the new oil prices would force his government to adopt measures similar to those of a "war economy." He pointed out that the 1979 OPEC oil price rises would aggravate Brazil's worsening inflation, slow economic growth, raise the cost and hence the price of manufactured imports, and reduce exports by causing a world economic slowdown. Projected oil imports would now cost $2 billion more than anticipated, raising the specter of a $2 billion balance-of-trade deficit. Brazil already owed nearly $50 billion to foreign creditors, and Figueiredo pointed out that the additional, oil-induced debt threatened "to compromise our international credibility and the stability of our development." Calling for sacrifices and warning that Brazilians would have to work harder merely to stand still economically, he promised reductions in government spending and "appealed to the patriotism" of his development-oriented ministers "to contain their entrepreneurial drive" during this reordering of priorities.

Figueiredo's program, like that of governments in nearly all the oil-importing nations, sought to increase domestic petroleum production, to develop other domestic energy resources, and to slow the growth of petroleum consumption. Specifically, he announced that domestic oil production would rise from some 170,000 barrels per day to 500,000 in 1985, by intensifying Petrobrás' exploration efforts and by bringing into production its offshore finds of the 1970s. To substitute domestic resources for imported oil, he proposed to increase the production of ethyl alcohol from sugarcane, so that by 1985 Brazil would produce 10.7 billion liters per year, or the equivalent of 170,000 barrels per day of crude oil. Finally, he called for a major increase in coal production, so it could replace fuel oil, particularly in such industries as cement production. Emphasis on domestic oil, alcohol, and coal constituted the heart of his near-term recommendations, although he urged continued research in solar heating, shale and vegetable oils, and other still-experimental technologies. The nation would press ahead in harnessing its rich hydroelectric potential. Nuclear generation, until recently a top-priority item, would be held back until hydropower is exhausted. [21]

Policies to conserve energy proved thornier politically than policies to produce it. Figueiredo promised "more energetic measures to contain fuel consumption" in his speech, but he did not spell these out. Immediately after the speech, the cabinet-level Council on Economic Development approved a freeze on oil imports (at 960,000 barrels per day) and urged that the National Energy Commission, which the president had just created, come up with specific measures to hold down consumption.

Council members discussed only such weak measures as gas-station closings on weekends and at night, however, suggesting that they felt that public policy could do relatively little to contain consumption. They, like the president, had apparently been influenced by Planning Minister Mário Henrique Simonsen, whose supply-oriented suggestions formed the core of Figueiredo's speech. In an extensive memo to the president, Simonsen had argued that, except for gasoline, fuel consumption could not be reduced by raising prices and that for diesel fuel and fuel oil, "the only effective rationing is through economic stagnation (or semistagnation)."[22]

Figueiredo therefore faced a tough policy dilemma, for he desired both to cut oil consumption and to maintain full employment. Indeed, a study done for the labor minister showed that to keep unemployment from worsening, Brazil would have to create a minimum of 1.3 million jobs in 1979 alone, a task requiring a GDP growth rate of at least 6.5 percent.[23] If Simonsen were right, the president could reduce energy consumption only at the risk of building up potentially explosive social tensions. These tensions would almost certainly abort the gradual liberalization that had already loosened controls over political and trade-union activity, a liberalization that Figueiredo strongly supported.

A way out of the dilemma emerged later, when the president was apprised of data that not only called into question Simonsen's theory on the price elasticity of fuel products but even made the planning minister personally responsible for increasing Brazil's appetite for imported petroleum. Simonsen had argued that only gasoline consumption responded to price changes. To be sure, official figures show that gasoline consumption per car registered fell from 3,700 liters in 1973 to 2,020 in 1978, a period in which gasoline prices increased in real (constant cruzeiro) terms by 116 percent.[24]

Brazil's recent history demonstrates, however, that demand for all petroleum products, and not just for gasoline, is price elastic. From 1970 to 1973, the four years prior to the first OPEC price shock, Brazil's gross domestic product grew by 43 percent and petroleum consumption by 50 percent. In other words, petroleum consumption rose by about 1.16 percent for each 1 percent of GDP increase. From 1974 through 1977, however, Brazilian output grew by 25 percent and petroleum consumption by only 16 percent. That is, after the price of imported oil had quadrupled, petroleum consumption rose only about 0.64 percent for each 1 percent of GDP increase. In 1978, however, the relationship reverted to its preembargo pattern, as increases in petroleum consumption outstripped GDP increases. In 1978 GDP grew by 6.5 percent and petroleum consumption by 9.7 percent (or 1.49 times the GDP increase), and in the first half of 1979 GDP rose at an annual rate of about 6 percent and petroleum consumption by 10.5 percent (or 1.75 times the GDP increase).[25]

To solve the mystery of the 1978-79 reversal, one need only check the price structures for petroleum derivatives in these three phases. The sharp rises in international petroleum prices after 1973 soon forced similar increases in Brazil, leading not only auto owners but also consumers of diesel fuel and fuel oil to conserve or substitute. At this time, the National Petroleum Council set prices, with the goal of reflecting the costs of imports. This council marked up gasoline faster than diesel fuel and fuel oil, in order to hold down unnecessary use of private cars, a policy that obviously worked.

In 1976, however, the war on inflation assumed top priority, with Simonsen as chief strategist. He now had the last word on national pricing policy for all products. In order to hold down the inflation index, he abandoned the policy of pricing refined petroleum at international levels. Simonsen viewed gasoline as nonessential and allowed its price to rise at the pace of inflation. Then he used its high price to subsidize fuel oil and diesel fuel, the mainstays of industry and transportation. Indeed, fully 20 percent of the price motorists paid for their gasoline went to subsidize low prices for other petroleum derivatives.

By 1978 the fuel-oil subsidy had distorted the energy-price structure to the point that imported oil was much cheaper than domestically produced energy resources. Firewood cost twice as much as fuel oil for the same caloric output, steam coal cost two and one-half times as much, and hydroelectricity (at the low, industrial rate) nearly six times as much. Fuel oil sold for a lower price in Brazil than in all but the oil-exporting nations. One critic pointed out in a data-filled newspaper article that the subsidy for fuel oil was roughly equivalent to the amount Petrobrás spent on exploration and production in 1978, and O Estado de São Paulo pointed out in an editorial that, taken together, the various subsidies for fuels totalled over $1 billion and made electricity from oil-fired power plants cheaper than from hydroelectric dams—in a country with rich hydroelectric potential and relatively little petroleum![26]

The data brought out during the national debate in mid-1979 surely demonstrate the fallacy of Simonsen's argument that demand for petroleum derivatives (other than gasoline) is not price elastic. Indeed, when Simonsen's pricing policy was conditioning demand, consumption of fuel oil in 1978 shot ahead by 10.4 percent, almost half again as fast as gasoline, and fuel oil overtook gasoline as the culprit responsible for rising imports.[27] Not one single barrel of imported crude could be backed out, therefore, until the demand for fuel oil could be reduced. This fact, of course, lay behind Figueiredo's emphasis on alcohol and coal to substitute for fuel oil and diesel fuel. In July the government announced plans to cascade lighter derivatives into the heavier categories in order to slow the growth of demand for fuel oil, and thus to slow imports. Brazil already mixes 20 percent ethyl alcohol into motor gasoline and is developing a fleet of vehicles to run on pure alcohol. Now the government proposed to mix 20 percent gasoline and 7 percent alcohol into diesel fuel, creating an excess of diesel fuel that could then be mixed with fuel oil, thereby lowering the total demand for imported crude.[28] It is too soon to see how this will work in practice.

Government officials briefly tried a shock treatment immediately after Figueiredo's "war economy" speech. They froze crude-oil imports and then made reductions in monthly delivery quotas for large customers and distributors. They chose quotas, of course, because they believed that price rationing would not work and that its inflationary side effect was unacceptable. Within two weeks, the press reported from the highly productive agricultural areas of southern Minas Gerais, northern São Paulo, and Mato Grosso do Sul that hundreds of out-of-gas trucks lined the highways or were stuck in endless lines at filling stations that had also run dry. Even harvesting vehicles were running out, and by this time the area had already begun using its October allocation. Other regions were also hard hit, and fuel oil used by industry was also running short.[29] By mid-August no doubt remained that rationing by quota would create economic chaos and, with it, great political liabilities for the government. The National Petroleum Council therefore abandoned the quota system and placed its hopes on the price mechanism.[30] Its pricing actions of August and September began chipping away at the subsidy for fuel oil and diesel fuel. Since these two clearly were responsible for the rise in imports, the government raised their prices by 50 percent at the beginning of August. In September it raised gasoline by 40 percent. The higher rate for the heavier products can be only a first step, however, if the government is serious about slowing their growth. The gasoline hike put its price at $1.90 per gallon, or in the upper range of oil-importing West European countries. The new price for fuel oil, however, remained not only far below the international average but just half the cost of importing, refining, and distributing the product![31]

Modest as the fuel-oil and diesel-fuel hikes may seem when viewed in this light, they could not easily have been effected as long as Simonsen directed Brazil's economic course. In early August 1979, however, Figueiredo provoked Simonsen's resignation and replaced him with Antônio Delfim Netto, a former finance minister who is generally considered the author of the "economic miracle" from 1968 to 1974. One should not conclude that Simonsen's energy subsidies provided the only reason for Figueiredo to sack him. His obsession to cut inflation through restrictive monetary policies and cutbacks in government spending created discontent among industrialists, and his advocacy of repressive force against striking workers raised the voice

of Brazil's new restive and militant labor movement against him. In
this context, leaders of the government's ARENA party demonstrated
to the president that Simonsen was a political liability whose policies
could ultimately generate a social explosion.

Newly appointed Delfim and his aides argued that to solve Brazil's inflation, the government should foster development expansion
that would produce more goods and services, not provoke a recession.
This optimistic view contrasted dramatically with the pessimism and
bitterness of Simonsen, who commented, the day after Delfim replaced
him, "Only God can save the Brazilian economy."[32]

Figueiredo, following Simonsen's advice in his July energy address, had called upon his ministers to restrain their developmental
and entrepreneurial drive, but now he could reverse himself and urge
them to employ their talents to the fullest. The energy sector, now
top priority, is destined to gain the lion's share of official attention,
as will be made clear later in the chapter. Moreover, while Simonsen
had cut the federal budget by 40 billion cruzeiros, postponing power
dams and many other energy projects, Delfim can expand spending.
He took office just at the moment that government coffers began receiving a windfall of resources. The fuel-price hikes were expected
to bring in about 50 billion cruzeiros (about $1.8 billion) in the remain-
ing four months of 1979 alone. Thus the funds now exist for the ministers to put their talents to work. Funding will remain assured, because
recent laws earmark 12.5 percent of the value of all refined petroleum
for an investment fund in alcohol, coal, and rail and mass-transit projects.[33]

One question is whether Delfim's expansion plans will work, and
another is whether they will be allowed to work. Here, Brazil's dependency upon the bankers of the world's dominant capitalist nations
becomes a key variable. Brazil, with the largest foreign debt in the
Third World—it passed $50 billion in late 1979—needs the goodwill of
the bankers to service its debt and pay for foreign oil. For 1979, interest and amortization are estimated at $11.2 billion and oil imports
at $7 billion, while exports will reach only about $15 billion, at least
$3 billion short. Simonsen, once selected by Institutional Investor as
one of "the world's five best finance ministers," was a favorite of the
bankers, for he put monetary stability above all else and did not shy
away from conservative recessionary measures to control inflation.
Delfim's developmental expansionism, on the other hand, is seen by
bankers as unorthodox and risky. His presumed political ambitions,
to become the first civilian president since 1964, also raise bankers'
fears that he will let political rather than strictly economic criteria
guide his decisions.[34] A key variable to watch, therefore, is the role
of foreign banks in limiting or expanding Brazil's options as it seeks
to lessen its energy dependence.

The previous pages have discussed presidential policy initiatives to the oil shocks of the late 1970s. The remainder of this section details efforts to implement these policies, that is, to produce more energy domestically and to use it more effectively.

Domestic production of crude petroleum began to rise significantly only after the establishment of Petrobrás in 1954. Annual output increased nearly 15 times between 1955 and 1960, from 5,500 to 81,000 barrels per day. These years of success fed the illusory belief that Brazil would soon become self-sufficient in oil, although the latter figure represented only about 40 percent of the petroleum consumed in the nation. [35] Output doubled again by 1970, to 167,000 barrels per day, but overall economic expansion and surging car and truck output in the 1960s caused consumption to rise more rapidly than production. The 1970 output now represented 34 percent of consumption. Domestic production peaked in 1974, with output at 177,000 barrels per day, after which it began slipping. By 1977 it had fallen below the 1970 level. In those seven years, however, national consumption had doubled, so that 167,000 barrels per day now amounted to only 17 percent of Brazil's petroleum needs. [36]

This dependence on imports, of course, was not alarming as long as oil remained cheap and plentiful on the world market—a situation Petrobrás officials expected to continue indefinitely, as Ernesto Geisel, then president of the oil monopoly, testified to a congressional committee in the late 1960s. [37] Relying on this projection, Petrobrás continued business as usual, searching for oil at home while buying ever-increasing quantities of foreign crude to refine in Brazil.

The oil crisis was therefore totally unexpected, and it forced Geisel, now president of the Republic, to order major policy shifts. As a result, Petrobrás dramatically increased domestic exploration, with a new emphasis on offshore areas; it began drilling for oil on contract in other countries; and, in a dramatic break with the past, it invited foreign oil companies to explore in Brazilian territory.

Offshore exploration soon brought discoveries. In 1974 Petrobrás announced a major find in the Campos basin off the state of Rio de Janeiro. This not only lent momentum to intensified drilling offshore, but it also gave the rulers of authoritarian Brazil an accomplishment to set before public opinion. Billboards went up showing gushers erupting while crude-bespattered Petrobrás workers looked on with satisfaction. Minister of Planning Simonsen claimed that the Campos basin could make Brazilian domestic crude production rocket to 1 million barrels per day by 1980. Even the more cautious petroleum professionals expected big and quick returns. The head of Petrobrás, for example, projected that output would rise to 225,000 barrels per day in 1975. [38]

Even the most guarded of these estimates, however, was far off target. Brazil's output has varied relatively little during the 1970s, and even the small variations since 1974 have represented declines, not rises. Simply put, the rising offshore production of the new finds has been more than offset by falling output in Brazil's now-aging onshore fields, particularly in Bahia. The latest available figures, for the first quarter of 1979, show that daily production has continued its decline, down to 164,000 barrels. [39]

Despite disappointing production so far, the Campos basin finds are truly significant. In addition to the Garoupa field, announced with such fanfare in 1974, ten other fields have been discovered and much of the basin remains to be explored. By mid-1979, estimated reserves were put at 1 billion barrels, with most fields still being delineated. One already-delineated field lends support to the optimists; it counts proved reserves of 250 million barrels. [40]

Through the late 1970s, haste, problems associated with novel technologies, and just plain bad luck have slowed the development of the Campos basin. Because imported oil drains so much foreign exchange, Brazilian officials decided to rush the area into production by installing a temporary subsea gathering manifold to collect the crude from nine wells in Garoupa and an adjoining field. This temporary system was to pump 45,000 barrels per day, beginning in mid-1977. In this way the Garoupa field could begin producing only three years after discovery, while the company went ahead with the eight-year process of erecting a permanent system of production towers and pipelines to shore. [41] Accidents, along with normal suppliers' delays, dashed these hopes. The provisional system finally began pumping from one well in February 1979, and by July, output had reached 9,500 barrels per day. Ironically, by then Petrobrás officials expected that full production from this temporary system, which had already cost $225 million and was sure to swallow at least $25 million more, would be reached in 1982, just about the time that the permanent system comes on-line. [42]

The oil-price pinch also led Brazil, in a major foreign policy departure in 1976, to invite foreign companies to explore for oil on "risk contract." This is the first time that foreign companies have been allowed to prospect on Brazilian territory since the 1920s. Three years later the Petrobrás director for exploration said that the results justified the decision. Not only was Brazil able to hasten the exploration of new areas, but, with the world's leaders in research and development now drilling in Brazil, Petrobrás has had the opportunity to observe the newest technologies and thus to reduce its technological lag, which this director estimated at only four years. [43]

The risk contracts are so named because they force the foreign companies to take specified economic risks. This distinguishes them from the old-time concession contracts of the Middle East and elsewhere, which gave the concessionaires long-term oil rights in major portions of a country's territory. The companies could then either work the concession or "bank" it, as suited their private interests. The Brazilian contracts, whose terms are probably the stiffest in the world, require a company to begin drilling by a fixed date, usually a year or less from signing, and to spend a stipulated minimum dollar figure while drilling a minimum number of wells. If oil is not discovered in a fixed time period, the contract area reverts to Petrobrás, thereby preventing the companies from hiding or postponing discoveries. If a contractee strikes oil in commercial quantities, it prepares the well for production and then turns it over to Petrobrás, in return for which it will recover all its expenses plus a previously negotiated but confidential percentage of the revenues. Crude-short Brazil has thus written these contracts to provide for revenue sharing, not production sharing. [44]

The first two rounds of bidding, from 1976 through 1978, yielded 17 contracts for 19 offshore blocks, mainly in the Santos basin off São Paulo state and at the mouth of the Amazon. So far, none of the contractees has struck oil in commercially viable quantities, a fact that helps explain the decline in interest in the third round in 1979, when only 11 groups submitted bids for 9 of 42 blocks. In this round, Petrobrás for the first time offered onshore blocks and four, near Manaus on the Amazon River, were taken. [45] In a fourth round, begun in late 1979, the government for the first time opened risk contracts to domestic private and state enterprises, and a number submitted preliminary bids. They will be bucking considerable odds, for even the Petrobrás director for exploration has described Brazil's onshore geology as "not among the most promising" for crude petroleum deposits. [46] If no oil is discovered, Brazil's supply will be limited to the considerable new reserves of the offshore areas Petrobrás itself has been exploring. These, of course, will at best provide less than half the nation's demand for crude in the mid- to late 1980s.

As Table 5.1 made clear, natural gas plays a negligible role in the Brazilian energy profile. The nation has few natural gas wells, and natural gas occurring in association with petroleum is usually separated from the oil and reinjected into the well to maintain the pressure. [47] There is no network of long-distance pipelines comparable to that of the United States. Urban "street gas" pipelines are very limited and in most cases supply gas manufactured from naphtha, a petroleum distillate, or coal. In 1979 a municipal gas company in São Paulo successfully produced street gas from alcohol and from garbage and began distributing it to its customers, part of the continuing effort to reduce reliance on petroleum derivatives. [48] Most urban Brazilians, however, are not served by street gas systems. They cook over bottled propane (liquid petroleum gas).

Completion of the permanent production system in the Campos basin will provide gas as well as oil pipelines to shore, and that gas will be used mainly in refinery operations and as petrochemical feedstocks. Brazil's recent agreement to import Bolivian natural gas to the São Paulo area accounts for the rising but still small consumption of natural gas that Table 5.2 projected for the 1980s.

Since Brazil, even under the best of circumstances, will not be able to produce enough petroleum to meet half of its liquid-fuel needs, policy makers and scientists have turned to alcohol as an alternative. This is not the first time, for sugar-growing areas used alcohol as motor fuel in the early decades of this century. During the Great Depression and World War II, when domestic petroleum production was nonexistent or negligible and shipping precarious, official regulations required that alcohol be mixed in gasoline. [49]

In late 1975 President Geisel established a National Alcohol Program to supply anhydrous ethanol for mixing with gasoline. After a transition period of several years, ordinary motor fuel would be gasohol containing 20 percent ethanol. The program also aimed to substitute alcohol for petroleum feedstocks in the chemical industry.

Geisel highlighted many benefits for the nation. On the economic ledger, the alcohol program would hold down oil imports and save foreign exchange, while giving a boost to the domestic capital-goods industry that would produce the new refinery equipment. Sociopolitically, it would reduce individual and regional income disparities, because the low-income regions of the country are capable of producing the raw materials, such as manioc, and because labor-intensive agricultural projects would increase employment and income among the poorest strata. Once the program is in full swing, Geisel affirmed, national income would rise because idle or underutilized factors of production, particularly land and labor, would be put to use. [50] Figueiredo, succeeding Giesel in 1979, increased the nation's commitment to alcohol.

The alcohol program is fraught with risk, of course. This is always true of crash programs to incorporate new technologies and productive processes into a complex economy whose parts are closely interdependent. Timing presents a major challenge, for government credibility and national confidence will be badly eroded if the program's elements get out of phase. Not only must a significant portion of national agricultural output (mainly sugarcane at this moment) be devoted to alcohol production, but distilling equipment must be installed at the same pace that sugar output increases. And finally, the motor vehicle fleet should evolve to consume the increasing alcohol output. Farming, distilling, and the capital goods and vehicle industries are all private-sector activities, and each subsector has sought assurances that it will be compensated if government planners miscallate and cause losses.

Ethanol production has thus far exceeded Geisel's target figures, a source of pride to many Brazilians. Indeed, the nation's mills and refineries produced 664 million liters of ethanol during the 1976/77 harvest year, the last before the National Alcohol Program took effect. In the 1977/78 harvest, production rose to 1.5 billion liters, then to 2.6 billion in 1978/79, and it is estimated at 3.8 billion liters for the current 1979/80 harvest. This greatly exceeds the 2.5 billion liters proposed in the original alcohol program.[51] Officials project that production will rise to 10.7 billion liters in 1984/85.[52] Moreover, highly placed figures in both public and private sectors assert that the ethanol target could realistically be doubled by 1985, to 20.5 billion liters.[53] Fuel-alcohol use reflects this rise in production. In 1978 the 15.6 billion liters of "gasoline" consumed in Brazil contained 1.5 billion liters of ethanol. This figure is 2.4 times that of 1977 and 9 times that of 1976.[54]

One should be cautious, however, about basing projections on the recent ethanol production figures. World sugar prices have been very low during the late 1970s, so Brazil has simply reduced sugar sales on the world market and converted a greater proportion of the crop into alcohol. If the world price rises sufficiently, it will necessarily force up the cost of alcohol, probably above the cost of gasoline refined from imported oil. This may in turn lead Brazilian policy makers to opt to export more sugar and hence put the squeeze on alcohol output. Finally, when considering the security of alcohol output, one must not neglect the potential impact of a serious crop disease. The government has also run into difficulty matching sugar production to refinery capacity.[55]

The expanded alcohol program announced in 1979 calls for service stations to carry straight alcohol as well as gasoline. This requires the development of new automobiles, designed to run on hydrated ethanol. Hydrated ethanol costs nearly 20 percent less to manufacture than anhydrous ethanol, but it cannot be used in gasohol because the 5 percent water in solution with it will, in the presence of gasoline, precipitate out and collect in the fuel tank.

The Brazilian auto industry is prepared to produce alcohol-propelled vehicles as soon as a stable supply of motor alcohol is offered to the public. The commercial development of alcohol engines marks an important step in Brazil's efforts to free itself from technological dependency. The government has financed alcohol-combustion research at the Air Force Research Center since 1974, and as a result, the technical norms to be met by the multinational auto firms producing the engines are norms devised in Brazil. When the alcohol engines began coming off the assembly line in late 1979, they were the first automotive engines produced in Brazil for which a royalty did not have to be paid to a foreign parent company.[56]

In August 1979 representatives of the auto industry reached agreement with the government on production targets for all-alcohol cars. By 1985 there should be 1.7 million such vehicles on the road, of which 1.2 million will be new and 475,000 will be older cars with converted engines. Public policy provides incentives for buyers of the new alcohol vehicles. The sales tax, annual highway-use tax, and fuel price at the pump will be lower, and financing terms will be easier. Experiments are also under way to develop diesel engines that run on alcohol. [57]

Brazilian scientists and engineers are now working on technologies to substitute alcohol for petroleum distillates in industrial processes. The president of Brazil's National Council on Scientific and Technological Development insists that "alcohol is more important as an industrial feedstock than as fuel." The Brazilian chemical industry has begun developing technologies to supplement or replace naphtha as its principal feedstock. Already one major chemical firm is producing ammonia, the key compound from which nitrogen fertilizer is made, from anhydrous alcohol instead of from naphtha, though still on a small scale. [58]

Although the alcohol program, as currently implemented, is not without negative environmental and social side effects, it nevertheless represents one feasible way of harnessing solar energy on a large scale. Hence the slogan of the alcohol-technology unit of the Air Force Research Center: "Alcohol is Solar Energy in Liquid Form." [59]

Sugarcane may be efficient at capturing solar energy, but unfortunately it competes with food crops for the best land. An alternative crop that grows on poor lands such as the acidic soils of Brazil's central plateau is manioc, a root crop. One technical consultant to the National Alcohol Program claims that manioc can already compete favorably with sugarcane in terms of alcohol output per hectare. In light of Australia's success at increasing manioc yields (up to 50 tons per hectare, compared with the Brazilian average of 14 tons), manioc is likely to be significantly more productive in the future as selection and cultivation techniques improve. Petrobrás has been refining 30,000 liters per day of ethanol from manioc since 1977 in an experimental distillery in Minas Gerais, and in 1981 a 150,000-liter-per-day plant will be completed. The main emphasis remains on sugarcane, however: as of June 1979, the nation was building 208 sugarcane distilleries and only 10 for manioc. The technical consultant cited above lamented that "manioc is being forgotten in the discussion of energy sources, because, unlike sugar cane, it lacks a godfather." Manioc, a crop traditionally planted by poor peasants on small plots, cannot compete politically with the "sugar sheiks" who control large plantations and wield great political power. [60]

The São Paulo Energy Company is currently testing a process

to produce methanol (wood alcohol) from eucalyptus. If this small-scale test shows promising results, the company will try it on an industrial scale in the early 1980s. José Goldemberg, a prominent nuclear physicist who has studied many aspects of the energy question, reports that the process of converting eucalyptus to methanol is more energy efficient than that for converting either eucalyptus or sugarcane to ethanol. According to his figures, energy output is 11.4 times energy input for methanol, 8.1 times energy input for ethanol from eucalyptus, and only 3.6 times energy input for ethanol from sugarcane. [61] Government-sponsored research is testing still other products as a source for alcohol, among them potatoes, yams, pineapples, babassu nuts, and sweet sorghum. Because sorghum is harvested during the dead season for sugarcane, it may prove an ideal complement to cane, enabling the distilleries to run year round. Researchers have also discovered a tree that produces virtually pure diesel fuel. [62]

In addition to renewable liquid fuels, Brazil possesses vast deposits of oil shale, a fossil resource that, when retorted, produces an oil similar to the lightest grade of crude petroleum. Brazil's oil shale formations, containing an estimated 800 billion barrels of oil, are second only to those of the United States. The energy potential of this shale has long been known in Brazil, and before the turn of the century gas extracted from shale lighted the gaslights of Taubaté, in São Paulo state. [63]

In 1951 Brazilian technicians began studying the shale deposits, and in 1959 they developed and patented the Petroxis process for retorting shale to obtain oil. By the early 1970s Petrobrás was producing about 1,000 barrels per day from 2.2 tons of shale at a small pilot plant in São Mateus do Sul, Paraná. From then until the second oil shock, however, high capital requirements and low petroleum prices on the world market delayed plans for an industrial-scale plant that would produce 50,000 barrels per day. In mid-1979 President Figueiredo accorded priority to the project, which should be completed by 1985. [64] Some observers still doubt, however, that those implementing this policy will be able to take enough capital from competing projects so that this one can be completed on time.

Coal has been selected to take the place of fuel oil, the main energy source used in Brazil's rapidly expanding industrial sector. After the second oil shock, the government proposed a program to increase annual coal production from 10 million tons to 35 million by 1985. Brazil's southern states have sizable coal reserves, estimated at 22 billion tons. These were not considered for extensive development until the crisis of 1979, however, because most of this coal is relatively low-grade steam coal with high ash content. This is why the steel industry has been importing over 3 million tons per year of high-grade metullurgical coal. [65]

The task for public policy, therefore, is to get Brazilian industries to accept the coal, and in September 1979 officials took a major step to achieve this goal. The ministers of industry and commerce, of mines and energy, and of transportation signed an agreement with the presidents of the sindicatos (associations) of cement firms and of coal mining firms to begin substituting coal for fuel oil in the cement industry, an industry that consumes at least 20 percent of the fuel oil used in Brazil. By 1984 the cement industry will burn 5.6 million tons of coal, saving the nation 2.8 million tons of fuel oil. To this end, the signatories agreed to perform the following tasks: the coal industry will raise its output to meet the target; the cement industry will progressively increase its coal consumption, according to an agreed-upon timetable, and, during the transition, government subsidies will neutralize fuel-price differences among firms in the sector; the Ministry of Mines and Energy will install coal depots in five cities and improve existing depots in six other places; and the Ministry of Transports will improve eight ports, enhance the navigability of three rivers in Rio Grande do Sul, build feeder rail spurs to the coal fields, and upgrade the rail lines between coal mining areas in the South and major cement-producing areas of the Southeast. [66] Experiments into other uses of coal are under way.

Traditional solid fuels—such as firewood, charcoal, and sugarcane bagasse—until recently supplied nearly one-half the energy used in Brazil (see Table 5.1), and they still supply one-quarter. Bagasse fuels the nation's sugar refineries and distilleries. Firewood, until 1968 the leading source of energy in Brazil, heats homes in rural areas and small towns of the temperate South, and throughout the interior fuels cooking stoves and small industries such as bakeries and brick works. The single biggest user of firewood and charcoal is the Brazilian steel industry, which owns, harvests, and reforests giant tracts of timberland to feed its furnaces. Other uses for these fuels are small-scale, often domestic, and sometimes picturesque. In recent years, for example, mechanics have been installing producer-gas generating equipment on cars and trucks, converting them to run on coal or charcoal. The concept of producer-gas generators is more than a century old; these units even propelled 20,000 cars in the city of São Paulo alone during the gasoline crisis of World War II. One mechanic boasts that his 1931 Ford gets 12 kilometers to a kilogram of coal. [67]

Far and away the most important future energy use of trees and other biomass will lie in their conversion to liquid fuels such as alcohol, not in burning them directly. To burn biomass is to waste by-products that could be put to advantage, such as coking charcoal, protein-rich animal feed, carbonic gas, and furfural. [68] Public policy is supporting research and development of hydrolysis techniques to obtain liquid fuels from biomass while conserving the by-products.

Brazil's greatest domestic energy resource is hydroelectricity. Up through the nineteenth century, Brazilians had good reason to feel that geography had played a cruel trick on them. In those early centuries, when settlers in the coastal cities struggled to penetrate and develop the interior, the nation's rivers were of little use. Nearly all of them cascaded over waterfalls as they descended from the central plateau, and, moreover, most flowed away from the coast, to feed into the River Plate or Amazon basins.[69]

In the twentieth century, however, the geographic curse became a blessing, for power-company engineers took advantage of the waterfalls to generate electricity. They even reversed the flow of some rivers on the plateau, forcing them over the coastal escarpment and into turbines at the bottom. By the late 1960s the best power-dam sites had been occupied near Rio de Janeiro and São Paulo, and Eletrobrás and the state power companies began building projects farther and farther afield, both to serve this industrializing area and to expand electric-power service in other areas of the country. By 1976 hydroelectric projects provided 92 percent of the nation's electricity, and, as Table 5.1 shows, this accounted for 24 percent of all energy used in Brazil.[70] The remaining electricity was generated thermally, by coal- and oil-fired peak-demand plants in large cities, by small diesel generators in interior villages not yet reached by the expanding power grid, and by factories generating their own power.

Eletrobrás in 1979 completed Plan 95, a 15-year projection of electric-power development from 1980 to 1995. Brazil already counts 22,000 megawatts of installed hydroelectric capacity, and this is expected to rise to 69,000 by 1990. This figure represents only one-third of the 209,000 megawatts of potential hydroelectric capacity in Brazil (compared with a total potential of 158,000 in all Western Europe). Included in the new capacity are two giant projects already under construction, the 12,600-megawatt Itaipu dam (to be the world's largest) being built jointly with Paraguay, and the 3,800 megawatt installation at Tucuruí in the Amazon basin. The share of electricity in total energy consumption should rise to 35 percent in 1985, as some industrial users switch from fuel oil to electricity, and new energy-intensive aluminum and steel projects are completed.[71] The aluminum and steel projects represent a Brazilian effort to produce and export energy-intensive manufactured products, thereby turning hydroelectric energy into hard foreign currency.

To extend national autonomy in applied technology and to supplement hydroelectricity (which many erroneously believed to be nearly fully exploited), the Brazilian government announced a mammoth nuclear power program in 1975. Based on an accord with West Germany, the program was highly touted as ensuring Brazil's entry into the club of nuclear nations and guaranteeing plentiful energy for continued rapid economic growth. For the first two years after the accord, expectations seemed limitless. The president of Nuclebrás, the national nuclear holding company, asserted that nuclear reactors would provide half the nation's electricity by 2000, and some officials claimed that Brazil would then have 63 nuclear power plants and would export nuclear technology and equipment to other parts of Latin America and the Third World. [72] By mid-1979 the bubble had burst, and most Brazilians, excluding a few die-hards in the nuclear sector, were demanding that Brazil repudiate the accord with West Germany or, at least, revise and downscale it.

When the accord, the largest nuclear transaction ever signed, was announced in 1975, it triggered vigorous opposition from the U.S. government, because it would transfer to Brazil the equipment and technology of the entire uranium fuel cycle. That is, it not only offered Brazil, a country that refused to sign the Nuclear Nonproliferation Treaty, two to eight 1,245-megawatt reactors, but it also provided for the establishment of enterprises to prospect for and mine uranium, to enrich uranium, to fabricate fuel elements, and to reprocess spent fuel. Reprocessing involves isolating and collecting plutonium, the key ingredient in atom bombs. [73]

By 1977 the Carter administration's persistent efforts to block the accord's implementation caused U.S. relations with Brazil to deteriorate significantly, but, if anything, the diplomatic flap merely reinforced Brazil's determination. Indeed, this case merely reminded Brazilians that for three decades the U.S. government has systematically frustrated their efforts to develop a measure of independence in nuclear science and technology. [74]

If U.S. pressure initially stiffened Brazilian resolve behind the accord with West Germany, why were its fortunes so quick to sink? By 1979 nearly all relevant domestic groups concluded, on the basis of compelling evidence, that the nuclear program will not serve either their group interests or the national interest. The public debate over nuclear power in Brazil has not turned on reactor safety and radiation danger, as it has in the industrial nations. The major issues have been cost, decisional competence, and the real extent of technology transfer. Nuclear electricity was shown to be at least twice as expensive as hydroelectricity in Brazil, engendering opposition from industrialists and economic planners. [75] Brazilian capital-goods producers complained that all the good contracts went to German firms, because Brazilian negotiators of the accord, astonishingly, had put into German hands the controlling votes on the project's executive bodies. When documents proving this were leaked to the press in 1979, it rallied broad nationalist support against the accord. Brazilian scientists argued that the deal would almost certainly retard rather than accelerate the effort to reduce Brazil's scientific and technological dependence on foreign nations. [76]

By 1979 the accord had come to symbolize the decisional bank-
ruptcy of a closed authoritarian system that could create a nuclear
program "without debates, without consultation with the scientific and
university communities, and without consideration of public opinion."[77]
By September, one federal deputy at last declared on the floor of the
Chamber what many had thought but none dared say in public in the
early years of the accord: "The men in power who signed the accord
were more interested in making an atom bomb than in producing electric power!"[78]

It was clear by late 1979, therefore, that Brazil would have to
cut its losses by revising the accord, but it was not clear how. Probably the best solution for the fundamentally flawed Brazilian nuclear
program is the simplest, and in the opinion of José Goldemberg, it
is almost inevitable: a delay in construction "will ultimately be beneficial, for it will allow time to evaluate and correct the errors."[79]

The delay might even give decision makers time enough to cut
off the nuclear-fission program with the second German reactor. The
nation could then rely on hydroelectricity while waiting the several
decades necessary to see if thermonuclear fusion, a much more benign nuclear technology, will become reality. In late 1979, researchers at the University of São Paulo made an important breakthrough by
confining plasma in a magnetic field, using a Tokamak device they designed and built themselves. Their achievement attracted official attention, buoying their hopes for significant financial support to expand
and intensify their research.[80] Whether they get a long-term commitment will depend in part on the fate of the West German accord, for
there is likely to be a tradeoff between fission and fusion. As one advocate of fusion power for the United States cautions, the probability
that a nation will be able to switch from fission to fusion in the early
twenty-first century "depends in large part on decisions to be made
within the next decade—especially on decisions regarding the funding
of fusion research and the extent to which fission development between
now and the year 2000 should be permitted to preempt the potential
fusion market of the more distant future."[81]

Following the first oil shock, the government charged a task
force with examining alternative energy technologies. The task force
prepared a national solar-energy research and development plan, the
most fully developed and productive aspect of which is the alcohol program described above. Other projects—including work on solar collectors for drying crops, heating space or water, running refrigerators or air conditioners, and distilling water—are in very early stages
of development. Although the government has funded research into applied solar technology at many universities and research institutes,
widespread industrial production of solar appliances has not occurred,
mainly because these devices lack a "godfather." In other words, the

researchers do not possess the political connections necessary to induce production. For example, the federal housing agency that finances low-income dwellings could have changed building regulations to permit solar water heaters. One analyst noted that this measure "could have made economically feasible the large-scale production of solar water heaters to replace the widely used, inefficient, and electricity-hungry electric resistance water heaters."[82]

In the near future, one solar application, crop drying, may get the kind of political backing that can link up financial support, industrial-scale production, and a guaranteed market. Following the second oil shock, the federal government ruled that, after 1980, drying activities must do without their current energy source, fuel oil. Since agriculture contributes a major portion of the nation's foreign exchange and is being geared up to contribute more, the government cannot risk cutting off fuel oil until some other means of drying proves practicable. Prototype solar dryers are being developed at a number of universities, and, if they test out successfully, there may be pressure for a crash industrial program to produce them.

Research funds are enabling scientists and technicians to develop equipment to harness the sun's energy trapped in winds, tides, and tropical waters. Researchers are measuring regional wind potential and designing windmills, preparing an experimental generator run by the tides on the north coast, and constructing a heat engine to generate electricity from the thermal gradient off the coast of Rio de Janeiro state, where cold water from the deep rises to meet warm tropical waters at the surface.[83] Apparently, the biogas digesters used extensively in China and India have received little or no official attention in Brazil. Perhaps Brazilian policy makers and technocrats, with their bias for bigness, do not believe support for biogas would enhance their reputations or advance their careers. After all, biogas units are small, they are not subject to easy central control, and, moreover, they would be dispersed among a rural population that carries little political importance.

Conservation, or the performance of a given task with a smaller amount of energy than has previously been the case, offers Brazil a potentially effective way to slow the growth in oil imports. Public policy gave uneven support to conservation in the mid-1970s, particularly through inconsistent pricing policies, but major decisions in 1979 unequivocally back it. Policies supporting conservation have been most systematically developed for the transport sector. To reduce gasoline consumption, the government not only raised prices, but it also banned cars in certain downtown areas of big cities, established exclusive bus lanes on some commuter roads, set a 50-mile-per-hour speed limit, and closed gas stations on Saturdays, Sundays, and at night during the week. Though of greater symbolic than material importance, bike paths have been established in many cities.[84]

After the first oil shock, the government committed massive funds to revitalizing rail and maritime freight hauling. The share of freight handled by the railroads dropped from 23 percent in 1952 to 17 percent in the late 1960s and early 1970s, but by 1975 it had rebounded to 20 percent and government planners hoped to raise it above 30 percent by 1980. Maritime shipping carried 25 percent of Brazil's freight in 1952, but only 10 percent in 1970. By 1975 this had risen only marginally, to 11 percent. [85]

Architecture and building standards have not received much attention from energy policy makers. Goldemberg, observing the energy inefficiency of the air-conditioned, sealed-window buildings being constructed nowadays in major cities, commented: "The Portuguese colonizers some 200 years ago constructed simple but functional buildings, and with some effort it should be possible for Brazilians today to design a new type of 'tropical civilization' that is appropriate to the environment."[86] A visit to the modern, beautiful, and architecturally distinguished Ministry of Education building in Rio de Janeiro, built in the 1930s, will convince any observer that Brazilian architects have, for more than a generation, been able to design such buildings. It is not yet clear whether the rising costs of energy will force Brazilian architects to return to the lessons of their heritage.

Despite the determination of Brazilian policy makers to increase sharply the domestic energy output, it is clear that for the foreseeable future, Brazil will not become self-sufficient in energy production. Foreign sources will continue to supply a large segment of Brazil's energy demand.

INTERNATIONAL ENERGY POLICY

Brazil's international energy policy is of pivotal importance to the bid for great-power status. The principal aspects of this policy include: intensive, pragmatic diplomatic activity, particularly through bilateral negotiations, to insure that the flow of oil will not be cut off and that cooperative projects with neighboring countries will increase Brazil's supplies of other forms of energy, particularly hydroelectric power and natural gas; aggressive promotion of Brazilian exports around the world so that Brazil can pay for energy imports, or can directly swap Brazilian products for coal and oil; direct exploration and production by Petrobrás in foreign countries, through risk or concession contracts; reversal of Brazil's long-standing refusal to allow foreign companies to search for oil on Brazilian territory; and development of nuclear power, through the agreement with West Germany.

The Arab oil embargo of 1973-74 posed two distinct policy challenges to Brazil: that Brazil's oil supplies might be cut off, causing an immediate, devastating economic dislocation; and that at the new prices, oil imports, even if unimpeded, would cause a slower but equally disastrous economic crisis, through a permanent hemorrhage in the balance of payments.

Brazilian fears of an oil cutoff were great enough in the embargo period that the Foreign Ministry gave in to Arab pressures and realigned the nation's Mideast policy. [87] Prior to 1974, Brazil had described its policy in the Middle East as "equidistant" or "evenhanded." The "evenhanded" policy ceased to be tenable, however, once the Arab nations employed the oil weapon after the October war. By January 1974, Arab ambassadors in Brasília demanded that Brazil make an explicit policy statement tilting toward their side. On January 31, at a reception for the representatives of the Arab League, the Brazilian foreign minister acceded. His address omitted the customary word "neutrality," condemned the conquest of territory by force, called for rapid Israeli withdrawal from all occupied territories, and urged that the rights of the Palestinian refugees be acknowledged. Soon afterward, the Egyptian consul general in Rio pronounced Brazil "friendly to the Arabs." The high point in Brazil's tilt toward the Arabs came at the UN General Assembly in November 1975, when Brazil voted in favor of a resolution defining Zionism as racism.

The degree of tilt diminished between 1976 and 1978. In the first place, the initial realignment kept Brazil off the embargo list, assuring economic stability in the short run. The degree of tilt also diminished because Brazilian diplomats and economic policy makers found that it produced fewer material benefits than they had expected. They had hoped that their nation's Third World status would gain them less onerous terms from the Arab oil dealers, as well as significant investments of petrodollars. These hopes proved groundless. By late 1976, therefore, Brazil chose not to support several Arab-sponsored UN resolutions, leading observers to conclude "that a warning was being flashed to the Arabs that Brasília's support (which the Arabs in fact do court) was not to be taken for granted in all political issues nor as gratuitous in any of them. "[88]

In 1979 the Iranian crisis again reduced Brazil's supply of crude, cutting the nation's normal three-month stock to little more than one month's. Arab influence naturally increased as Brazil began negotiating increased Iraqi deliveries to make up for the Iranian shortfall. It is thus hardly coincidental that in May 1979, in a joint statement with Iraq, Brazil recognized the Palestine Liberation Organization as the legitimate representative of the Palestinian people. [89]

Of longer-term importance than Brazil's activity in such traditional diplomatic domains as the United Nations are its aggressive and innovative departures in economic diplomacy. Indeed, both Brazil's drive for greatness and the obstacles to it are fundamentally economic

in nature, compelling this approach. Chief among the obstacles is
the energy crisis.

International Monetary Fund statistics reveal the magnitude of
the problem. Oil imports, which had cost Brazil $573 million in 1972,
soared to $3.2 billion in 1974 and $4.5 billion by 1978. Brazil faced
a hitherto inconceivable trade deficit of $4.7 billion in 1974. The gov-
ernment—by aggressively marketing Brazilian products, developing
new exports, and benefiting from unusually high international commodity prices—all but closed the gap by 1977, but in 1979 the second shock
caused the nation's oil bill to hit $7 billion, raising the trade gap to
about $2.5 billion and causing a current account deficit of $10.5 billion. In 1980 oil imports are likely to come to between $10 and $12.5
billion. [90]

Brazil's new economic diplomacy has shifted an important share
of responsibility in foreign policy making away from the Ministry of
Foreign Relations, most of whose high-level officials have little training in economics, to other institutions, notably the Ministry of Mines
and Energy, under whose supervision fall four giant state companies:
Petrobrás, Nuclebrás, Eletrobrás, and the Vale do Rio Doce Company, whose iron ore is sometimes bartered directly for energy supplies; Petrobrás itself; the Ministry of Industry and Commerce; and
the Ministry of Finance. [91]

To cope with its new role, Petrobrás spun off two internationally
oriented subsidiaries in the early 1970s. Brazilian officials decided
in 1970 to seek contracts to explore and produce oil in foreign countries; to this end they created a subsidiary, Braspetro, in 1972. Braspetro's contracts generally involve joint ventures with transnational
majors, independents, or host-country national companies, and the
oil to which Braspetro is entitled is usually an earmarked portion of
production. In Iraq, for example, Brazil will receive 25 percent of
Braspetro's output up to 350,000 barrels per day. It will have to pay
for this, but at a rate 25 percent below the prevailing OPEC price.
The discount declines progressively for purchases in excess of that
figure. [92]

Braspetro has been drilling for less than a decade and has several successes as well as the possibility of more good strikes. It was
producing 10,000 barrels a day in Colombia by 1977, with further exploration under way. Exploration in Algeria yielded one of the few new
fields of good size found there in recent years; by mid-1978 the discovery well was producing 3,000 barrels per day while Braspetro
drilled step-out wells to delineate the field and prepare for full production several years hence. [93] In Iraq, Braspetro prospectors found
a field with reserves estimated to exceed 7 billion barrels. The company planned to invest nearly $2 billion to bring it into production in
1982, with ultimate output anticipated at 750,000 barrels per day. [94]

By late 1979, Braspetro had also contracted to prospect in Angola, Egypt, Iran, Libya, the Philippines, Madagascar, Guatemala, and Indonesia. And Norway, whose offshore drilling equipment industries had made several big sales to Brazil and hoped for more, invited Braspetro to bid for a risk contract there. [95]

Braspetro originally handled international trade for Petrobrás, but in 1976 it transferred this function to a new subsidiary, Interbrás. This is a trading company that markets Brazilian manufactures and commodities, and sometimes exchanges them directly for energy resources. In conjunction with officials from the Ministry of Industry and Commerce, Interbrás officials travel the globe in pursuit of major deals.

Brazil, now at middle-level economic development, has a wide variety of goods and services needed by many of the oil-producing states. Possessing good water resources for agriculture, for example, it can supply sugar, coffee, soybeans, cotton, and other foods and fibers to the arid nations of the Middle East. Brazilian agricultural planners have been sensitive to market opportunities, as when they turned the country from an occasional importer of soybean products in the 1960s to a major exporter, second only to the United States, in the 1970s. The government that was inaururated in 1979 has promised to foster the expansion of export agriculture.

During the sustained economic growth of the postwar decades, Brazilian architects and civil construction firms gained a wealth of experience in designing and building residential, commercial, and industrial establishments, as well as highways, airports, power dams, irrigation systems, and the like. This experience in an "underdeveloped" milieu with tropical and subtropical conditions provides the Brazilians with a selling point when negotiating with nonindustrialized oil producers. One survey for 1979 forecast $3 billion in Brazilian service exports, mainly from construction contracts, with the possibility that these would rise to $5 billion by 1982. [96] Finally, because Brazil's level of industrialization is several steps ahead of that of the OPEC countries, it offers diverse products such as shoes and textiles, simple instruments and machine tools, automobiles and airplanes.

Since the Brazilian military are seeking major-power status, it is not surprising that the nation has developed the largest armaments industry in the Third World. Brazilian weapons salesmen aggressively pursue export contracts around the world, with particular emphasis on the Middle East. The Brazilian weapons sales catalog, as of 1979, included fighter and transport planes, armored combat vehicles, light tanks, boats, surface-to-air missiles, rocket launchers, mortars, cannons, laser-directed targeting devices, machine guns, automatic pistols and rifles, grenades, ammunition, and communication equipment. [97]

Let us now look at some specific examples of Brazilian sales and barter deals. The Middle East is one focal point for Interbrás efforts, because the region drains such a great share of Brazil's foreign exchange. In 1974, for example, Brazil made $2.4 billion worth of purchases from Mid-East countries, while selling them only $331 million, or 7 percent. [98] One of the prime targets of Interbrás negotiators, at least through the overthrow of the shah, was Iran. In 1977 the two nations signed a five-year, $6.5 billion trade agreement that moved Iran back into top-supplier position by increasing Brazil's oil purchases from 75,000 to 275,000 barrels per day, or a little less than one-third of Brazil's then 800,000 barrel-per-day import appetite. More significant than the amount of this agreement was the stipulation that the Iranian government buy products from Brazil worth at least 30 percent of the value of its oil sales. Iran did not fulfill its commitment, however, for the shah's fascination with the most sophisticated gadgetry and his close affinity for the United States caused his government to order from the United States or Western Europe, even when ordering food that Brazil could supply. [99] When the revolution in Iran halved oil output in 1979, Brazil was forced to scramble on the world market to find replacement oil. Iraq, now delivering 400,000 barrels per day, is the new top supplier.

Barter deals in the Middle East include a 1978 agreement by Brazil and Algeria to exchange 24,000 Brazilian-made Volkswagens for $62 million in crude oil. In early 1979, Prince Abdul al Saoud, the head of Saudi Arabia's national investment company, took steps to create a jointly funded Brazilian-Arab trading company to market Brazilian goods and services in the Middle East and to channel Arab capital investments to Brazil. [100]

Brazil has actively pursued Nigeria, Black Africa's major oil producer. Nigeria is now Brazil's largest trading partner on that continent, and when a Nigerian trade mission visited Brazil in January 1979, the negotiators discussed Brazilian participation in agriculture, food processing, and the construction of ports, highways, and railroads. Some 25 Brazilian consulting and service companies established themselves in Nigeria to seek contracts, and a Brazil-Nigeria Committee was created to coordinate the formation of joint ventures and technology transfers. By late 1979 Nigeria had doubled its oil deliveries to Brazil, to 40,000 barrels per day. [101] Brazil has sought similar contracts with Angola, though at a lower level.

Toward the socialist nations, the Brazilian military regime maintained correct, low-level ties in the 1960s. As economic diplomacy came to overshadow traditional diplomatic concerns in the 1970s, however, the government actively expanded relations with them. Indeed, in 1972 the minister of finance announced that next on the agenda of Brazil's export drive would be "the conquest of communist markets." Braspetro and, later, Interbrás were conceived as means to this end. [102] To be sure, Brazil's large state agencies help smooth the way, for they are the type of organizations that negotiators from the socialist world are used to dealing with. During the Arab oil embargo, the Soviets sold Petrobrás 30 million barrels of diesel fuel. This opened the door to significantly expanded trade between the two countries, mainly Soviet oil for Brazilian coffee, other commodities, and manufactures such as textiles and shoes. Brazil has also established beneficial trade relations with Poland. In 1976 the two signed a five-year, $3.2 billion agreement to exchange Brazilian iron ore for Polish coal. [103]

In post-Mao China, relatively minor levels of trade ballooned to new dimensions. Brazil now exports iron ore, iron and steel products including oil field equipment, and sugar to China, and Brazilian negotiators appear likely to win hydroelectric power plant contracts. In return, China has begun selling oil to Brazil, with 1980 deliveries at 30,000 barrels per day. [104]

Brazil has courted, and in some cases coerced, its Latin American neighbors as it expands its sphere of influence, its export markets, and its appetite for energy. Brazil is building with Paraguay the world's largest hydroelectric complex at Itaipu on the Paraná River. With the signing of the Itaipu accord in 1973, Brazil definitively displaced Argentina as the dominant power over Paraguay. Some analysts now feel that Paraguay will evolve under Brazilian hegemony in a fashion similar to Puerto Rico under the United States. [105] Itaipu strengthens Brazil's economic hold over Paraguay, because Paraguay's electricity needs are so modest that nearly all of its share of the power output will be sold to Brazil.

Bolivia, another former dependency of Argentina and one from which Brazil took 73,000 square miles as it expanded westward at the turn of the century, has been drawn increasingly into the Brazilian sphere of influence in recent years. [106] By 1978 Brazil's energy crisis and Bolivia's economic difficulties led the energy ministers of the two countries to reactivate a dormant 1974 agreement under which Brazil would import 400 million cubic feet of Bolivian natural gas per day over a 20-year period. Brazil would finance a $1 billion pipeline to carry the gas from Santa Cruz, Bolivia, to the São Paulo area. To pay for the gas, whose price will be renegotiated every six months, Brazil will finance an industrial complex in Bolivia and will later buy from it such products as rolled steel and urea. [107] Implementation of the accord in 1979 merely meant preparing studies and drafting plans, so political instability in Bolivia posed no threat. When construction activities approach, however, debate over its implications for Bolivian national sovereignty will surely begin again there.

Mexico's recent oil discoveries and growing exports have not gone unnoticed in Brazil, but the Brazilians refused to pay Mexico's high price until December 1978. Then, with the shah tottering in Iran, Brazil finally agreed to buy Mexican oil, as part of a large petroleum, petrochemical, and mining agreement. Although the oil component of of this deal is relatively small—20,000 barrels per day beginning in 1980—it opens the door for expanded Brazilian purchases once Mexico completes its offshore barge-storage system to handle supertankers. [108]

Brazilian hydroelectric diplomacy has raised oil imports from Venezuela. The Brazilian minister of mines and energy and the president of Petrobrás visited Venezuela in 1978 and won a $1.2 billion power-dam contract, for which Venezuela will pay with increasing crude deliveries. By January 1980, Brazil's crude imports from Venezuela had risen from 14,000 to 50,000 barrels per day, and the Brazilians were hoping to raise that to 150,000. When President Figueiredo paid a state visit to Caracas in November 1979, his party included 100 Brazilian private businessmen, and the deals discussed, if consummated, could amount to between $2 and $3 billion. [109]

Venezuela is also the target of a Brazilian effort to make its costly nuclear deal with West Germany generate some foreign exchange. In August 1979, Brazil agreed to supply its northern neighbor with some low-level nuclear technology. The Venezueleans are hoping that the waste heat from nuclear power plants could be used to recover oil from heavy-crude fields and the Orinoco tar belt, now that its light crudes are nearly exhausted. In November, the Brazilian minister of mines and energy announced that Brazil intended to transfer nuclear technology to Iraq. Neither of these deals involved sensitive technology such as reprocessing or enrichment. [110]

ENERGY AND THE ENVIRONMENT

In Brazil, as elsewhere, policies to expand energy production and consumption necessarily affect the environment. By the early 1970s, for example, soaring petroleum consumption was creating annoying air pollution in many of the major cities, including extremely unhealthy smogs in São Paulo, where winter temperature inversions concentrate the pollutants in chill, misty air.

Brazil's recent efforts to increase domestic energy production, particularly through untried or innovative technologies, have clearly aggravated environmental abuse. For the most part the nation possesses the technology to minimize the dangers, but the nub of the issue is whether policy makers possess the will and ability to enforce ecological responsibility upon energy producers and users.

This section discusses some of the dangers inherent in Brazil's massive energy-production programs. Because offshore petroleum production presents pollution problems similar to those in other countries discussed in this book, we will omit treatment of it. In like manner, the dangers of radioactive emissions and waste disposal associated with nuclear power are not peculiar to Brazil and can be inferred from discussions of other countries where nuclear energy is further developed. Nevertheless, let us record the comment of a German consultant who was appalled by practices at the nuclear plant construction site: "After seeing the way the Brazilians resolve their problems, I won't go in there again without a lead suit."[111]

Coal presents major environmental challenges wherever it is mined and used. A report from one Santa Catarina coal-mining company to the Ministry of Mines and Energy in June 1979 pointedly warned of dangers from expanding mining activity there: "At present levels of production in Santa Catarina, pollution of the land, air, and water has already reached 'alert' levels in the south of the state. The groundwater is completely polluted by the wastes from coal beneficiation. When coal production levels double or quadruple in this area, the pollution problem will reach unbearable levels. It is urgent, in our view, to apply existing regulations rigorously and to improve them." There is, unfortunately, reason to doubt that this recommendation will be implemented, for the official cost projections for expanded coal mining do not include pollution clean-up. [112]

Coal presents other dangers, too. Coal mining has been a dangerous, often fatal occupation everywhere. The report cited above points out that Brazilian mine-safety standards lack many safeguards required in the United States, making the occupation in Brazil all the more dangerous. Inferior dust control in the mines, for example, has aggravated lung disease among miners. Finally, if the nation fudges on environmental controls for coal-burning industries, it will cause a massive deterioration in health, because most cement plants and other major industries are near the population centers.

Brazil's shale-oil program is too new to allow an empirical assessment of its environmental impact, but Petrobrás officials claim they have designed it so it will not damage the environment. The company plans to strip-mine the shale, a technique that allows easier disposal of the spent shale than the underground mining used in the American West. In the United States, not all of the spent shale will fit back underground, raising a variety of so-far unsatisfactory suggestions for its disposal. In Brazil, however, all the residue can be replaced in the original cut before the topsoil is bulldozed back into place. Petrobrás is experimenting with fast-growing eucalyptus, fruit orchards, and cereal plantings on the restored land. [113] If large-scale restoration of the strip-mined land exceeds cost estimates, only a policy decision from the highest officials will secure continued funding for it. One cannot, therefore, take the idyllic forecasts for granted.

The alcohol program has the potential to be relatively benign ecologically and socially. Its balance sheet in Brazil, however, is at best mixed. Combustion of ethanol generates fewer short- and long-range atmospheric hazards than combustion of gasoline. It contains no lead, and its emissions of hydrocarbons, carbon monoxide, and nitrogen oxides are much lower than from gasoline. However, it produces acetaldehyde, an irritant of the eyes and mucous membranes. Research into acetaldehyde control is under way, but no breakthroughs have been reported. [114] Methanol (wood alcohol) has more serious toxic and corrosive characteristics, but technologies to produce and use it are still experimental, permitting us to pass over it here. [115]

Ethanol combustion does not threaten the earth's temperature balance the way burning fossil fuels does, via the greenhouse effect. In this, renewable biomass fuels differ dramatically from fossil fuels. Burning fossil fuels increases atmospheric carbon dioxide by oxidizing the carbon stored in animals and plants in prehistoric times. Burning alcohol, on the other hand, does not change the carbon-dioxide balance, for it oxidizes only the carbon that has recently been present in the atmosphere. Then the next season's crops of sugarcane and manioc reduce that carbon dioxide by fixing, through the solar-driven process of photosynthesis, an equivalent amount of carbon in their tissues. [116]

The production of ethanol has seriously affected the quality of water. In producing each liter of alcohol, Brazilian distilleries produce an average of 12 liters of fluid wastes. This they routinely dump into adjacent rivers, destroying animal and plant life. Since early 1979 the practice has been illegal, but it nonetheless continues in many areas. Recent research indicates that these liquid wastes could simply be flooded back into the fields from which they came, replacing many of the nutrients and thus holding down fertilizer costs for the planter. They can even be dried and used as cattle feed or, through anaerobic fermentation, converted to methane gas and fertilizer. [117] Ecologically sound disposal of these wastes is, therefore, not beyond the technological capacity of Brazilian scientists. The key unknown is whether the nation can summon the political will to enforce the law. Given plans to produce more than 10 billion liters of ethanol per year by 1985, action will have to be taken soon. In particular, lending agencies must refuse to finance new distilleries unless they include the mandated waste-conversion facilities.

Despite President Geisel's upbeat predictions in 1975, the alcohol program has not fulfilled its social promise to create jobs and raise incomes in the poorest regions of the country. It has created some 500,000 jobs in rural São Paulo, where 90 percent of the ethanol will be distilled. São Paulo is, however, the wealthiest state in the union and many of these jobs are only technically new, because most of this cane is being grown on lands formerly used for soybeans and coffee, and many canefield workers worked on those previous crops. [118] It has also driven food crops farther from the cities or displaced them altogether, leading to the following caustic assessment in 1979: "the $2 billion that Brazil will spend this year importing food-stuffs is probably due as much to the displacement of food crops by cane as to the regional droughts and floods."[119]

The alcohol program raises a serious land-use issue. Less than 10 percent of Brazil's land surface is currently arable, and the pro-duction of 10.7 billion liters of ethanol by 1985 will require one-tenth of that. Rising costs of commercial fuel may also lead to deforestation of other large areas. Many observers fear that the poor will cut trees in parks, forests, greenbelts, and hillsides near population centers, creating eyesores and provoking serious erosion and water-supply problems. Larger enterprises may seek to exploit the Amazon basin's fuel potential. Consider the dangers inherent in this assertion: "the vegetation of Amazonia will be sufficient to supply energy for Brazil, at present consumption levels, for 1,000 years, even if there were no renewal."[120] Unfortunately, the robust exuberance of the tropical rain forest is something of a sleight-of-hand trick by Mother Nature. Ecologists note that this ecosystem is in fact "among the most vulner-able on the planet." Deforestation in the Amazon, an area covering more than half of Brazil's territory, would be irreversible for "the soil supporting the forest, and the water in the streams are both sin-gularly deficient in nutrients." Astonishingly, "the tropical wet forest is ecologically a desert covered by trees!"[121] Without the trees, it would return to desert, with unpredictable climatological consequences.

Even the apparently benign hydroelectric projects pose some environmental problems, although the Brazilian electric-power sector has for the most part behaved responsibly toward the environment. One ecologist observed that "Eletrobrás' environmental assessment department, now several years old, is in the vanguard of environmen-tal planning," and when he prepared a major environmental-impact study on the giant Tucuruí dam project in the Amazon basin, he de-scribed it as "an environmentally rational development." He nonethe-less offered many suggestions and warned that "the costs associated with not implementing environmental considerations could, at worst, exceed the total cost of the project.!"[122] The great economic risks of not being ecologically responsible, therefore, have motivated the power sector's sound behavior.

Electricity from the completed Tucuruí project in the early 1980s will attract iron and aluminum mining and refining industries into the area. The attendant economic expansion, particularly if it takes place in an uncontrolled boom-town atmosphere, will irrepar-ably damage this ecosystem. Myriad individuals and companies pur-

suing their own short-term profit simply cannot be expected to restrain themselves so as to protect the long-term interest of the ecosystem. And the threats can only multiply, for more and more hydroelectric projects will be undertaken in the Amazon in the future.
Eletrobrás may, in its own interest, be self-regulating and ecologically responsible, but the new arrivals will not. Only the strict exercise of enlightened political will can prevent the worst abuses, but so
far the government, through its development and road-building agencies, has done more to foster than to prevent abusive conditions. [123]

Whatever benefits the hydro projects may provide for urban
residents, they are sure to further damage the interests of Brazil's
dwindling Indian population. One survey cites 11 Indian areas known
to be threatened by proposed dams, and it observes of the Itaipu site:

> Although [the Indians] have occupied this area for genera-
> tions, they have no 'legal' title to it. The policy of Itaipu
> Binacional regarding this seems to be that it will only re-
> imburse those landowners who can show legal title to the
> lands they claim. Hence, the Indians are liable to receive
> nothing, outside of the establishment of a museum for the
> "archeological and cultural preservation" of the flooded
> areas. . . . The survival of their living culture, however,
> remains problematic. [124]

What are the prospects for ecologically rational energy production in Brazil? Responsible behavior varies directly with the size of
the enterprise, the complexity and hence the vulnerability of its technology, the degree to which it sells in a monopoly market, and the
degree of state ownership or control over the sector. When these conditions hold, the "externalities" (costs of production that are external
to the market relationship between the seller and buyer, because they
can be imposed upon other parties, such as the victims of industrial
pollution) can be internalized in the price the consumer pays or the
state subsidizes.

It is thus no accident that Eletrobrás and its subsidiaries get
the highest marks for ecological responsibility. These giant state-
owned firms hold a monopoly, and they surely wish to prevent ecologically irrational behavior from silting up their reservoirs, damaging
their turbines, or even changing rainfall patterns—misfortunes that
have resulted from unsound practices elsewhere. Petrobrás, in petroleum and shale, is likely to be next in responsibility. The success
of its offshore operations depends almost as greatly upon ecologically
responsible behavior as do hydroelectric facilities. Its onshore operations, particularly in shale, do not, so the risk of environmental degradation is greater there.

The greatest risks occur in the competitive domain of the private sector, notably in alcohol and coal. To maintain ecologically responsible energy exploitation in these fields, there must be a public-policy commitment to enforce the letter and spirit of environmental laws. Such enforcement is at best spotty as of 1979.

The Brazilian federal government created a Special Secretariat for the Environment in 1973, and the industrial states also possess analogous bodies. Implementation of environmental norms has nevertheless remained weak, so Brazil's industries, seeking to keep their prices low on the export markets, have externalized the social and health costs of polluted air and water upon nearby populations. In the São Paulo area, however, pollution is so intense that the state and local governments have begun putting teeth into the regulations. [125]

More rigorous enforcement of environmental standards may develop in the 1980s. A growing but still small environmental movement has emerged in Brazil in recent years. If the significant political liberalization of the late 1970s continues, it may create conditions that will increase the movement's political leverage.

FUTURE PROSPECTS

To describe the Brazilian energy profile in an era of crisis is necessarily to shoot at a moving target, one that may again move and change even before this book is published. Nevertheless, the principal components of Brazil's energy policy for the 1980s are now clear. Changes over the next decade, therefore, are likely to be modifications of emphasis among components, not new departures.

Policy makers, spurred by the spiraling foreign-oil bill, are investing massive resources in domestic energy production—most notably domestic oil and such petroleum substitutes as alcohol, coal, and hydroelectricity. These efforts will, by 1985, raise domestic production to at least 350,000 and perhaps even to 500,000 barrels per day while producing significant quantities of energy from other sources. Even so, the nation will continue to depend heavily on foreign oil. Rapid economic development will raise overall energy demand, keeping oil imports at the current level of about 1 million barrels per day. This import burden will imperil the Brazilian economic model unless the nation can increase exports even more dramatically than it did during the 1970s.

The policies to achieve greater energy self-sufficiency can only loosely be called a national energy policy, at least up to 1979. As in most other capitalist nations, Brazilian energy policy has really been a series of discrete policies, often decided in isolation one from the other, rather than a coherent, integrated policy. By 1979, however,

the second oil shock on top of Brazil's worsening economic crisis
forced the government to treat its energy policies as parts of a whole,
thereby acknowledging the tradeoffs involved in each alternative. In
this integrated evaluation, conservation for the first time won serious
official attention, and it also became apparent that the nuclear agreement with West Germany, at least on its original terms, was too costly
to merit priority.

A coherent policy to assure adequate energy supplies has become bound up with national security policy, particularly in light of
Brazilian aspirations for great-power status. As the minister of
mines and energy put it in 1979, "Today those who rule are those who
have energy. . . . " Confidently looking to the future, he continued,
". . . Brazil, fortunately for us, has the medium- and long-range
energy potential to enable it to sit at the table where the great international decisions are made."[126]

Especially significant to the bid for great-power status is the
development of new technologies. No nation can lay claim to such
status without first creating conditions for independent inquiry in experimental science and then building upon it an important degree of
self-sufficiency in applied technology.[127] The energy crisis has forced
Brazil to intensify its efforts in this area. Indeed, it is moving into
technologically uncharted terrain, where Brazilian scientists and engineers themselves are developing the new technologies, rather than
importing them from already industrialized nations. Brazilian programs to harness the photosynthetic energy of plants, to produce shale
oil commercially, to develop fusion energy, and even to design and
export weapons systems therefore have an importance that transcends
their short-term purpose of alleviating the balance-of-payments deficit.

The very term "great power" evokes an image of bigness, and
Brazil's energy policies, with their reliance on complex, large-scale,
centralized technologies, surely fits the image. While Brazil is, of
course, a large-scale nation with large-scale problems, two key characteristics of its political system also give it a bias for bigness. The
first is the primacy of the technocracy in economic policy making. To
the entrepreneurial mind of top-level technocrats, a system of strip
mines, a 10-billion-liter alcohol program, or a massive power dam
is a far more impressive and prestigious accomplishment than the
production and distribution of a large number of decentralized biogas
digesters or solar plates, each of which would be under the control
of its owner rather than under the direction of the manager of some
giant organization. The second characteristic is the corporative nature of interest representation in Brazil. In a political system where
officially recognized sindicatos represent the interests of each industrial sector and give it a voice in the policy-making bodies, big organ-

izations have far greater access to the technocrats than small ones, and individuals have none at all. The coal and alcohol programs illustrate the way technocrats interact with major economic groups to create giant, complex schemes.

Paradoxically, the very success of these massive programs may impose a serious cost on the nation, for it may obscure the complementary benefits Brazil could obtain from simultaneously pursuing small-scale technologies. Policy support for the commercial production of solar water heaters (and space heaters in the temperate South) could save a good deal of the commercial fuels and electricity that currently heat space and water. The savings would increase in the future as more and more people move into income brackets that can afford hot water or heated homes. Small-scale, decentralized biogas and solar units could also provide millions of rural Brazilians with modern energy sources for the first time, while saving the energy and materials that would, at some future point, have to go into creating a long-distance transmission and distribution system to these areas of low population density.

The economic benefits of a small-scale strategy to complement the current large-scale approach are significant and can be measured in barrels of oil saved. However, the social and political benefits may be even more significant. If policy makers facilitate the production and distribution of small-scale units for installation at the point of use, they would enable many Brazilian citizens to lead more rewarding and productive lives, and that surely is one of the functions of government. If the political liberalization of the 1970s continues into the 1980s, it may well create conditions where individuals, the long-forgotten members of the Brazilian political system, can use the political process to force public-policy attention onto smaller-scale applied technology. Combined with energy-efficient architecture, this would at last help create the new type of tropical civilization that Goldemberg has advocated.

The importance of such a policy departure would go far beyond the tropics, for it would serve as an example for the entire world. Brazil, an industrial power to be, is rather uniquely positioned to set such an example, for most of its energy-producing, -distributing, and -consuming infrastructure remains to be created. It thus has more of a choice than the already established industrial nations, which have invested enormous sums of capital into an infrastructure requiring fossil fuels. Short of investing in a second costly infrastructure, these industrial nations have no choice until either the first infrastructure wears out or fossil-fuel prices rise so high that an investment in an alternate infrastructure becomes economically feasible.

The energy policies pursued in Brazil through the 1970s suggest, unfortunately, that policy makers have no predisposition toward renewable, point-of-use energy technologies. The fortunes of these technologies may improve during the next few years, however. If solar crop-drying devices prove effective during the early 1980s, a solar-technology industry may develop, and the citizen pressures hypothesized earlier may lead to official support for its application in housing and other sectors. If so, Brazil could strike out on its own, providing a model for the industrial world in this era of transition.

NOTES

1. The notion that Brazil is between First and Third Worlds is Wayne A. Selcher's, in his Brazil's Multilateral Relations: Between First and Third Worlds (Boulder, Colo.: Westview, 1978).

2. For a brief analytic overview of the Brazilian political system and principal policy issues, see Kenneth Paul Erickson, "Brazil: Corporatism in Theory and Practice," in Howard J. Wiarda and Harvey F. Kline, Latin American Politics and Development (Boston: Houghton-Mifflin, 1979), pp. 144-81.

3. Brazil, Ministério das Minas e Energia, Balanço Energético Nacional, 1977 (Brasília, 1977), pp. 12, 20. (Hereafter cited as Balanço Energético, 1977.)

4. Norman Gall, "Noah's Ark: Energy from Biomass in Brazil," American Universities Field Staff Reports, No. 30 (1978), p. 2.

5. Luciano Martins, Pouvoir et développement économique: formation et évolution des strutures politiques au Brésil (Paris: Editions Anthropos, 1976), pp. 407-43.

6. Kenneth Paul Erickson and Patrick V. Peppe, "Dependent Capitalist Development, U.S. Foreign Policy, and Repression of the Working Class in Chile and Brazil," Latin American Perspectives 3 (Winter 1976): 19-44.

7. Brazil, Fundação Instituto Brasileiro de Geografia e Estatística, Anuário Estatístico do Brasil, 1956, p. 171; 1960, p. 119; 1966, p. 239; and 1977, p. 573.

8. Gall, "Noah's Ark," p. 4.

9. Kenneth Paul Erickson, "The Political Economy of Energy Consumption in Industrial Societies," in Martin O. Heisler and Robert Lawrence, eds., International Energy Policy (Lexington, Mass.: Lexington Books, forthcoming, 1980).

10. Data on cement from O Estado de São Paulo, July 7, 1979; Anuário Estatístico do Brasil, 1963, p. 129, and 1978, p. 453. Data on steel from Werner Baer, The Development of the Brazilian Steel Industry (Nashville: Vanderbilt University Press, 1969), p. 87; Ronald M. Schneider, Brazil: Foreign Policy of a Future World Power (Boulder, Colo.: Westview, 1976), p. 201; Janet D. Henshall and

Richard P. Momsen, Jr. , A Geography of Brazilian Development (London: G. Bell, 1976), pp. 161-66.

11. Henshall and Momsen, A Geography, pp. 134-36, 171, 173; Peter Evans, Dependent Development: The Alliance of Multi-national, State and Local Capital in Brazil (Princeton, N.J.: Princeton University Press, 1979), pp. 228-49.

12. Thomas F. Kelsey, "Brazil," in Donald R. Kelley, ed. , The Energy Crisis and the Environment: An International Perspective (New York: Praeger, 1977), pp. 198-99; Henshall and Momsen, A Geography, p. 202.

13. Balanço Energético, 1977, p. 100.

14. T. L. Sankar, "Alternative Development Strategies with a Low Energy Profile for a Low GNP/Capita Energy-Poor Country: The Case of India," in Leon M. Lindberg, ed. , The Energy Syndrome: Comparing National Responses to the Energy Crisis (Lexington, Mass.: Lexington Books, 1977), pp. 235-36.

15. Sylvia Ann Hewlett, Cruel Dilemmas of Development: Twentieth-Century Brazil (New York: Basic Books, forthcoming, 1980), Chapter 8.

16. Henshall and Momsen, A Geography, p. 110.

17. Peter Seaborn Smith, Oil and Politics in Modern Brazil (Toronto: Macmillan of Canada, 1976), pp. 1-102; Judith Tendler, Electric Power in Brazil: Entrepreneurship in the Public Sector (Cambridge, Mass.: Harvard University Press, 1968), pp. 1-42; The Economist Intelligence Unit, Quarterly Economic Review for Brazil, 1978, No. 1, p. 21.

18. Cited in Pedro Ricardo Dória, Energia no Brasil e dilemas do desenvolvimento (Petrópolis: Vozes, 1976), p. 7.

19. The Economist Intelligence Unit, Quarterly Economic Review for Brazil, 1974, No. 4, pp. 4-8.

20. O Estado de São Paulo, July 7, 1979, p. 24.

21. Text of speech in Jornal do Brasil, July 5, 1979, p. 17.

22. Text of memo in Jornal do Brasil, July 7, 1979, p. 13.

23. Joelmir Beting in Folha de São Paulo, July 22, 1979.

24. Data from ibid. ; O Estado de São Paulo, September 9, 1979, p. 49; Conjuntura Econômica, August 1975, p. 250; August 1979, p. 67.

25. Joelmir Beting in Folha de São Paulo, July 22, 1979.

26. The critic is M. F. Thompson Motta in O Estado de São Paulo, June 17, 1979; editorial in O Estado de São Paulo, May 20, 1979; other data drawn from Joelmir Beting in Folha de São Paulo, July 22, 1979.

27. O Estado de São Paulo, January 3, 1979, p. 20.

28. O Estado de São Paulo, July 7, 1979, p. 24.

29. Jornal do Brasil, July 5, 1979, p. 17; O Globo, July 21, 1979.

30. O Estado de São Paulo, August 18, 1979, p. 25.

31. Joelmir Beting in O Globo, August 1, 1979; O Estado de São Paulo, September 9, 1979, p. 49.
32. Latin America Political Report, August 17, 1979, p. 249; Latin America Economic Report, August 17, 1979, pp. 249, 256; August 24, 1979, p. 257; August 31, 1979, p. 267.
33. O Estado de São Paulo, August 2, 1979, p. 32; August 15, 1979, p. 26.
34. Latin America Economic Report, August 31, 1979, p. 266; Institutional Investor quoted therein.
35. Instituto Brasileiro de Geografia e Estatística, O Brasil em numeros (Rio de Janeiro: IBGE, 1966), p. 64; Smith, Oil and Politics, pp. 131, 135.
36. Anuário Estatístico do Brasil, 1973, p. 202; Frank Niering, Jr., "Brazil: Offshore Search at Crucial State," Petroleum Economist, June 1978, p. 232.
37. Latin America Economic Report, July 20, 1979, p. 220.
38. Petroleum Economist, January 1975, p. 28.
39. Petroleum Economist, June 1979, p. 253.
40. Oil and Gas Journal, June 18, 1979, p. 172.
41. Petroleum Economist, October 1976, p. 393; Platt's Oilgram News, June 29, 1978, p. 3; Oil and Gas Journal, May 8, 1978, p. 140; O Estado de São Paulo, April 6, 1979, p. 27.
42. O Estado de São Paulo, April 4, 1979, p. 27; April 6, 1979, p. 27; July 8, 1979; Petroleum Economist, February 1979, p. 78; March 1979, p. 130; June 1979, p. 251; Platt's Oilgram News, March 29, 1978, p. 5; June 29, 1978, p. 3.
43. Folha de São Paulo, July 22, 1979.
44. Petroleum Economist, January 1977, pp. 29-30.
45. Platt's Oilgram News, September 22, 1978, p. 4; Oil and Gas Journal, November 27, 1978, p. 5; Petroleum Economist, June 1979, p. 251.
46. Folha de São Paulo, July 22, 1979.
47. Balanço Energético, 1978, pp. 60-62.
48. O Estado de São Paulo, April 4, 1979, p. 17; Jornal do Brasil, August 9, 1979.
49. Gall, "Noah's Ark," pp. 9-10.
50. Ibid., pp. 10-11.
51. Folha de São Paulo, February 1, 1979, p. 35; O Estado de São Paulo, August 10, 1979, p. 29.
52. O Estado de São Paulo, August 22, 1979, p. 24.
53. O Estado de São Paulo, August 11, 1979, p. 29; Folha de São Paulo, August 14, 1979, p. 22.
54. O Estado de São Paulo, April 11, 1979.
55. O Globo, May 13, 1979, p. 36; Jornal do Brasil, June 10, 1979; O Estado de São Paulo, August 10, 1979, p. 29; August 22, 1979, p. 24; Folha de São Paulo, August 28, 1979, p. 20.

56. Latin America Economic Report,., June 29, 1979, p. 199; September 14, 1979, p. 286; O Estado de São Paulo, August 2, 1979, p. 34; O Globo, August 28, 1979, p. 25.

57. O Estado de São Paulo, August 29, 1979; Latin America Economic Report, October 12, 1979, p. 315.

58. O Globo, December 12, 1978; Luiz Straunard Pimentel, "The Brazilian Ethanol Program," paper presented to the September 1979 meeting of the American Chemical Society, Washington, D. C. , pp. 9, 26. See also Melvin Calvin, "Petroleum Plantations for Fuel and Materials" (Berkeley: Lawrence Berkeley Laboratory, Preprint LBL-9013, April 1979), p. 7.

59. New York Times, June 14, 1979, p. D-5.

60. O Globo, September 10, 1979, p. 15; O Estado de São Paulo, June 15, 1979; Pimentel, "The Brazilian Ethanol Program," pp. 24-25.

61. José Goldemberg in O Estado de São Paulo, August 26, 1979.

62. Folha de São Paulo, August 14, 1979, p. 22; New York Times, June 14, 1979, p. D-5; Thomas H. Maugh II, "Unlike Money, Diesel Fuel Grows on Trees," Science 206 (October 26, 1979): 436.

63. O Estado de São Paulo, September 16, 1979, p. 51.

64. O Estado de São Paulo, April 6, 1979, p. 27; O Globo, February 11, 1979, p. 31; September 12, 1979, p. 25; Brazilian Business 59, no. 10 (October 1979): 17-18.

65. O Estado de São Paulo, August 26, 1979, p. 51.

66. O Estado de São Paulo, September 20, 1979, p. 35.

67. O Globo, August 8, 1979, p. 31; O Estado de São Paulo, September 9, 1979, p. 36; Business Week, January 9, 1943, pp. 54-59; April 17, 1943, pp. 84-87.

68. Visão, August 1, 1977; O Estado de São Paulo, August 16, 1979, p. 41.

69. Vianna Moog, Bandeirantes and Pioneers (New York: Braziller, 1964), pp. 19-20.

70. Anuário Estatístico do Brasil, 1978 (Rio de Janeiro: IBGE, 1979), p. 498.

71. Latin America Regional Reports: Brazil, November 9, 1979, p. 4.

72. O Globo, October 12, 1977; Norman Gall, "Atoms for Brazil, Dangers for All," Foreign Policy, No. 23 (Summer 1976), p. 194; Schneider, Brazil: Foreign Policy, p. 115.

73. On the international aspects of the Brazilian-West German accord, see Gall, "Atoms for Brazil," pp. 155-201; H. Jon Rosenbaum, "Brazil's Nuclear Aspirations," in Onkar Marwah and Ann Schulz, eds. , Nuclear Proliferation and the Near-Nuclear Countries (Cambridge, Mass.: Ballinger, 1975), pp. 255-77; Edward Wonder, "Nuclear Commerce and Nuclear Proliferation: Germany and Brazil, 1975," Orbis 21, no. 2 (Summer 1977): 277-306; Marshall C. Eakin,

266 / NATIONAL ENERGY PROFILES

266 / NATIONAL ENERGY PROFILES

"The Politics of Energy: The 1975 Brazilian-West German Accord" (University of California at Los Angeles, April 1979).

74. Allen L. Hammond, "Brazil's Nuclear Program: Carter's Non-proliferation Policy Backfires," Science 195 (February 10, 1977): 657-59; Schneider, Brazil: Foreign Policy, pp. 47-50, quote, p. 48. Gall, "Atoms for Brazil," pp. 181-82.

75. O Estado de São Paulo, April 8, 1979, pp. 7-9.

76. O Estado de São Paulo, August 23, 1979, pp. 6, 21; August 25, 1979, p. 5; September 9, 1979, p. 21; September 20, 1979, p. 6; September 29, 1979, p. 6; Jornal do Brasil, May 10, 1979. See also José Goldemberg, Energia nuclear no Brasil: as origens das decisões (São Paulo: Editora Hucitec, 1978).

77. O Estado de São Paulo, April 6, 1979, p. 9.

78. O Estado de São Paulo, September 12, 1979, p. 5.

79. O Estado de São Paulo, April 17, 1979, p. 5.

80. Folha de São Paulo, October 7, 1979, p. 7.

81. Gerald Garvey, Nuclear Power and Social Planning: The City of the Second Sun (Lexington, Mass.: Lexington Books, 1977), p. xix.

82. José M. Miccolis, "Alternative Energy Technologies in Brazil," in Norman L. Brown, ed., Renewable Energy Resources and Rural Applications in the Developing World (Boulder, Colo.: Westview, 1978), pp. 56-60; quote, pp. 59-60.

83. Ibid., pp. 58-59; O Globo, March 13, 1977.

84. Miccolis, "Alternative Energy Technologies in Brazil," pp. 71-72; José Goldemberg, "Brazil: Energy Options and Current Outlook," Science 200 (April 14, 1978): 163; O Globo, August 26, 1979, pp. 26-28.

85. Data on rail and maritime transport from Conjuntura Econômica, April 1978, pp. 88-91; May 1978, pp. 67-70; May 1979, p. 65; August 1979, pp. 71-72; The Economist Intelligence Unit, Quarterly Economic Review for Brazil, 1974, No. 4, p. 8; O Estado de São Paulo, April 5, 1979, p. 35; Latin America Economic Report, October 19, 1979, p. 327.

86. Goldemberg, "Brazil: Energy Options," p. 164; see also Richard G. Stein, Architecture and Energy (Garden City, N.Y.: Anchor, 1977), pp. 23-47, 291-93.

87. The following account is drawn from Selcher, Brazil's Multilateral Relations, pp. 108-17.

88. Ibid., p. 117.

89. Platt's Oilgram News, May 23, 1979, pp. 3-4.

90. Balance-of-trade data from International Financial Statistics 19 (April 1966): 62; 24 (January 1971): 64; 28 (October 1975): 72-73; 32 (July 1979): 80-81; Latin America Weekly Report, November 30, 1979, p. 57; Latin America Regional Reports: Brazil, January 4, 1980, p. 8.

91. Schneider, Brazil: Foreign Policy, pp. 107-25.

92. Peter Seaborn Smith, "Brazilian Oil: From Myth to Reality?" Inter-American Economic Affairs 30 (Spring 1977): 49-50; O Estado de São Paulo, September 22, 1979, p. 26.

93. Petroleum Economist, October 1977, p. 400; Platt's Oilgram News, June 12, 1978, p. 5; Middle East Economic Survey, June 12, 1978, p. 5.

94. Petroleum Economist, October 1978, p. 442; Platt's Oilgram News, May 23, 1979, p. 304; O Estado de São Paulo, September 22, 1979, p. 26.

95. Petroleum Economist, October 1977, p. 400; Platt's Oilgram News, September 5, 1978, p. 2; Smith, "Brazilian Oil," p. 49; O Globo, August 21, 1979, p. 18.

96. The Economist Intelligence Unit, Quarterly Economic Review for Brazil, 1979, No. 1, p. 20.

97. O Estado de São Paulo, May 27, 1979, pp. 6-8; Latin America Political Report, March 4, 1977, p. 66; Latin America Regional Reports: Brazil, November 9, 1979, p. 3.

98. Anuário Estatistico do Brasil, 1976 (Rio de Janeiro: IBGE 1977), pp. 246, 264.

99. Petroleum Economist, June 1978, p. 235; Platt's Oilgram News, September 25, 1978, p. 3; O Estado de São Paulo, August 22, 1979.

100. Petroleum Intelligence Weekly, July 24, 1978, p. 6; The Economist Intelligence Unit, Quarterly Economic Review for Brazil, 1979, No. 2, p. 23.

101. The Economist Intelligence Unit, Quarterly Economic Review for Brazil, 1979, No. 1, pp. 1, 20; Latin America Weekly Report, November 30, 1979, p. 58; Selcher, Brazil's Multilateral Relations, pp. 222-24.

102. Stanley E. Hilton, "Brazil's Bilateral Relations" (ms., January 1976), pp. V6-V18, quote, p. V11; Jornal do Brasil, February 19, 1976, pp. 1, 23.

103. Hilton, "Brazil's Bilateral Relations," pp. V16-V18; Schneider, Brazil: Foreign Policy, pp. 195-220.

104. Platt's Oilgram News, July 7, 1978, p. 4; November 14, 1978, p. 3; January 2, 1979, p. 4; Petroleum Intelligence Weekly, November 20, 1978, p. 7; Oil and Gas Journal, November 20, 1978, p. 5.

105. Daniel Fretes Ventres, "Evolución y Perspectivas de la estructura social y económica en Paraguay," Estudios Paraguayos 3 (October 1975): 5-30, as cited in Riordan Roett and Amparo Menéndez-Carrión, "Authoritarian Paraguay: The Personalist Tradition," in Wiarda and Kline, Latin American Politics and Development, p. 142.

106. Lewis A. Tambs, "The Influence of Brazil: A Historical

and Geopolitical Survey," paper presented to a conference on "Modern Day Bolivia," Center for Latin American Studies, Arizona State University, March 14-17, 1978, pp. 60-75.

107. Platt's Oilgram News, October 31, 1978, p. 4; Petroleum Intelligence Weekly, November 13, 1978, p. 10.

108. Platt's Oilgram News, December 14, 1978, p. 3; Oil and Gas Journal, January 30, 1978, p. 5.

109. Petroleum Intelligence Weekly, May 8, 1978, p. 9; Platt's Oilgram News, April 19, 1978, p. 2; April 24, 1978, p. 4; November 27, 1978, p. 3; November 28, 1978, p. 6; Latin America Weekly Report, November 16, 1979, p. 29.

110. Latin America Economic Report, August 3, 1979, p. 237; O Estado de São Paulo, November 25, 1979, p. 70.

111. Isto É, September 27, 1978, p. 81, citing an interview in Der Spiegel.

112. O Estado de São Paulo, August 26, 1979, p. 51; report cited therein.

113. O Globo, September 12, 1979, p. 25.

114. Pimentel, "The Brazilian Ethanol Program," p. 15; O Globo, April 29, 1979, p. 38.

115. O Globo, June 11, 1979.

116. Melvin Calvin, "Petroleum Plantations," in Richard R. Hautala, R. Bruce King, and Charles Kutal, eds., Solar Energy: Chemical Conversion and Storage (Clifton, N.J.: Humana Press, 1979), pp. 5-9.

117. Pimentel, "The Brazilian Ethanol Program," p. 25; O Globo, April 29, 1979, p. 38; May 13, 1979, p. 36; June 25, 1979; O Estado de São Paulo, August 9, 1979, p. 46; August 10, 1979, p. 30.

118. O Globo, August 28, 1977; August 8, 1979; O Estado de São Paulo, May 24, 1979.

119. Mark Sonnenblick, "Brazil Embarks on Alternate-Energy-Based 'War Economy,'" Executive Intelligence Review, August 21, 1979, p. 48.

120. Rogério Cézar de Cerqueira Leite, in O Estado de São Paulo, December 10, 1978, Cultural Supplement, p. 7. Cerqueira Leite, a scientist involved in developing alternate-energy projects, would surely not exploit the Amazon's resources irresponsibly, but real doubts arise about the ability or willingness of private-sector firewood and charcoal companies to submit their pursuit of profit to the dictates of sound ecological management.

121. Robert J. A. Goodland and Howard S. Irwin, Amazon Jungle: Green Hell to Red Desert? (New York: Elsevier, 1975), pp. vi, 28, 47; first quote from preface by Harald Sioli.

122. Robert J. A. Goodland, "Environmental Assessment of the Tucuruí Hydroelectric Project, Rio Tocantins, Amazônia" (Bra-

síia: Centrais Elétricas do Norte do Brasil, S. A. , Eletronorte, 1977), p. 205.

123. Shelton H. Davis, <u>Victims of the Miracle: Development</u> <u>and the Indians of Brazil</u> (New York: Cambridge University Press, 1977), pp. 135-57.

124. Paul L. Aspelin and Sílvio Coelho dos Santos, "Indian Areas threatened by Hydroelectric Projects in Brazil" (Florianopolis: Federal University of Santa Catarina, April 1979), mimeo., p. 40.

125. Kelsey, "Brazil," pp. 211-14; <u>Latin America Economic</u> <u>Report</u>, August 10, 1979, p. 243.

126. <u>O Globo</u>, August 1, 1979, p. 24.

127. See Caryl P. Haskins, <u>The Scientific Revolution and World</u> <u>Politics</u> (New York: Harper and Row, 1964), pp. 10-12, 29-46.

6

MEXICO
Laura Regina Rosenbaum Randall

THE PATTERN OF ENERGY USE

Through much of Mexico's history, human and animal power
were the sources of energy for both labor and the transportation of
people and products, while heat was provided by wood and charcoal.
Transformation of the economy in more recent times has seen the
limited use of water power and coal, limited because Mexico has fewer rivers suitable for power generation than most industrial nations
and coal supplies that are of low quality and high cost. As a result,
the economy of Mexico in the twentieth century has been dependent
increasingly on the use of petroleum and is expected to make use of
atomic and other nontraditional sources of energy. 1
The first commercially important oil production is attributed
to Edward Doheny, who brought in his first well in 1901 at a level of
10,000 barrels. Weetman Pearson started production in 1906 with an
initial output of 502,000 barrels for that year. Despite fighting entailed
by the Mexican Revolution, production climbed steadily to 193.4 million barrels in 1921, a level not to be reached again until 1974. The
output of 1921 made Mexico the world's second largest producer, an
achievement that was not sustained. The Pearson holdings were sold
to Royal Dutch in 1923 and those of Doheny went to Standard Oil of
Indiana in 1925, events accompanied by the entrance of other foreign
producers. The decline of oil production was blamed by oil company
economists on adverse policies of the Mexican government, world

---

Research for this chapter was done mostly by mid-1979.

depression, and salt deposits in some of the oil fields. Mexican economists claim that the new owners held Mexican oil lands in reserve
while they exploited more profitable Venezuelan holdings. More recently it has been claimed that early overproduction ruined the oil
fields. Although new fields were brought into production in 1933, labor difficulties and the extraordinary discourtesy of the companies
resulted in the nationalization of most foreign-owned oil property in
1938. The companies unaffected by nationalization produced less than
10 percent of the oil and were purchased by the government after
World War II.

Since 1938, Mexican oil development has been carried out predominantly by Petróleos Méxicanos (Pemex)[2] The precedent set by
Pemex for government involvement in energy matters has encompassed
the electricity sector, which was nationalized in 1960, and a new
draft law, published in January 1979, reaffirmed government ownership of uranium and reserved to the nation its development and use.
Government, therefore, exercises substantial control over present
and future sources of energy.

The degree of government control over energy production has
increased as the traditional sector of the economy has declined. This
traditional sector relies on firewood, charcoal, and dung. In 1940
the consumption of these fuels was 15 percent of total energy use; the
share fell to 8 percent in 1950 and to 4 percent in 1972. In 1960 an
estimated 63 percent of the population used traditional fuels; by 1970
the figure had fallen to 41 percent, suggesting that many in this group
were using commercial energy. [3]

Within the commercial sector, the relative importance of Mexico's primary energy sources in 1972 can be described as follows: 90. 6
percent came from hydrocarbons, 5. 3 percent from coal, and 4. 1 percent from hydraulic sources. [4] From 1965 to 1975 hydrocarbons provided 88. 3 percent of commercial energy, growing annually at 6. 9
percent, with coal providing 4. 8 percent, growing annually at 10. 6
percent. Electric power provided 6. 8 percent of commercial energy,
growing annually at 11. 3 percent. Within the hydrocarbon group, avia-
tion and diesel fuel use grew well above average rates, while much
coal was absorbed by the steel industry. [5] The use of electric power
underwent a striking shift between 1962 and 1975. Nationalization of
electricity was followed by a rapid residential increase in its use,
which rose from 16 to 27 percent of gross energy production. From
1965 to 1975, above-average rates of growth for electric power occurred in the transport, government, and domestic sectors. [6]

Shifts in the supply and use of energy in Mexico have not been
solely a result of free market forces, but have been influenced strongly
by government policy. Indeed, the development of an adequate energy
policy is perhaps the foremost problem facing the country, not only

because sufficient energy supplies are crucial to economic development, but also because the energy sector is becoming a significant part of the economy. It contributed over 7 percent of the gross domestic product, 23 percent of the economic growth rate, and 31.8 percent of projected government expenditures in 1978.[7] Moreover, energy policy is fraught with domestic problems, such as the recent operating losses of the energy sector due to underpricing and high cost.

An adequate energy policy is also of importance to the United States, for Mexico, which did not export crude oil from 1968 to 1974, provided 5.8 percent of U.S. oil imports toward the close of 1978 and is likely to supply increasing amounts of oil, natural gas, and electric power to satisfy U.S. and world energy needs in the mid-1980s.[8] For example, Mexican average daily production in 1978 is estimated at 1.4 million barrels, with exports of 300,000 to 400,000 barrels of crude a day.[9] Finally, the size of Mexican energy reserves, which are said to be as great as those of Saudi Arabia, make it likely that Mexican energy policy will affect world energy output and pricing decisions in the near future.[10]

The upshot is that energy policy is too vital a matter to be left to the energy-producing sector. Instead, policy is derived from the general development goals of the nation and is revealed in the actions of many government organs. In practice, this means that decisions on prices, exports, and production must be approved by the president; such decisions are not left to the "free play of market forces," nor are they reached by any one ministry or agency of the government, such as the Comisión de Energéticos.[11]

The significance of energy decisions is suggested by the size of Mexican energy reserves. While estimates vary, they appear to indicate sufficient reserves to last into the twenty-first century, providing that actual production justifies the reserve estimates (see Table 6.1). As of the end of 1978 there were an estimated 29 billion barrels of proved oil reserves, 20 trillion cubic feet of natural gas, 8 billion tons of coal, 83 million kilowatt hours a year of hydroelectric energy, 10,000 tons of uranium, 13 billion kilowatt hours a year of geothermal energy, as well as potential reserves well in excess of these amounts. Solar potential is as yet unknown.

The relative importance of each source of energy for the generation of electricity is shown in Table 6.2. Installed electric capacity in 1977 was 11,821 MW. Of this amount, 61 percent was thermoelectric, which relied on diesel and fuel oil and gas; 39 percent was hydroelectric; 0.6 percent was geothermal; and 0.3 percent was thermoelectric, which relied on coal. Yet the relative importance of these sources is expected to shift, reflecting both resource availability and projected costs per kilowatt hour for each energy source. Nuclear power advo-

TABLE 6.1

Estimates of Proved and Potential Reserves in Mexico, 1978

Category	Unit of Measurement	Size of Proved Reserves[a]	Adequate to Year	Potential Reserves[b]
Petroleum[c]	10^9 barrels	40	2025	200
Natural gas	10^{12} cubic feet	20	2020	
Coal[d]	10^9 tons	8	centuries	
Hydroelectric	10^9 kWh/year	83	60 will have been used by 2000	
Uranium[e]	10^3 tons	10	1990+	100
Geothermal	10^9 kWh/year	13		100
Solar			Not yet known for nontraditional sources of energy	

[a] 40×10^9.

[b] 200×10^9.

[c] Pemex claims most recently that it has 60 years of reserves for oil consumption at the present rate.

[d] In Coahuila Province, the reserve potential of coal, as it comes from the mine, reached 8.1 billion tons, of which 1.4 billion correspond to coal resources in veins greater than 1.4 meters, 324 million to veins less than 1.4 meters, 462 million to probable, and the rest, 6.0 million, to inferior or geologic resources. (Secretaria del Patrimonio Nacional, Memoria, 75/76, p. 37.) Proved reserve estimates increased to 2 billion tons in 1978. Of these reserves, 300 million tons will be used at Río Escondido. 1.5 billion tons are reserved for the siderurgy industry, and 200 are of a quality that, for the moment, prevents their use for generation. (Boletín Energéticos, March 1978, p. 12.)

[e] Of the 8,000-ton 1976 estimates, "uranium reserves are some 6,123 tons, the remainder, 1,850 tons, being subject to confirmation; . . ." the on-site tonnage which the nation can actually count on for its short-term energy development policy is 4,750 tons of uranium, which, as a result of recuperation factors, will be substantially reduced; however, it is estimated that in the future the nation can count on a potential of great magnitude, if an aggressive exploration program is developed. (Secretaria de Patrimonio Nacional, Memoria 75/76, p. 37.) Morales Amado places 1978 reserves at 10,000 tons. Interview, July 1978.

Sources: Excelsior, September 2, 1978; Boletín Energéticos, March 1978; Boletín IIE, July 1977; Secretaria de Patrimonio Nacional, Memoria 75/76; Arnulfo Morales Amado, "Uranio," El Economista Mexicano 12, no. 2 (March-April 1978); Latin American Political Report, October 20, 1978; New York Times, January 6, 1979.

TABLE 6.2

Two Estimates of Sources of Electricity in Mexico, 1977–2000 (installed effective capacity)

Type of Plant	1977		1982		1986		2000	
	MW	Percent	MW	Percent	MW	Percent	MW	Percent
Hydroelectric	4,547	39	6,587	37	8,555	35	21,900	27
Thermoelectric (fuel oil, diesel and gas)	7,169	61	9,795	55	11,576	48	23,000	28
Thermoelectric (coal)	30	0.3	630	4	2,430	10	9,500	12
Geothermal	75	0.6	180	1	290	1	5,400	7
Nuclear-electric	—	—	654	4	1,308	5	21,400	26
Total	11,821	100	17,846	100	24,158	100	81,200	100

	1980		1990		2000	
	MW	Percent	MW	Percent	MW	Percent
Hydroelectric	6,900	43	8,600	22	10,600	12
Conventional	8,600	53	9,100	23	9,400	11
Nuclear	670	4	21,600	55	68,000	77
Total	16,170	100	39,300	100	88,000	100

Sources: Comisión Federal de Electricidad, Gerencia General de Estudios e Ingeniería Preliminar, cited in Energéticos, Boletín Informativo del Sector Energético 2, no. 3 (March 1978): 4; Arnulfo Morales Amado, "Uranio," El Economista Mexicano 12, no. 2 (March–April 1978): 52.

274

cates argue that heavy water nuclear plants have a cost advantage compared to hydrocarbons in the generation of electric power. Since the cost of decommissioning plants is not included, this claim has been challenged. Coal may turn out to be Mexico's cheapest source of energy. Moreover, hydroelectric and geothermal power are often cheap when available.

In addition to traditional factors bearing on the choice of energy sources, many Mexicans believe that in the 1980s the United States will use economic, political, or even military means to force Mexico to sell its oil in far greater volume than Mexico might wish. Thus it is held that Mexico must develop new reserves of existing energy sources to secure its own energy supplies and to achieve a measure of flexibility.

The use of Mexico's varied energy sources is an indication of its economic development. Thus in 1977 the distribution of energy use reflected both economic structure and pricing policy (the latter is discussed below).

Agriculture used only 0.8 percent of the nation's energy while producing 9.1 percent of the GDP, which reflects the limited mechanization of agriculture, a conclusion open to slight modification if one includes energy-related products such as chemical fertilizer.

Manufacturing, which has benefited the most from government pricing policy, absorbed 25.8 percent of the energy while contributing 28.8 percent of GDP. From 1960 to 1973 the industrial sector increased its use of energy at an annual rate of 7.0 percent, while its output went up at a rate of 8.4 percent. Within this sector, the rates of growth for subgroups may be tabulated as follows, for energy use and output, respectively. [12]

Chemicals	16.0 and 12.2 percent
Food products	11.9 and 5.3
Automotive industries	10.7 and 14.6
Construction	7.4 and 11.0
Other manufactures	7.7 and 9.1
Other industries	7.4 and 8.1
Beverages and tobacco	7.0 and 7.0
Basic metals	7.0 and 6.3

Transport required large energy inputs for which few substitutes were available and benefited from substantial gasoline subsidies; it accounted for 30.4 percent of energy use and, if one adds communications, 4.1 percent of the GDP. The generation of electricity consumed 16.7 percent of the energy pie, including amounts lost in the processes of conversion and transmission, petroleum production used 13.4 percent, the residential sector 8.2 percent, and 4.7 percent went to non-energy and other uses. [13]

As expected, the most rapidly growing regions of the country accounted for 56.9 percent of the energy used in 1975, regions with below-average rates of growth for 16.2 percent, while the rest of the country accounted for the balance. On the other hand, the least developed regions had the highest rates of growth in energy use, which reflects the low base against which their energy growth rates were measured, while the most rapidly growing regions had the lowest rate.[14]

With regard to the impact of economic structure on energy use, as Mexico industrialized, the amount of energy required to produce a unit of GDP increased by 14 percent from 1965 to 1977. The ratio of energy to GDP was typical of a country at Mexico's level of income and the 1.12 income elasticity of demand for energy during those years was the same as the Latin American average for 1950-74. Yet it would be misleading to judge the efficiency of energy use from these figures, since energy/GDP ratios reflect geography, composition of output, and the price of energy.

Mexico's mountainous terrain entails greater energy use than is the case with flat countries nearer to sea level. Also, the relationship between industrial structure and the use of energy is complex. On the surface the structure of industry is related to the composition and levels of energy use, but these in turn are largely reflections of government economic development policy. The most visible of government policies is subsidized pricing for energy and power. The subsidy on refined petroleum products, excluding natural gas, is 146 percent; other subsidies are 10 percent on basic petrochemicals, 526.5 percent of Mexican sales receipts on natural gas, all for a total subsidy of 75 billion pesos. * The most generous subsidy is for natural gas, followed by liquid gas, paraffin, fuel oil, diesel, asphalt, and gasoline among refined products.

The petrochemical subsidy pattern is varied, some products selling for more and others for less than world prices. The overall hydrocarbon subsidy program is for the most part helpful to those who own cars, buy modern products, or own industries. One notable exception is the larger-than-average subsidy on kerosene, which is sold at 9 cents (U.S.) a gallon, compared to 62 cents in the United States. The cost of the subsidy program is borne by those affected by inflation, which hits the most poor the hardest.

The regional pattern for the sharpest fall in real wages was Yucatan, Guadalajara, Comarea, Lagunera, Guerrero, Distrito Fed-

---

*Equal to 4.5 percent of the 1977 GDP. Subsidies vary among the products.

eral, and Guanajuato. If Pemex were to control its costs, it could begin to make an appropriate allowance for the depletion of natural resources. If it were to end its subsidy program, it could finance its own expansion, instead of depending upon a combination of foreign borrowing and inflationary domestic financing. All of this influences the distribution of income, demand, and derived demand for energy.

Until recently the electric power sector, like the petroleum sector, had to "subsidize the nation" with low electricity rates, which were intended to spur industrialization and raise employment levels. Relationships underlying this policy of underpricing electricity to subsidize industrial development are complex. Electricity accounts for only a small share of industrial costs and its price is not likely to be crucial in determining investment patterns. However, it does influence the choice between capital-intensive and labor-intensive equipment. The lower the price of electricity, the more likely that energy-rich capital-intensive equipment will be selected. The demand for electricity would be greater than full-cost pricing warranted, and would lead to greater than necessary depletion of nonrenewable fuel sources. Furthermore, pricing subsidies limit the ability of the electric power sector to finance its own expansion. In 1978 rate increases were granted in an attempt to place electric power on a self-financing basis.

The question of rate structure is complicated by regional variations in rates. Within each region tariffs for electric power are fixed for each type of customer, the average regional rate reflecting the composition of economic activity. The widening gap between regional rates in 1976, combined with the fact that differences in rates among regions did not strongly reflect differences in average generating costs, led the Federal Electricity Commission (CFE) to reduce the disparity in rates charged from 61 percent of the average in 1976 to 43 percent in 1977. Similarly, if we exclude the rate for temporary electric service, the difference in rates for varied uses was reduced from 154 to 138 percent of the average rate. [15] To some extent the regional pattern appears to encourage decentralization of development. The long-awaited regional development incentives were announced in March 1979 and included a 30 percent discount in electricity, gas, fuel oil, and petrochemical prices for firms meeting specified development goals.

The structure of rates according to category of user has been protested by those who consume small amounts of energy; that is, small-scale industry and households end up "subsidizing" large-scale industry by paying higher rates. On the other hand, they also end up "subsidizing" agricultural production and corn mills, which results in lower food prices when rates alone are considered. When the total volume of electric sales is examined, however, the largest "subsidy" is received by the 400 largest firms, irrigation projects, and house-

holds in areas with hot summers. [16] Electricity subsidies have been
extended to powerful established industries rather than to potential
new consumers in rural areas (industry in the United States claims
that changing the rate structure would increase small-user demand
so much that this would offset reductions in use by large industry).
Rural areas are likely to benefit from the development of nontraditional energy and sources of electricity such as solar, applications
of new technology for small-scale use of hydropower, and the use of
combustion engines to generate electricity.

In the absence of strong, continuing, and effective regulation
by the government, questions of pollution and energy-saving devices
have not seriously influenced the level and distribution of energy use.

DOMESTIC ENERGY POLICY

Mexican economic growth policy is shaped by domestic needs
and by Mexico's relations with the rest of the world, above all with
the United States. The complex interactions of industrial, population,
and energy growth are understood, debated, and partly underlie the
bargaining that goes on over energy questions. The government has
manipulated newly discovered oil deposits to obtain enough foreign
funds so that it might be able to announce future limits on energy
growth rates, both to preserve Mexican resources for future generations and to control the inflation and undesirable surge of imports
that might be triggered by Mexico's oil wealth. It is well known in
Mexico that some countries, from sixteenth-century Spain to modern
Iran and Venezuela, have had difficulty transforming great influxes
of foreign exchange into productive goods. Within the context of its
development policy, the federal government, at least in theory, is
taking first steps to achieve a coordinated energy policy.

In the end, decisions rest with the president. Directly under
the president is the Secretaria de Patrimonio y Fomento de Industria
(SEPAFIN), which sets prices and serves as a consulting commission.
Within SEPAFIN, the Subsecretariat of Mines and Energy contains in
turn the Comisión de Energéticos, which is composed of the directors
of SEPAFIN, the ministries of Commerce, Agriculture, and Hydraulic
Resources, of Pemex, the Federal Electricity Commission, the National Nuclear Energy Institute (INEN), and the National Council of
Science and Technology (CONACYT). The Comisión de Energéticos
is served by an executive and his technical staff.

The strength of the various ministries influences energy policy,
which is coordinated by the Comisión de Energéticos. The analysis of
investment decisions is the responsibility of SEPAFIN, but they have
to be negotiated with the Secretaría de Programación y Presupuesto

(Secretariet of Programming and Budget). Energy policy in Mexico
emphasizes applied design and engineering problems because their
solution is essential to effective implementation of energy policy in
a developing nation. The CFE and Pemex, therefore, have established
technical institutes—the Instituto de Investigaciones Eléctricas (IIE)
and the Instituto Mexicano del Petróleo (IMP)—which help with the
analysis and solution of problems related to electricity and petroleum.
Similarly, INEN is being reorganized and the creation of coal and
solar institutes is under consideration. The pattern of decision making is in flux, for the Comisión de Energéticos wants to formulate
rather than just coordinate policy. Since energy policy, broadly conceived, affects operational details of the entire economy, the Comisión de Energéticos probably cannot determine energy policy unless
it is converted into the most important planning agency of the Mexican government. Thus energy policy with respect to oil affects petrochemicals as well. Arturo del Castillo of IMP, which advises on oil
and petrochemicals, has stated that no other nation has an energy
policy, so it is odd to expect Mexico to have one. [17]

Basic questions concerning the scope, criteria, priorities, and
techniques of Mexican energy policy are determined against this de-
cision-making background and reflect the concern of Jose López Portillo's administration with government reform, foreign debt, and Mexico's limited ability to absorb new resources in ways consistent with
an emphasis on development of small, decentralized projects capable
of generating employment. The two broadest areas to which energy
policy decisions are applied are sources of energy and the demand
for various sources of energy, by both the electric power sector and
other users.

Decision making in energy policy was strongly affected by the
1973 oil embargo and its aftermath, which were different for Mexico
than for most developing nations. The embargo had two strikingly different effects. On the one hand, the price rise, which came while
Mexico was importing oil, damaged the Mexican balance of payments
and contributed to the 1976 devaluation and subsequent recession. On
the other hand, the price rise also increased the value of Mexican oil
reserves, and thus made possible the massive inflow of foreign funds
subsequently attracted by the new wealth of petroleum. At the same
time, there were fears that the United States would succumb to the
temptation to pressure Mexico for oil, which has led to an effort to
develop substitutes and to forge links with other nations to strengthen
Mexico's bargaining position with the United States. However, the international price rise had a limited impact on the domestic consumption of oil, since the Portillo administration has subsidized the internal price of oil and its derivatives. Nevertheless, it is likely that
Mexican domestic energy prices will rise eventually to international

levels. The probable effects of future oil price rises on economic activity and the cost of products have not yet been completely analyzed.

The oil embargo led to complex relations between Mexico and other producing nations. Mexico has not joined OPEC; as a result, no harm was done by the U. S. Trade Act of 1974, which excluded OPEC countries from tariff preferences contained in the act. Mexico also has ignored suggestions of a producer's alliance from Venezuela. On the other hand, Mexico has followed OPEC pricing policies and has announced a limit to its oil production after 1980.

The OPEC price rise also has permitted Mexico to expand the range of oil sales. Crude oil and its derivatives are sold to 14 countries and agreements for the refining of Mexican oil have been signed with Japan and Spain. Deep-water oil terminals to accommodate supertankers, reduce oil transport costs, and increase the number of customers for Mexican oil are being built at Pajaritos on the Gulf Coast and at Salina Cruz on the Pacific Coast. Nonetheless, the United States is still Mexico's chief oil trading partner and purchases 90 percent of its oil exports. Swap agreements are being discussed so that Mexican oil might be processed in Caribbean refineries and sold to the United States in exchange for Middle Eastern oil, which would then be refined and sold to Mexico's European customers. [18]

The energy embargo has led to limited conservation and recycling measures, primarily because Mexico wishes to promote both rapid economic development and a brisk export trade. It is feared that measures adding to production costs will result in reduced investment, higher sales prices, and lower export earnings. The most ingenious conservation measure was the 1977 phase-out of the long midday lunch break for 80 percent of the government employees in the Distrito Federal (Federal District), which cut the use of transportation in half and saved a good deal of energy. [19] On the other hand, the government is unwilling to promulgate unenforceable regulations in the private sector, such as those that would require use of the most energy-efficient system for each task to be performed. [20] Instead, pricing policy is used to stimulate efficiency and frugality. Conservation is discouraged because oil and energy prices are below international levels.

The government is not unaware of the need to conserve energy. The flaring of natural gas, which was worth 126 billion pesos at export prices from 1956 to 1976, was supposed to come to an end in March 1979, but did not, because of unexpectedly high gas-to-oil ratios. [21] The construction of a gas pipeline to transport gas produced in conjunction with oil permits complete utilization of gas from this source and the removal from production of wells that produce gas but not oil. Research in the coal industry is being given to the use of by-products of energy production such as ash. [22] The geothermal industry is pursuing a method for the use of geothermal fluid to produce potassium fertilizer both for domestic consumption and for export. [23] Solar energy development will make possible the better use of existing resources. Atomic generation of electric power does not yet permit recycling; enriched uranium yields plutonium, which must be returned to those who provide enrichment services.

Conservation does have a role in decisions of government about the uses and production of energy. In order to reduce problems of shifting power from one part of the electric system to another, Mexico recently spent 500 million pesos to unify cycles (at 60 cycles, 110 volts), and established fines for the installation of electric generators for private use. [24] Pemex found recently that some pumps could be moved by steam turbines in place of electric motors, and it plans to design and install steam equipment. [25] The fact that electric traction is twice as efficient as diesel-electric traction has led to an increased emphasis on the former. [26] Moreover, IIE managed to locate sources of the greatest loss in the electric system, such as the breakdown of distribution transformers, which had resulted in a loss of 436 million pesos in equipment investment and 119 million pesos annually to produce the energy wasted in breakdowns. Several projects have been undertaken to aid in design, manufacture, and quality control of transformers. [27] Trolleybuses provide another example, where 10 percent of energy was lost in the form of heat, costing about 5 million pesos in the Distrito Federal alone. IIE is developing a new system of speed control intended to avoid such losses. [28] In general, it can be expected that continuing government attention to design and quality control will improve energy efficiency and promote conservation.

With regard to domestic energy sources and their development, Mexico will, in the short run, conserve and develop its traditional energy supplies and explore new methods of producing energy. In the long run, when the view shifts from 1984 to the twenty-first century, when known oil and gas reserves will be virtually depleted, Mexico can be expected to rely on other sources of energy. In both periods, however, the most important source of energy is petroleum.

The most prominent aspect of Mexican energy policy, therefore, is the development of the oil and petrochemical industries. Crude oil provided 65 percent of the country's primary energy in 1977. [29] It supplied feedstock to the petrochemical industry, accounted for 22.4 percent of goods exported, and 8.7 percent of foreign exchange income. [30] Oil and petrochemicals accounted for 6.7 percent of GDP, 22 percent of its growth rate, and 22.5 percent of projected federal sectoral expenditures for 1978. [31] In the first half of 1978, a sum of 126.2 billion pesos was spent on energy out of total government spending amounting to 913.1 billion pesos. [32] (Demand for hydrocarbons, coal, and electricity from 1965 to 1985 is shown in Table 6.3.)

TABLE 6.3

Energy and Electricity Demand in Mexico, 1965–85
(cubic meters of crude petroleum equivalent; See note 5)

	Hydrocarbon Demand	
Year	For Energy Use Not Including Electricity Generation	For Production of Electric Energy
1965	23,947,003	2,489,000
1975	48,081,005	9,892,000
1985	85,178,355	25,805,000

	Per Capita Demand			
	Hydrocarbons	Electric Energy	Coal	Total
1965	0.588	0.032	0.025	0.645
1975	0.799	0.062	0.043	0.904
1985	1.020	0.136	0.092	1.248

Source: Instituto Mexicano del Petróleo, Energéticos, Demanda Regional Analisis y Perspectivas, 1977, pp. 77, 207, and C.1.

Still, the size of the oil and petrochemical industries does not suggest fully its historical importance. Growth of the oil industry has been necessary to provide electric power for industrial development, to asphalt roads, and to produce fertilizer for the expansion of agriculture. When oil growth has been rapid, anticipated oil exports have enabled Mexico to maintain the exchange rate and to obtain favorable interest rates in international markets. When oil growth has been slow, as in the late 1960s and 1970s, imports of oil contributed to Mexico's balance-of-payments deficit and to subsequent devaluation. Thus Mexican energy policy, as it affects oil, has an impact on a wider range of issues than the provision of electric and motor power. It must take into account the size of oil reserves, the profitability of various forms of oil development, and the implications of oil development for the nation's finances, employment level, and international relations.

The spectacular increase in Mexican oil reserves is based on the giant oil and gas fields in the Reforma area of Chiapas-Tabasco and in the Gulf of Mexico. Although some North American oil geologists had been rumoring their existence in 1973, there was no official announcement of new reserves until President Portillo took office in 1976. The actual size of reserves is somewhat speculative, as Pemex

was far too busy bringing in the oil for the first few years to find time
for the detailed statistical work necessary to estimate the profile of
each oil deposit and build up reserve estimates. Nevertheless, both
Mexican and foreign estimates have driven up the proved reserves to
16 billion barrels of crude oil and gas liquids (estimate is based on
wells in production and includes the possibility of secondary and tertiary recovery) at the end of 1977. In January 1979 President Portillo
announced proved reserves of 40 billion barrels, probable reserves
of 44.6 billion, and potential reserves (including proved and probable
reserves) of 200 billion barrels of oil and 20 trillion cubic feet of gas,
which, if proven, may equal the holdings of Saudi Arabia. [33]

Oil reserves are said to be essential for faith in Mexico's future.
Consequently oil production is seen as one of the main features of Mexico's current development plans. One of the most pressing questions
with which Mexico, and Pemex, is faced is just how much oil to produce. According to Jorgé Díaz Serrano, director-general of Pemex,
the basic criteria for oil development are:

(1) satisfying immediate, mid-term and future domestic
demand; (2) rational use and integral development of all
hydrocarbons; and (3) exportation of surpluses, the latter
with a view to financing an accessible price for domestic
consumption and also to obtaining the economic means
needed to give substance to the general development of
the nation, the only road and solid basis of economic independence and of national prosperity. [34]

The relative weight given to each criterion in choosing among projects
is not mentioned. These criteria are broader than the technical criterion that oil production should be greater than a tenth of reserves,
which is recommended for optimal secondary and tertiary recovery. [35]

Díaz Serrano recently stated that once the production of crude
oil reaches 2.25 million barrels a day, it will be maintained at that
rate. [36] This production decision may have been compelled by the fact
that gas yields in Reforma are said to be very high, perhaps more
than can be used or exported should rapid oil production occur. [37]
Nonetheless, Mexico's $34 to $40 billion foreign debt suggests that
rapid production will be the goal through the next few years. Thus the
task of Pemex is "to generate foreign exchange and contribute to national development."[38] It is anticipated that oil export earnings will
be over 46 billion pesos in 1978 and will continue to rise thereafter,
and that Pemex will contribute 20 billion pesos to reducing the foreign
debt by 1982. Until the foreign debt is reduced, however, oil is likely
to be produced for export even if the costs exceed what it brings on
foreign markets, a policy that Pemex is said to have followed in the

early 1970s. Once the foreign debt is substantially reduced, the government should be able to sustain a lower rate of production, thus reserving oil for Mexicans, although a change in presidents every six years makes planning for a longer period of time unlikely. [39] Mexico recently announced that it would reconsider its oil production level once 2.25 million barrels a day are reached.

In theory, decisions on how much oil to lift should reflect technically optimal production rates, the cost of producing oil, the rate of return on producing oil, the rate of return that could be earned on investment in activities other than oil, and the indirect benefits and costs to the nation of oil production. One must also take into account, in addition to the foreign debt and world oil supplies, various factors touching on production and conditions surrounding their exploitation. Of these, the least certain item is the cost of producing oil. Since Mexico owns oil, and does not have to purchase it from a supplier, Pemex does not subtract the natural resource cost of oil from its pretax profit of 10,027 pesos to obtain a net profit statement. [40] If Pemex had to purchase all the crude oil it processed, then the 293.1 million barrels of crude it produced in 1976 would have cost 63.5 billion pesos, implying a loss of 53.5 million pesos.

For purposes of comparison, a typical U.S. refinery on the East Coast purchased only 30 percent of its crude. For Pemex, the comparable cash outlay (not resource value) would have been 19 million pesos and a loss of almost 9 million pesos. An alternate way of looking at the problem is to use the markup between crude oil and the refined product price. Crude costs about 75 percent of the price that wholesalers/jobbers pay for refined products. If we apply this figure to Pemex's 46.5 million pesos of sales income, that leaves 34.8 million pesos as the cost of crude, with a loss of 24.8 million pesos. [41]

The problem of losses does not end with the question of imputed resource costs. In 1976, Pemex used 9.6 million pesos of its pretax profits to pay income taxes. It had an investment program of 24 million pesos that could not be covered by its own resources; the financial deficit was 9.3 million pesos. Various estimates of loss range from .63 to 3.74 percent of gross domestic product (the fall of GDP, before accounting for Pemex's loss, was 2.91 percent).

Losses stem from two severe difficulties: high operating costs and the underpricing of internal sales. An indication of high operating costs can be derived from certain facts. Field operating costs per barrel were 64.8 pesos in 1976. This figure should be compared with the current Alaska North Slope wellhead price of 136.62 pesos per barrel, of which 9.20 pesos are field operating costs, 5.05 pesos are field financing costs, and 19.55 pesos are for amortization and depletion, for a total of 33.81 pesos. Pemex's higher costs cannot be attributed to differences in the depth of wells drilled in the Alaska North

Slope and Mexico. It is said that Pemex is overstaffed, and that its salary and loan bill came to some 119,482 pesos a worker in 1976, which is high by Mexican standards. Permanent jobs with Pemex are said to cost workers about 70,000 to 80,000 pesos in payment to labor leaders. [42] Agreements are negotiated so that competent workers can be assigned to key jobs, but featherbedding problems are enormous and led in 1977 to a government requirement that employment growth be restricted to 4.8 percent. At the same time, a reduction in overtime cut 255 million pesos from payrolls. In a further step to solve this problem, both management and labor were represented on productivity committees. New training centers will cost 180 million pesos, compared to 80 million pesos that were spent on worker's housing. [43] A related problem is that Pemex signed a contract with the labor union according to which Pemex is obliged to bring 40 percent of the work to the union for it to accept or contract to third parties. [44]

The losses of Pemex may have stemmed also from a determination to bring in oil despite escalating costs. In the last year of President Luis Echevarría Alvarez's term, Pemex's investment was 45 percent higher than that authorized, which provoked questions about the ability of the government to control Pemex.

The inability of Pemex to finance its own expansion is a problem of national importance. Pemex's debt at the end of 1977 came to $2.9 billion, or 12.7 percent of Mexico's $23 billion public foreign debt. [45] Currently the growth of Pemex requires a combination of foreign borrowing, the creation of funds, and transfers of funds from other quarters. This practice has led to strong inflationary tendencies and to the weakening of development in the short run of other sectors of the economy. [46]

All of this probably contributed to the government's recent decision to finance additional Pemex expansion by using Mexican financial sources and limiting the rate of oil growth. In 1977, for example, $175 million of Pemex's financing was placed with the Bank of Mexico. [47] It seems likely that this was as inflationary as Mexico's earlier requirement that the Bank of Mexico purchase Government Development Bank (Nacional Financiera, or NAFIN) obligations. Since inflation is a notably regressive tax, and the real income of Mexico's poorest citizens has been falling, the question that cries out for an answer is whether or not there exists a nonregressive system of financing that does not have unfortunate implications for Mexico's external position.

Such a system clearly does exist and now operates along two lines. First, Pemex's indebtedness is placed with nonbank sources of credit, begun in July 1978. The second aspect entails a major reversal of Pemex's pricing policy. Subsidies on domestic sales are provided by charging below-market prices, with generally lower rates for industry than for domestic users. The subsidies are as follows:

Oil and refined products	6.9 billion pesos
Natural gas	14.2 billion pesos
Petrochemicals	940.4 million pesos

The total subsidy comes to about 22 billion pesos, for a lesser figure of 715.2 million pesos for petrochemicals has also been cited. Commentators as divergent as Juan Eibenschutz and Heberto Castillo argue for pricing at international market rates; in March 1978 the government announced that energy prices will gradually approach, but not equal, them. Castillo stresses that the present rate structure favors the rich and the transnationals and leads to decapitalization of the energy sector. Eibenschutz emphasizes the need for prices that will lead to efficient distribution and conservation of resources and that will cover operating costs, interest on the debt, and a margin to finance an expansion program. [48]

Even if it were possible to put Pemex on a paying basis, there would be questions as to whether rapid development of the oil industry is the best strategy for Mexican economic development. Issues most often raised concern implications of oil development for regional development and the creation of employment. Also of concern is the use to which Mexican oil is put.

Petroleum is one of the most capital-intensive industries in a modern economy. Much development of oil fields is transitory. Hence there are serious doubts that skills needed for oil work can be transferred easily to other activities, and that employment in an oil region can be sustained after the basic construction phase is over. In this respect, oil development resembles any other mining "boom town" pattern. Money goes for bars and groceries, schools are few, and inflation is rampant. Complaints about the absence of adequate social development in Chiapas and Tabasco persist, while the contamination of fisheries from Pemex's wastes continues. [49] Pemex is said to be slow in paying for the land that it expropriates, but it is also viewed as a source of wealth that can be used to deal with regional problems. Yet an attempt by the governor of oil-rich Tabasco to tax Pemex for the development of his state failed. President Portillo declared subsequently that social development of Mexico's oil regions is the responsibility of the federal government. [50] Such development is to be accomplished, ostensibly, by a 30 percent discount on energy and raw materials used for industrialization in Tabasco, as part of a government decree to stimulate development of the Gulf-Southeast zone. [51]

These considerations, plus the growing excess of refining capacity around the world, [52] have contributed to Pemex's recent decision not to increase the rate of production after 1980, and to sell crude to be refined abroad* rather than to insist on selling refined products or petrochemicals. [53] Further considerations governing the best way of developing the petroleum industry are the uses to which petroleum is put, and the cost of developing substitutes for petroleum. According to Heberto Castillo, a critic of the government,

> policy should seek to satisfy the internal needs of development with all energy sources . . . we should examine all aspects of Mexican economic development and should not consider oil as foreign exchange to get foreign goods or money. In this sense, oil is like land, and one should not sell the country or see it as fuel. Burning it is the worst use. Society has been developed by the industrial revolution thanks to mineral fuel, but for all these years, its use in this respect has been destroying the environment. [54]

Some Mexicans also feel that because there is more value in refined petroleum and petrochemicals than in crude oil, and because oil is a high-cost electricity producer, Mexico should develop its petrochemical industry. Current plans are nearly to double the use of 4 percent of Mexico's oil and gas as feedstock by 1985, and to substitute other materials in the production of electricity. [55]

That part of oil policy involving the use of foreign technicians has come under attack, in spite of the fact that they may be needed for specialized tasks such as offshore oil development and the containment of drilling disasters. So as to reduce dependence of this sort, the Mexican Petroleum Institute has negotiated joint technology-development agreements with foreign firms. [56] Similarly, Pemex has signed an accord with Ortloff International, whereby Ortloff will supply technology and complete engineering designs and provide all equipment for two new cryogenic plants, while Japan will supply capital and technology in exchange for oil. [57]

Areas in which dependence on foreign suppliers is likely to be reduced are capital goods for the oil industry, and the development of petrochemicals. † Mexico supplied 85.5 percent of its own apparent consumption in 1975 and plans to continue refining at least half of its

---

*Pemex cut its third quarter 1978 export price by 15-20 cents per barrel for 34 gravity isthmus crude to aid its competitive position.

†Nevertheless, the petroleum industry is growing so rapidly that Pemex may have to import 75 percent of its machinery. Joint Latin American oil and petrochemical development was therefore proposed,

own oil output. [58] The production of basic petrochemicals is reserved to Pemex; it will participate as a minority partner in tripartite association and anticipates that it will export 16 percent of its refined products and 17 percent of its basic petrochemicals by 1982.

Mexican natural gas policy follows oil policy in importance. On September 1, 1978, President Portillo said: "Having surpluses, we can sell or reserve them, but never underprice them, which would be equal to burning them." Natural gas provided 20 percent of Mexican primary energy in 1977. [59] Relatively small proposed sales of Mexican gas were the focus of Mexican-U. S. energy relations in 1977-78, despite Mexico's great oil wealth.

Gas policy takes into account the kinds that are available, ability to transport them within Mexico and to other nations, nonenergy uses of gas, the effect of gas policy on the rest of the economy, and the political implications of pricing policy.

Mexican natural gas policy is tied in part to petroleum policy, since 57. 4 percent of Mexico's natural gas is associated with petroleum; the remainder comes from "dry" wells. Gas must be produced so long as the petroleum with which it is associated is lifted, and Mexico has a limited capacity to store, liquify, or reinject gas into the ground. Dry gas, however, does not have to be produced until it is needed. The question—How much gas should be produced and from which source?—has various answers depending on whether the gas is intended for domestic use or for export, and on whether it can be shifted readily from one part of the country to another.

The North and Northeast have dry gas with a low sulfur content that requires little processing. Although dry gas is easier to sell, many northern fields are small and exact higher drilling and collection expenses than is the case in southern fields; however, the latter entail greater processing expenditures. Until recently, gas was flared because other options were not yet open. Pemex wanted to end flaring losses, making a convincing case that liquifaction of gas for export to most markets would be far less profitable than export of gas to the United States. Pemex believed that it could obtain $2. 60 per 1, 000 cubic feet of gas* and that at such a price the export of gas to the United States would earn 43. 7 billion pesos a year within two years. The cost of a gasduct needed for export (23 billion pesos) and the as-

---

with the hope that a self-sufficient capital goods industry for this sector would result. Excelsior, January 4, 1979, pp. 1, 16, and editorial.

*A price equal to $13. 00 a barrel for the equivalent heating-power amount of No. 2 fuel oil in New York, the price to escalate with that of No. 2 fuel oil.

sociated nonduct costs (19. 5 billion pesos) would be earned back rapidly. [60] The United States, however, was trying to limit gas price increases as part of its proposed energy bill, and therefore denied U. S.
companies permission to purchase Mexican gas. Subsequently work
on connecting the gasduct to the United States was suspended.

The alternatives to selling gas to the United States were viewed
at first as either liquifaction or the production of ammonia. [61] The
trouble with the latter option is that if all the gas that otherwise would
have been sold to the United States were processed into ammonia, the
resulting amount would equal more than half of world consumption.
Such an amount could not be absorbed or produced profitably at a natural gas cost of $2. 60.

A recently suggested alternative use of natural gas as a substitute for oil in Mexican markets is being implemented by connecting
Southern, Central, and Northern Mexico with a gasduct, which permits the shutting down of dry gas fields and the shifting of "associated
gas," after processing, to the rest of Mexico. As a result, petroleum
is released from other uses to the petrochemical industry, and contamination, as a by-product of energy production, is reduced. The
outcome of this strategy is that gas flaring was less than 10 percent
of gas production in mid-1978, but had not ended, as hoped by March
1979.

Although firm data on the demand for natural gas do not exist,
the impression is that industrial demand both in Monterrey and Mexico would absorb gas if there were an adequate distribution network.
The decision on whether to consume gas or oil is to some extent up
to the energy sector. Between them, Pemex and the Federal Electricity Commission (Consejo Federal de Electricidad) use more than one-
half of Mexico's natural gas and can shift between sources in most instances. The decision about which fuel source to use domestically
rests on the fact that it pays to export gas rather than fuel oil once
prices rise above $2. 20 per 1, 000 cubic feet at current prices for
fuel oil No. 5 or 6 in New York. President Portillo underscored flexibility in decision making by his September 1978 announcement that
"practically, use of all southeast gas is assured. "[62]

In the next few years, increasing attention is likely to be paid
to anomalies in international and domestic pricing policy. European
prices for No. 2 fuel oil are 38 percent higher than those in New York,
while gas prices are similar. [63] As it becomes possible to sell fuel oil
to Europe, a larger share may be exported than is used domestically,
with gas filling the gap at home. [64] Domestic prices are said to be
strikingly different for natural gas as between users; the industrial
use of gas costs five times less than domestic use. [65] Although distribution costs are higher to many homes than to single industrial sites,
the differential is greater than that usually associated with distribution

costs. There is a conflict of goals between a desire for prices that
reflect distribution costs and that also reflect the need to conserve
scarce resources. A third conflicting goal is the desire to use prices
to promote industrial decentralization. SEPAFIN will charge different
gas rates in different regions, in accordance with the national development plan. It is believed that Mexico can absorb its entire gas output. On the other hand, the high distribution cost of natural gas to distant areas suggests that this may not be economical for thinly settled
regions; in such locations, the development of biogas for heating and
power is in its first testing phase.

Mexico's hydroelectric reserves are estimated at 83 million
kilowatt hours per year. Some 60 million of these hours will have
been used by the year 2000, a development that will cost around 150-
200 billion pesos. [66]

Hydroelectric plants provided 38.5 percent of Mexican electric
capacity in 1977, which is expected to decline to 35 percent in 1986
and 27 percent in 2000, unless a massive nuclear power program is
undertaken for the production of electricity, in which case the share
of hydroelectric would be only 12 percent (see Table 6.2). The CFE
is oriented toward the development of large hydroelectric plants and
has paid little attention to the development of profitable secondary hydroelectric sites, but these should become more attractive since hydroelectric power is cheap and renewable. Reductions in transmission
costs should also encourage development of secondary sites.

At the present time, 40 percent of the hydroelectric potential is
in the South, in the Grijalva-Usumacinta complex, and 34 percent in
the Balsas and Papaloapán areas. Some of the projects will affect
Guatemala and require international agreement. [67] Five major hydroelectric projects are included in Mexico's ten-year plan for electric
development: Chicoasen, Peñitas, Caracol, Aguamilpa, and the expansion of Malpaso and Angostura. Of these the most important is the
1,500 MW Chicoasen dam, which has an underground plant. Supplies
for this partially completed project have come from France, Japan,
Switzerland, and Mexico, with all civil engineering designs being carried out by Mexico. [68] Many specialists have been trained, but there
is still a shortage of mechanics at Chicoasen. As in the case of other
hydroelectric plants, Chicoasen will be important for generating electricity during periods of peak demand. As future development of hydroelectric power shifts necessarily to smaller sites as the larger ones
are used up, secondary hydroelectric potential and optimal design for
5- to 20-megawatt hydraulic plants probably will be looked into, at a
cost of about 1.5 percent of the energy sector's research and development budget (see Table 6.4). [69]

Coal can be viewed as a new resource with respect to electric
power generation. [70] Present consumption of coal is estimated at 6 to

TABLE 6.4

Estimated IIE Budget, 1978 and 1979
(in percentages)

	1978	1979
Geothermy	17	15
General fossil fuel	16	17
Hydroelectric	5	8
Nuclear and nonconventional	8*	13
Transmission and distribution	27	19
Technical assistance to equipment suppliers	21	24
Transportation	7	7
Million pesos	162	280

*Includes 20.0 for rural communities.
Source: Estimated from various sources.

7 million tons per year, about 98 percent of which is used by the side-rurgic and minerometallic industries. [71] The remaining 2 percent is used in the generation of 37.5 MWe of electricity in the "Venustiano Carranza" plant in Nava, Coahuila. Coal-fired electric plants will produce 630 MW by 1983,* and will consume an estimated 1.8 million tons of coal, equal to at least 15 percent of projected coal production. [72] Coal development has been a relatively low priority area because of the pollution coal causes and its higher cost in relation to nuclear energy. At the same time, IIE has suggested that since there is relatively limited experience with coal-fired thermoelectric plants, Mexican research is given to solving problems often already dealt with in countries with longer experience. [73] For example, there are problems such as how to characterize coal, specify boilers, produce and manage coal, and separate, utilize, and store ash. Some IIE laboratories,

---

*A $158 million dollar loan has been announced, granted by the Inter-American Development Bank, to develop Rio Escondido (Coahuila) coal mines. The total investment will be $309.4 million. The coal produced will supply a 1,200 MW thermoelectric plant. Some 3.6 million metric tons will be produced, plus 600,000 tons from strip mines. Excelsior, December 16, 1978.

design facilities, and research have been focused on thermoelectric plants. The lack of efficient workers is a problem. Output per Mexican coal worker is about one-tenth of average worker output. Moreover, there are few engineers trained for coal-fired thermoelectric plants. Thus a significant part of the research and development budget for coal is devoted to training as well as to exploration, reduction of pollution, and minimization of costs. All of these developments are essential if Mexico is to supply all of the coal needed for steel as well as electric power production.

Geothermal energy is desirable for electric power generation, because under ideal circumstances it costs two-thirds of the bill to produce energy with nuclear or hydrocarbon fuels. Geothermal is economical at a scale of 3 MW and can be developed in as little as three years, which makes it possible to avoid investing in excess power capacity.[74] Geothermal sources provided 75 MWe (0.6 percent) of installed capacity in 1977, and are expected to reach 1.2 percent in 1986 and 6.6 percent in 2000. Growth will be slow because development of geothermal fields in Mexico takes 10 to 15 years. Estimates of reserves vary from 40 to 100 million kilowatt hours yearly. Reserves may prove to be greater as knowledge of Mexico's subsoil grows.

The geothermal plant in operation is Cerro Prieto, 30 kilometers from Mexicali, Baja California Norte. This plant's generating capacity will increase to 400 MW by 1986.[75] New test wells may confirm sufficient vapor to permit the generation of 1 million kilowatts. It should be noted that Cerro Prieto is only one of 130 known geothermal structures that have been identified.[76] Information from Skylab will enable Mexico to locate other geothermal sites.[77]

Geothermal development entails associated problems and benefits. On the one hand, contamination of the environment and sinking of soil require remedial action. On the other hand, in addition to providing energy, geothermal fluid contains salts useful for the fertilizer industry. Mexico imports 150 million pesos worth of potassium a year; potassium from geothermal fluid could replace these imports, with some left over for export.[78]

According to the Workshop on Alternative Sources report, "Mexico is chronically short of specialists in the exploration for and exploration of energy resources other than hydrocarbons," so that geothermal development will depend to some extent on borrowing foreign technology and technicians.[79] Estimates of geothermal reserves are 9,200 MW in known areas, and another 9,000 MW in areas requiring further exploration.[80] The location of geothermal energy sources, in relation to Mexico's transmission network, renders it necessary either to construct new transmission lines or to arrange for an exchange of energy with the United States. The complex needs of geothermal energy development explain the nearly 2 percent share that geothermal has in development plans and in the energy sector's research and development budget.

The most controversial aspect of Mexican energy policy is the development of nuclear electric plants, suggested by J. Poppendieck's remark, "Let them eat yellowcake." Two questions are relevant here: what should be the share of nuclear energy in the overall energy picture, and how should nuclear energy development be organized?

Mexico's first nuclear plant, Laguna Verde, is now being built. It includes two 650 MW units, and substations and transmission lines needed to connect them to the electric grid, which will permit Laguna Verde to supply energy to areas ranging from Tampico to Monterrey. [81] Mexico's ten-year energy development plan calls for nuclear to provide 5.4 percent of the nation's electricity by 1986 (see Table 6.2). Recent announcements by ex-Foreign Minister Santiago Roel suggest that there will be 20 reactors by the year 2000. Pronuclear advocates urge that nuclear plants supply 77 instead of 26 percent of electricity by 2000. * Mexico's ability to carry off a large nuclear power program has been questioned because of the large investment costs and real difficulties attending the creation of a new industry. [82] The construction of Laguna Verde has been plagued by delays, caused in part by the cancellation of a foreign contract. Nonetheless, because Laguna Verde is Mexico's first nuclear plant, and work experience and quality control are extremely important, some of the technical aspects of the construction project have been contracted to foreign companies. [83]

Perhaps the most important part of Laguna Verde is centered on INEN, which is the authority responsible for authorizing licenses for the construction and operation of the plant, and which acts as the chief judge of security-related aspects of the work. At the same time, it receives assistance and inspection services from the International Atomic Energy Organization.

The Laguna Verde plant will use a light water reactor (LWR), which was developed by the United States and is the dominant type used in the world. [84] Although Mexico has considerable uranium reserves, the uranium for the first charges of both reactors was bought in France. Enrichment of uranium is to be supplied by the U.S. Energy Research and Development Agency. The manufacture of fuel is to be provided by General Electric, which is also the manufacturer of the reactor. The dependence of Laguna Verde on foreign suppliers and technology has provoked a sharp debate on whether or not the share of Mexico

---

*According to the U.S. Department of Commerce, in Electric Energy (January 1977), p. 94, capital investment in nuclear was to reach $171 million in 1980 and $240 million in 1985.

in the nuclear industry is more important than its size. Mexico's limited share in the present nuclear power industry is blamed on lack of government budgetary support. Also, Mexico lacks the technology for enrichment of uranium and cannot afford the $6.7 billion that enrichment and associated electric development would cost; it is anticipated that centrifuge separation will require 10 percent of the electricity used by current diffusion techniques, so that enrichment may become practical in time.

Both choice of reactor and provisioning of the Laguna Verde plant are controversial. The alternative to an LWR is a heavy water reactor (HWR) using natural uranium and heavy water: "The principal advantage of the HWR over the LWR stems from the fact that heavy water is sufficiently more effective than light water as a moderator to permit the use of natural unenriched uranium as fuel. The commercially available HWR is of Canadian design, and is known as CANDU."[85] Nationalists are pleased that a heavy water reactor would permit the manufacture of the entire fuel cycle in Mexico, although there are few sources of heavy water. Both an LWR and an HWR will be used as Mexico's nuclear electric industry develops; moreover, many observers argue for an immediate escalation of the Mexican nuclear program, so that Mexicans will be able to develop technology and participate in the design, construction, testing, and operation of nuclear electric plants as they are needed.[86] Twenty-five percent of all nuclear plant parts already can be made in Mexico, but there are problems relating to quality. Mexican manufacturers will need assistance if their participation in the nuclear power industry is to reach a level of 85 percent, which would yield a domestic expenditure estimated at 15.2 million pesos for each plant.[87]

Since the problem of an independent Mexican nuclear industry ranges from invention of new technology and education of scientists to the provision of engineering services, the question is more complex than it first appears. The question, in its simplest form, is whether or not Mexico should reinvent the wheel, or perhaps provide a substitute. Education, research, and invention are the most capital-intensive activities known, all the more so in atomic physics, and Mexico is discouragingly short of capital. In this regard, it is interesting to note that, although Arturo Warman suggests that Mexican conditions are closer to those of China than to those of the United States, and that Mexico should study the Chinese model for solutions,[88] Vice Premier Deng Xiaoping states that adhering "to the policy of independence and self-reliance . . . does not mean shutting the door on the world, nor does self-reliance mean blind opposition to everything foreign."[89]*

---

*President Portillo says that Mexico does not aspire to scientific

Given the difficulty of obtaining enough trained personnel, and taking into account the experience gained from the first plant, which can be applied to succeeding ones, it is realistic to project the construction of 20 reactors the size of Laguna Verde. According to Arnulfo Morales Amado, the proposed budget for the next six years is 20 billion pesos. [90] This is a frighteningly small amount. Estimates of the capital cost for a new building for part of the City University of New York run between 1. 9 and 2. 3 billion pesos, while a laboratory for a Tokamak reactor costs 133. 4 million pesos and the reactor itself will cost at least 3. 2 million pesos. [91] Yet these figures for atomic industry are consistent with Mexico's small research and development expenditures, and with the education budget, which has risen of late to levels average for developing nations. In evaluating this budget, it should be kept in mind that Dr. Amado says no scientist unemployment should be created, which leads one to wonder what would happen if some were sick or did not turn out well. Most advanced work requires redundancy, for it is hard to run an efficient enterprise without spare parts or spare people. For this reason it is possible either that the budget is too small or that the use of foreigners will be required. The latter can come to pass in one of two ways, neither of which has been discussed very much in public.

The first possibility is the formation of a Latin American Common Market for the atomic energy industry. Economies of scale might result from this eventuality, but some scientists doubt that this would be a good strategy for Mexico in view of the fact that Argentina and Brazil are more advanced at present in relevant scientific and technological developments. The second is the use of technicians with nuclear experience from the United States or other countries. Interestingly enough, the two countries most often mentioned as collaborators with Mexico in nuclear industry development are India, because of its HWR reactor, and Canada, because of its willingness to sell Mexico technology soon after it sells Mexico a CANDU reactor. [92]

In a related development, on November 21, 1978, Mexico and Spain signed a treaty involving nuclear equipment and study and technique programs. Spain is to give Mexico's CFE a line of credit for work on nuclear energy. Also, on December 11, a ten-year agreement was signed with France under which, beginning in 1980, France will purchase 100, 000 barrels of oil a day, will pay for it with capital goods and technology, and will offer uranium enrichment technology

---

and technological self-sufficiency, but seeks self-determination, while other Latin Americans will be employed if sufficient Mexicans are not available.

for Laguna Verde. These arrangements imply that Mexico may follow Venezuela's lead in bartering oil for energy development technology as well as energy research and development capacity.

If, however, most reactors are to be of the light water type, then the question of uranium enrichment will be central and touches a very sore spot. Although plans for agreements covering uranium enrichment and its technology have been announced for the socialist bloc nations, Great Britain, and France, the only signed agreement is with the United States. [93] The problem with the latter agreement is that spent fuel from enriched uranium can be used to create an atomic bomb. The United States, nervous about what happens to enriched u-ranium, has insisted that it be allowed to inspect Mexican plants as a condition for uranium enrichment services. Mexico wishes to rely on international inspection teams of the International Atomic Energy Organization. A compromise was reached when an international in-spection team, which included an American, was sent to Mexico in August 1978. Nevertheless, the position of the United States is viewed as an infringement of Mexican sovereignty, thus it seems likely that alternate sources of enriched uranium will be sought when possible. [94]

Future organization of the nuclear industry has been a subject of sharp debate centering on the Law Regulating Article 27 of the Con-stitution on Nuclear Matters (Ley Reglamentaria de Articulo 27 Con-stitucional en Materia Nuclear). The debate was resolved when the law was published in January 1979. The Constitution itself establishes that in the

> treating of petroleum and solid, liquid or gaseous hydro-carbons, or of radioactive minerals, neither concessions nor contracts will be authorized, nor will there be substi-tution of those which, in their case, will have been author-ized, and the Nation will carry out the exploitation of these products, according to the terms which the respective regulating law indicates. The generation, transmission, transformation, distribution and supply of electric energy which have public service as an object are reserved ex-clusively to the Nation. Concessions will not be granted to individuals, and the Nation will make use of the goods and natural resources that will be required for said ends. The Nation also will make use of nuclear fuels for the generation of nuclear energy and will regulate its appli-cation for other ends. The use of nuclear energy can have only peaceful aims.

The law assures the exclusive position of the nation in the ex-ploration and exploitation of radioactive materials, regulates exports of uranium, and explicitly prohibits concessions or contracts.[95] The latter restriction is maintained despite the huge capital and technological requirements of the nuclear industry, which in the case of the petroleum industry led the government to welcome contracts and to suggest that where Mexican scientists are unavailable, those of other Latin American nations may be employed.[96] Surprisingly, the Mexican debate was missing the idea that Mexico adopt a law similar to the 1975 Venezuelan Organic Law governing petroleum, whose Article 5 provides that the Venezuelan government, after a special vote in Congress, can sign joint contracts in partnership with private firms for limited periods of time under the overall control of the state.

Debate over regulation of the nuclear industry was so intense that for the first time in recent Mexican history, a law submitted by the government was held over for the next session of the Chamber of Deputies rather than being passed immediately.[97] In the law adopted in January 1979, the government placed responsibility for nuclear policy in SEPAFIN, and created a National Atomic Energy Commission, which will coordinate the activities of the newly created URAMEX and ININ, and administer the various provisions of the law. URAMEX, the Mexican Uranium Company, will explore, mine, and process radioactive minerals, and develop the various stages of the nuclear fuel cycle, except for spent fuel, which is to be used for electric energy production by CFE; the Institute for Nuclear Research and Development (ININ) will plan and carry out research and development in nuclear science and technology, and will provide peaceful uses of nuclear energy linked to national development; the National Security and Safety Commission will revise, evaluate, and authorize the bases for design, construction, operation, modification, and documentation for nuclear plants and installations, establishing and carrying out the national system of responsibility and control for nuclear materials, and look after similar responsibilities for the nuclear industry.

The separation of the National Security and Safety Commission from the rest of the nuclear industry is generally accepted: "the judge and the judged cannot be the same person." However, the split of INEN into two groups, albeit under an umbrella organization, the National Atomic Energy Commission, was criticized because a going concern was broken up, and research was separated from practical applications, which could make communication between scientists in the two groups more difficult than would have been the case if they were under one roof.[98] The many scientific breakthroughs that occurred in the laboratories of giant corporations rather than in pure research laboratories were cited as reasons for maintaining URAMEX and INEN as one group. It is not clear whether the locus of innovation reflects the organization with which the innovation occurred, or the huge sums available for investment in giant corporations, compared to the small budgets of pure research laboratories in deficit-ridden universities. As a result, opinions differ in Mexico as to the respective merits of pure or applied research for nuclear electric development.

There are still other questions touching the future of the nuclear power industry. One is whether the importance of the nuclear industry will force the government to look upon it as not subject to strikes, like the defense industry, a topic with implications many find hard to discuss. The U. S. solution of a no-strike pact with unions in the nuclear industry, which leaves the industry outside of military jurisdiction, may be a useful model for the Mexican nuclear industry. The question of how to provide for the future development of activities yet to be invented is covered by requiring ININ to stimulate nuclear research and development in universities and institutes of higher education, coordinating these activities with its own programs, and to train personnel in the federal government and its agencies in nuclear energy and its various applications. ININ is also to promote national and international exchange to favor research and development in nuclear material, organize conferences, and carry out relations with the International Atomic Energy Commission and other groups. To the extent that Mexico must develop its own nuclear industry, a number of current projects now under consideration are important. These include a zero-potential reactor, the training of scientists for the nuclear sector, and research into, as well as experimental plants for, theoretical and applied nuclear energy. Of these the most interesting is the construction of a machine (Tokamak) to confine plasma, the key to future mastery of fusion energy. [99]

A remarkable innovation in the January law is the provision that the export of radioactive materials or minerals will not be approved until a national nuclear energy development plan is approved; the plan is to quantify the nation's annual needs for a period of at least 15 years. Export authorization will not be granted if it affects proved reserves of these materials that the nation needs. Authorization, if granted, will not exceed 5 percent of the nation's needs.

Solar energy includes energy obtained directly from sunlight and indirectly from biomass, biogas, or the wind. If applications of solar energy are expensive for Mexico City, where overcast skies make on-site installations of solar devices impractical, they are of prime importance for many rural areas at a costly distance from the main energy distribution network. In the countryside, most traditional energy sources are solar, providing at least 4 percent of Mexico's energy to 42 percent of the population in 1970. [100]

Much of Mexico's desert is thus receiving considerable solar radiation that could be converted into electricity and fed into the national power grid. In spite of this advantage, solar power is not included in Mexico's ten-year development plan. Predictions of falling costs in solar technology vary, but the solar alternative is expected to be commercially competitive somewhere between 2000 and 2030. Mexico plans to introduce solar technology as it is developed elsewhere, but apparently has not yet accepted suggestions that it develop a major solar research institute parallel to INEN and funded on the same scale.[101]

There are no plans for massive rural electrification, by means of either solar or conventional sources. According to Eibenschutz,

> I do not consider the sun as an energy source which could be massively applied, neither in Mexico, nor in any place in the world, at present. . . . It is one thing to electrify a few villages with advanced solar methods in a technological development plan, and another, very different thing, to propose a plan for solar plants in each isolated community of 200 people. The volume of resources needed is outside the reach of the country, even when we become very rich from petroleum.[102]

This view is not shared by all observers. For example, the co-directors of the Solar Energy Program of the IPN (National Polytechnic Institute) say that

> notwithstanding the current cost of photovoltaic generators in our country, there exist various applications for community use which are economically feasible, such as school television, rural radio-telephone, and, in the next few years, water pumps, as well as other applications which require an autonomous energy source easy to maintain, as is the case with systems operating in remote places. . . . Areas as small as one square meter will be needed to supply energy to electronic apparatus such as educational television and radio-telephone.[103]

Some recent studies have challenged the notion that nuclear electric power is necessarily cheaper than solar power, but Mexican estimates place nuclear-generated electric power at half the cost of the cheapest solar options, such as small-scale hydroelectric systems and cooking and lighting by means of biogas. The remaining solar options cost more than electricity delivered from a grid, while solar refrigeration costs less than diesel refrigeration.[104] Several solar techniques are competitive with diesel power, or will be in a few years.

One reason that Mexican estimates of solar energy costs are high is that they do not include conservation as part of the package.

In Eibenschutz's view, this is because conservation measures are often related to questions of conserving heat in winter, a problem far more severe for the United States than for Mexico, and also because it would be very difficult to secure compliance with conservation measures in Mexico. [105] It follows that the solar energy option may turn out to be less expensive than current Mexican estimates suggest, but not by as much as solar plus conservation estimates by the United States imply. Finally, Mexico seems capable of manufacturing its own equipment for solar power.

At present, the greatest attention among solar projects is aimed at integrated energy systems for rural communities. There are 80,000 communities that have not been electrified. [106] These villages are scattered throughout Mexico. The chief interest lies in those with 200 to 500 people. The condition of these villages differs according to their local resources, which can be used variously for biogas, solar, wind, and hydraulic microsystems as part of integrated village development. In the past year and a half, IIE has worked on process and system development and is now in the second phase of selecting three or four communities with different characteristics to use as pilot projects. [107] At the same time, the United Nations is negotiating to set up a rural energy center, which will be located in Vallecitos, Guerrero. [108] The government has a solar habitations project going jointly with West Germany that will cost an estimated 380 million pesos. [109]

Within the rural IIE project, only the simplest techniques are being used, such as flat plate collectors, water, or wind. No photovoltaic systems are contemplated due to their high cost. The integrated system should be able to provide a biogas digestor and water for washing. A problem not faced in other systems is how to get cows together in one place for the collection of their waste, which after processing is a better fertilizer than mere dung and also produces methane for biogas. The biogas can be used for internal combustion engines, water pumps, or electric generators. An advantage in using biogas is that it is substituted for charcoal and reduces the rate of deforestation. Methane is also less polluting than burning wood or charcoal. For the utilization of biogas the Volkswagen motor was chosen, because of its ease of maintenance, and converted to burn gas instead of gasoline.

A sociological/anthropological study will be made of the test village before and after the introduction of the integrated system. The community will be involved in planning so as to reduce resistance to the changes and to see how energy can be used to create additional employment, based perhaps on the establishment of an agroindustry that processes local resources. Villagers are to take part in building their own equipment so they can also learn how to maintain it.

Wind power is not a strong feature of this program because it is unevenly distributed in Mexico, but it has some potential for pumping water at cattle ranches. Related new measures, such as passive architecture,* are just beginning; it is expected to be of the greatest interest in Monterrey, where there is considerable variability of climate.

The importance of solar policy can be gauged to some extent by the fact that only $200,000 is currently spent on the rural energy project, and there are no plans for massive training of solar energy specialists. In Mexico there is a distinct lack of available jobs in solar energy. [110] In comparison, the United States spends an estimated $400 million annually on solar energy and has some 10,000 solar energy professionals. Of solar energy studies in the United States, one-quarter are devoted to development of inexpensive photovoltaic cells. Joint solar energy research and development between Mexico and the United States was proposed in December 1978 and is being explored. [111]

This assignment of priorities is not the necessary result of the comparative advantage of Mexico and the United States in nontraditional energy sources and sciences. Instead it reflects Mexican faith in abundant oil and in nuclear power. Brazil, in contrast, has insufficient oil reserves, a nuclear power program more advanced than Mexico's, and a major program in nonconventional energy sources, which includes research on crops and vegetable wastes that can be used in alcohol-burning engines, and the design of automobile engines that burn 100 percent alcohol.

INTERNATIONAL ENERGY POLICY

For most of its recent history, Mexico has been a net exporter of crude oil and an importer of some refined products. From 1968 to 1974 imports of crude oil were necessary because demand outstripped supply, but the recent oil discoveries have restored Mexico's position as an exporter of crude and exports of electricity to California are being discussed. Mexico is also able to export natural gas. The most important issue surrounding foreign energy sources for Mexico is energy technology. The best-known case is the enrichment of uranium rather than supplies of raw materials. At this time, Mexico cannot afford to enrich its uranium and relies on the United States for the service. There are other potential sources of enrichment as well. However, if Mexico were to use an HWR with natural uranium, it is unlikely that more than a few sources of heavy water would be available, and heavy water is not yet produced in sufficient quantity by Mexico for the operation of nuclear electric plants.

---

*Placement of buildings to take advantage of the sun.

Similarly, foreign technology and equipment are used in the petroleum industry, especially in offshore and cryogenic plants. Current attempts to limit dependence on a single foreign source have led to an agreement with Japan to exchange oil for technology and equipment, although there is some protest to the effect that Mexican suppliers should be used. There are attempts already to change the pattern of education so as to stress innovation and the adaptation of foreign technology, rather than relying simply on its use. It is possible that Mexico will follow the precedent of the United States and Venezuela in exchanging oil for investment in joint research and development of energy technology, part of which will be located in Mexico, rather than exchanging oil for technology that is already "embodied" in the equipment.

With regard to multilateral foreign relations, Mexico is not a member of OPEC and will not formally join in bilateral producer agreements, but nevertheless it follows OPEC pricing policy. Mexico has sponsored and signed the Nuclear Non-Proliferation Treaty. Bilateral foreign relations are determined by the fact that under the Constitution energy sources and their production are virtually a monopoly of the government; necessarily, foreign equity investment is precluded. Foreigners provide technology, finance, and services. They also act as subcontractors, especially in areas where Mexicans lack experience. In such cases, the foreign firm often trains workers in the new process. Recently, in response to nationalist attacks on the use of foreigners, there has been an indication of preference for Latin American over other foreign technicians in cases where there are not enough Mexicans available in a given specialty.

ENERGY AND THE ENVIRONMENT

Industrialization and energy use grow together, and the inevitable result is pollution. The problems caused by increased levels of pollution depend on geographic factors. In this respect, Mexico is in an unfortunate position. Pollutants from the United States contaminate Mexico. More than 50 percent of Mexican industrial production has been located in Mexico City, whose geographic setting aggravates the ill effects of pollution. The city is ringed by mountains and years ago it benefited from a shallow lake (Texcoco) as well as a system of canals. The lake has dried up, contributing now to dust pollution, and the mountains trap smog above the city in thermal inversions, especially during the dry season. Thus the atmospheric pollution of the Metropolitan Zone of Mexico City is considered one of the worst cases among the world's great cities. Serious respiratory illnesses and the deterioration of real estate are major social costs resulting from pollution. [112] The official position has been stated by Mario Alberto Chavez González: "Mexico endeavors to bring about a complete social and economic development, and simultaneously sustains a pitched battle against contamination and deterioration of the environment." [113] The government therefore advocates decentralization in order to solve the air pollution problem. In addition, the use of antipollution equipment is required in some cases and encouraged by fiscal techniques in others. [114]

Recent action to limit contamination of the environment stems from the 1971 amendment of Article 73 of the Constitution, which establishes the jurisdiction of the Council of Health in environmental matters, and subjects its environmental rulings to congressional review. The Federal Law for the Prevention and Control of Environmental Contamination, which was enacted March 23, 1971, defines pollutants as substances that alter the environment and includes smoke, dust, ashes, bacteria, wastes, residues, as well as all forms of energy, such as heat and radioactivity, and energy-related pollutants such as noise. Regulations issued on September 17, 1971, set standards for smoke and dust emission and pay special attention to refineries, thermoelectric plants, railways, and automotive vehicles, as well as waste incinerators, fertilizer plants, and concrete and asphalt plants. Motor vehicles cause 70 percent of the air pollution in Mexico City. [115] The government focused its antipollution measures on automotive vehicles and chose diesel- rather than gas-powered buses to reduce pollution; but insufficient attention has been paid to maintenance, so the diesel buses contribute to pollution. [116]

The possible effect of a given activity on pollution levels is one of many considerations in making siting decisions. For instance, it was argued that siting oil refining or petrochemical facilities at Tula should be reversed, because wind from Tula would blow smog into Mexico City. Moreover, the siting of hydroelectric plants in Tula and of tourist centers in Jalisco, Quintana Roo, and Baja California is being studied for its ecological impact. [117] Some commentators have argued that the administrative structure governing pollution matters makes it difficult to take pollution into account in decision making, for there are multiple jurisdictions over different aspects of pollution. The Ministries of Health and Welfare and of Agriculture and Water Resources are involved in formulating regulations, as are various advisory committees. Enforcement may prove difficult and fines were set at the low range of 50 to 100,000 pesos. [118] Critics point out that influential industrialists ignore the law with the complicity of mayors and governors. [119]

Specific action to reduce energy-related pollution has not yet been recommended by the National Commission on Energy (Comisión Nacional de Energéticos). [120] However, a number of actions have been announced to reduce air pollution that occurs as a result of Pemex operations. The 1977-82 plan, based on measurements of air and water contaminants, and evaluation of control systems, provides for increasing control of burners to reduce smoke emissions from refineries, while related studies are underway pertaining to the petrochemical industry. [121] Nonetheless, pollution considerations are apparently secondary to decisions for more growth. The 1979 budget for Pemex provides that Pemex will pay fines as required, but the most important decision affecting pollution was reached for other reasons. This was the decision to switch from oil to gas for electricity generation, which will reduce atmospheric contamination, but the decision was made because gas could not be exported. The single most important means of cutting down on pollution would be to raise the price of energy, a step that has been taken for electricity but not for oil or petrochemical products. Finally, it is estimated that antipollution devices to reduce air pollution from fossil fuel plants would increase electricity costs by 4.2 to 8.5 percent. [122]

Water pollution in Mexico results from many different actions: the dumping of "salt" water into the Colorado River in the United States, which flows later into Mexico; Mexican irrigation projects, which limit the availability and quality of water for shrimp and other forms of marine life; sewage disposal inadequacies, which affect agriculture and drinking water; and industrial, as well as other energy-related, operations.

Pemex has announced several strategies designed to limit water pollution. More control will be provided for waste water discharges, especially regarding grease and oil content, while relevant personnel have been trained to control spills from dams containing wastes. Ballast water from boats is to be treated, as well as waste water from petroleum refining operations. Plans to deal with related problems in the petrochemical industry are being studied. On the other hand, there are persistent complaints about the contamination of fisheries from Pemex's wastes. [123] With respect to electric energy and pollution, the cost of reducing water pollution from fossil fuel plants would increase electricity costs 1.8 percent, and the increase for nuclear plants would come to 2.7 percent. [124] In June 1974 the Undersecretariat of Water Resources announced that 45,000 factories had fulfilled their obligations to register and to inform authorities about the contents of their residual water. It was also announced that sanctions were to be applied to 25,000 more factories for noncompliance. [125]

The spectacular blowout of a drilling rig at Ixtoc I in the Bay of Campeche has led to a disastrous and long-continuing spill of oil into the Gulf of Mexico. Because of the expense of such rigs, and their scarcity in Mexico, the equipment in question had been rented by a Mexican firm from a U.S. firm to avoid direct foreign participation in the oil industry. The rig was 15 years old and the apparatus designed to cut off the oil flow failed to work. Equipment that could have limited the area of the spill was offered to Pemex before the Ixtoc I site blew, but Pemex declined to purchase it. Pemex's handling of the spill has been criticized by some foreign observers, who say that Mexican labor laws have prevented cleanup workers from staying on the job more than ten hours a day, that Pemex accepted insufficient amounts of cleanup equipment, and that the equipment was not properly used. The oil slick has killed dolphins, fish, crustaceans, and turtles, but fisherman have asked the government to minimize pollution stories lest their contaminated catches end up sold at a reduced price or not sold at all. There are conflicting reports about ecological damage. A Mexican government report evaluating such damage is expected soon. It is likely that offshore drilling will continue, but with increased attention to ecological and safety factors. In the meantime, the question of compensation to other nations for environmental harm is a touchy issue, whose resolution will take some time for these reasons: Pemex insures itself, so that the financial burden will have to be borne by Mexico; it is unclear whether tanker spills are precedents for blowouts; Mexico is likely to link the Ixtoc 1 spill to the issue of U. S. and Mexican pollution of one another's environments.

The nuclear plants at Laguna Verde are not yet in operation, so the discussion of atomic-related pollutants is centered, therefore, on potential effects. The difficulty one has in reaching valid comparisons of nuclear with other sources of power, with respect to health consequences, is that insufficient experience with nuclear is available to assess its long-range impact. One recent estimate states that under normal operating conditions the total health risks from nuclear power are from 0. 6 to 1. 0 deaths per reactor year, less than the . 04 to 25 deaths from a coal-fired 1000 MWe plant. [126] Another recent estimate cites 120 deaths per gigawatt hours (GWh) from coal and . 47 (GWh) from nuclear-fired plants. Both thermal and other nongenetic pollution are said to be less from nonfusion nuclear than from other sources of energy. The worst assumptions about nuclear accidents lead to estimates that they would be comparable to peacetime accidents such as hurricanes or earthquakes, which have not threatened the survival of societies. [127]

The most important factor in estimating the consequences of a nuclear accident is the location of the plant. Laguna Verde is mercifully far from Mexico City, thus reducing the consequences of a possible accident. However, should Mexico decide to use breeder reactors in the future, the problem must be faced that risks from the reprocessing of spent fuel are greater than those from current plant operations, and that there is considerable doubt over the effective storage and disposal of reprocessed fuel. Furthermore, permanent genetic damage from low-level radiation in routine nuclear plant operations is a continuing risk associated with any nuclear electric system. Underground facilities to control radioactivity would increase electricity costs by 1. 0 percent. [128] In the future, the development of fusion energy would overcome the radioactive waste disposal problem that might contribute to a greenhouse effect. On the whole, it appears that in Mexico the issue of national independence is more important than public health. [129]

Pemex maintains that lands in oil drilling areas are to be restored either to their natural state or to a state most favorable for further use. A similar policy is being followed with respect to the construction of the gasduct. If 90 percent of all electric wiring for power transmission were placed underground, the cost of electricity would increase by 3. 2 percent. [130]

For a developing nation, the most important aspect of the relation between energy and the quality of life is the availability of energy to the entire population. A shortage of natural fuel with the absence of electricity contributes to rural urban migration with its associated environmental problems. The first priority, therefore, should be the provision of energy to rural and poor neighborhoods. A second priority area is the relation between energy and employment; in some cases, energy availability encourages both investment and employment. In domestic services, however, electric-powered domestic appliances often replace maids without creating comparable employment opportunities in the appliance industry. Similarly, some agricultural mechanization reduces demand for labor while making possible the farming of larger areas. In the absence of a mechanism to transfer the increased output to those who are unemployed or underemployed, the availability of energy decreases the quality of life for some people while enhancing it for others. The government has been trying to help small businesses and to create employment. At the same time, population policy is aiming to slow the rate of population growth until a reasonable quality of life can be provided for all. There is a growing environmentalist movement, which may make possible strict enforcement of Mexican environmental and antipollution laws and avoidance of the intense environmental disruption caused by Japanese industrialization.

FUTURE PROSPECTS

Mexico has proclaimed a policy of economic growth with the emphasis on decentralization, aid to small businesses, the creation of employment, limitation of population growth, and pollution control. Since the details of plans by which these goals are to be reached are not available, it is difficult to assess either the priorities among these goals or the likelihood that they will be attained. There is some indication of rising interest in soft energy paths (solar, for instance), but efforts are modest, especially when one considers the ease with which capital can now be procured. The recent escalation in nuclear electric energy costs, and the difficulties in obtaining efficient coal production, make the development of alternate energy systems even more urgent than current government priorities indicate.

The problems in the next decade hinge on how much oil is there, and on how long investor confidence can be maintained. If the proved reserves are in fact as great as estimates say, and no new proved reserves are added to current totals, then Mexico has 45 years of oil reserves at 1980 levels of production. This guarantees enough oil for Mexican needs, plus exports. Massive capital inflows, based on these estimates, are helping to end Mexico's economic depression and to ensure that there will be sufficient capital available for economic growth. Bottlenecks will continue to exist for trained personnel, for the adaptation of existing technology to semitropical areas, and for technological innovation. Major priorities such as the achievement of agricultural self-sufficiency and decentralization in public administration and economic growth depend on factors other than an abundance of energy as does a successful population policy. Mexico's newly discovered energy wealth has made possible, but does not assure, an adequate level and equitable distribution of economic growth. 'If the ten-year strategy is successful, then Mexico will be more capable than today of providing for its own economic growth, even if surplus energy for export and foreign investment should disappear at that time.

NOTES

1. Laura Randall, A Comparative Economic History of Latin America, 1500-1914, Vol. 1 Mexico (Ann Arbor, Mich. : University Microfilms International, 1977).

2. Frank R. Brandenburg, The Making of Modern Mexico (Englewood Cliffs, N. J. : Prentice-Hall, 1964), pp. 272-73; Nacional Financiera, S. A. , Statistics on the Mexican Economy (México, D. F. , 1977), p. 52; Wendell Gordon, The Expropriation of Foreign Owned Property In Mexico (Washington, D. C. : American Council on Public Affairs, 1941); William D. Metz, "Mexico, the Premier Oil Discovery in the Western Hemisphere," Science, December 8, 1978.

3. Luis E. Gutierrez Santos, Algunas Hipótesis Sobre la Demanda de Energéticos Tradicionales (Mexico: CIDE, August 1975), pp. 11-12.

4. Instituto Méxicano del Petróleo, Energéticos: Panorama Actual y Perspectivas (México, D. F. , 1974), p. 2. 5.

5. Various energy sources are placed on a comparable basis by

use of cubic meters of crude petroleum equivalent (CMCPE). The definition of CMCPE used in Mexico is quite special: it is the energetic power of a cubic meter of oil measured according to the products obtained in the Mexican petroleum industry from 1960 to 1968. The calculation was made by converting to kilocalories the energy products obtained from refinery production during this period and dividing this figure between the barrels of crude processed in the same period. Energéticos: Panorama Actual, p. 5; Instituto Mexicano del Petróleo, Energéticos: Demanda Regional Analisis y Perspectivas (México, D. F. , 1977), Table A. 11.

6. Statistics on the Mexican Economy, pp. 73-74; Energéticos: Demanda Regional, p. 246.

7. Inter-American Development Bank, Economic and Social Progress in Latin America, 1977 (Washington, D.C. , 1978), pp. 302-11; Comercio Exterior de México 24, no. 2 (February 1978): 51-54.

8. U. S. Department of Energy, Energy Information Administration, Energy Data Reports, July 28, 1978; Joaquín Córdova F. et al. , La energía nuclear en México (México, D. F. , February 1978), p. 8.

9. Excelsior, September 2, 1978, p. 12A.

10. President Portillo announced petroleum reserves of 200 billion barrels of petroleum and 20 trillion cubic feet of natural gas on September 1, 1978.

11. Secretaría de Patrimonio Nacional, Comisión Nacional de Energéticos, Propuesta de Lineamientos de Política Energética (México, D. F. , 1976).

12. Boletín Energéticos, March 1978, p. 18; Inter-American Development Bank, op. cit. ; Instituto Mexicano del Petróleo, Energéticos: Demanda Sectorial Analisis y Perspectivas (México, D. F. , 1975), pp. 188 and 193.

13. Boletín Energéticos, March 1978, p. 18.

14. Energéticos: Demanda Regional Analisis y Perspectivas, Tables III. 1. 1, III. 2. 10. The most rapidly growing regions were region number I, which includes Baja California, Norte; Sonora; Baja California, Sur; Sinaloa; and Nayarit; Region number III, which includes Coahuila, Nuevo Leon, and Tamaulipas; and Region Number VIII, which includes Distrito Federal and Area Metropolitana. The regions of average growth were region number II, composed of Chihuahua and Durango; region number V, composed of Jalisco, Colima, and Michoacan; region number VI, which is Vera Cruz; and region number X, composed of Tabasco, Campeche, Yucatan, and Quintana Roo. The regions of below-average growth were number IV, which includes Zacatecas, Aguascalientes, and San Luis Potosi; region number VII, composed of Guanajuato, Queretaro, Mexico, Morelos, Tlaxcala, and Puebla; and region number IX, which includes Guerrero, Oaxaca, and Chiapas.

15. Spearman rank coefficient equals .74. Boletín Energéticos, March 1978, p. 18; and Carroll L. Wilson, Energy: Global Prospects 1985-2000. Report of the Workshop on Alternative Energy Strategies (New York: McGraw-Hill, 1977), p. 277.

16. Sector Eléctrico Nacional, Comisión Federal de Electricidad, Resultados de Explotación, México 1975-1976; Sector Eléctrico Nacional, Información Básica, 1977.

17. Statement of Lic. Arturo del Castillo, Instituto Mexicano del Petróleo, at Foro Nuclear, July 1978.

18. El Sol de México, April 5, 1978; Excelsior, August 11, 1978 and August 13, 1978; McNeil-Lehrer Report, September 14, 1978; Sevinc Carlson, Mexico's Oil: Trends and Prospects to 1985 (Washington, D.C.: Georgetown University, May 1978), p. 26.

19. Federal Reserve Bank of Atlanta, Caribbean Economic Survey 4, no. 5 (September/October 1978): 15.

20. Juan Eibenschutz, Interview, July 1978.

21. Comparacencia del Sr. Ing. Jorgé Díaz Serrano, Director General de Petróleos Mexicanos, Ante el H. Congreso de la Unión (Mexico, D. F., October 1977).

22. Boletín IIE, No. 2 (June 1977), p. 4.

23. Boletín IIE 1, no. 3 (July 1977): 4, 11; Boletín Energéticos, October 1977, p. 14.

24. Eibenschutz, Interview; Excelsior, July 25, 1978.

25. Eibenschutz, Interview.

26. C. González Ochoa and E. Bianchi, "Tracción eléctrica en México," Boletín IIE, No. 1 (May 1977), pp. 9-11.

27. Boletín IIE 1, no. 5 (September 1977): 5; and Boletín IIE, No. 5 (May 1978), pp. 1, 25-26.

28. Boletín IIE, No. 1 (May 1977), pp. 6-7; Boletín IIE 1, No. 6 (October 1977): 2, No. 2 (February 1978): 2, No. 3 (March 1978): 7-11.

29. Energéticos, Boletín Informativo del Sector Energético 2, no. 3 (March 1978): 17.

30. Banco de México, Informe Anual, 1977.

31. Inter-American Development Bank, Economic and Social Progress, pp. 302-11; Comercio Exterior de México, 24, no. 2 (February 1978): 51-54; Excelsior, February 7, 1978.

32. Excelsior, September 2, 1978, p. 10A.

33. Ibid., pp. 10A, 13A. Jorgé Diaz Serrano notes that Saudi Arabia's reserves may increase; the Middle East has the largest known reserves. McNeil-Lehrer Report, September 14, 1978. Carlson, Mexico's Oil, p. 3.

34. "La política de petróleos Mexicanos," El Economista Mexicano 12, no. 2 (March-April 1978): 13.

35. Wilson, Energy, p. 116. If the September reserve estimates

prove to be correct, the reserve/production ratio is 39. 2/1 and will fall to 22. 5/1.

36. Excelsior, August 15, 1978. See also Excelsior, September 2, 1978.

37. Carlson, Mexico's Oil, p. 12. 1982 gas estimate is 8-10 billion cubic feet, of which she estimates 4 billion will be used in Mexico, 2 billion exported, and 2 billion surplus.

38. Statement of José López Portillo, Excelsior, July 30, 1978.

39. The most recent statement of this is Jorgé Díaz Serrano, McNeil-Lehrer Report, September 14, 1978.

40. This oversimplifies the problem. Since gas and oil are often found together, it is not always clear what share of drilling costs should be assigned to each. Similarly, it is not always clear what share of refinery and petrochemical plant costs should be assigned to each product.

41. Pemex's financial data are taken from Adrián Lajous Vargas and Víctor Villa, "El sector petrolero Mexicano 1970-1977, Estadísticas Básicas," Foro Internacional 72, Table 2. (The first commercially recoverable oil in Chiapas-Tabasco was discovered in 1972.) Share of crude purchased by U. S. refineries from John M. Blair, The Control of Oil (New York: Pantheon, 1976), pp. 300-01. Markup from Federal Energy Administration, Office of Regulatory Programs, Preliminary Report: Preliminary Findings and Views Concerning the Exemption of Motor Gasoline From the Mandatory Petroleum Allocation and Price Regulations, August 1977, p. 44.

42. Heberto Castillo, Interview, July 1978.

43. Pemex, Informe, March 18, 1978, pp. 27-28; Pemex, Memoria, 1978.

44. Miguel Arroche Parra, "Cuatro Puntos Cómo Sanear a Pemex," Excelsior, August 5, 1978.

45. Lajous Vargas and Villa, "El sector petrolero. "

46. The inflation rate is currently 15 percent. Excelsior, September 2, 1978. The scarcity of supplies in some sectors, and 30 percent idle capacity in industry, are consistent with a crowding-out hypothesis. Excelsior, July 16, 1978; July 26, 1978.

47. Pemex, Memoria de Labores, 1977.

48. Interviews, Heberto Castillo, July 1978, and Juan Eibenschutz, July 1978; Secretaria de Patrimonio Nacional, Comisión de Energéticos, Propuesta de Lineamientos de Política Energética (México, D. F. , 1976).

49. Marco Antonio Michel and Leopoldo Allub, "Petróleo y Cambio Social en el Sureste de México," Foro Internacional 8, no. 4 (April-June 1978): 72; Excelsior, August 28, 1978.

50. Excelsior, July 12, 23, 30, 1978; August 1, 1978.

51. Excelsior, September 3, 1978.

52. John H. Lichtblau, "OPEC as Export Refiners," prepared for the OPEC Review, September 1978.

53. Petroleum Intelligence Weekly, July 17, 1978.

54. Heberto Castillo, Interview, July 1978.

55. Arturo del Castillo, Desarrollo y perspectivas de la industria petroquimica mexicana (México, D. F.: Instituto Mexicano del Petróleo, 1977), p. 486 and p. 85; and Lajous Varga and Villa, "El sector petrolero"; Excelsior, August 15, 1978.

56. El Sol de México, March 17, 1978.

57. Excelsior, August 5, 1978.

58. del Castillo, Desarrollo y perspectivas, p. 21, and Carlson Mexico's Oil, p. 24.

59. Boletín Energéticos, March 1978, p. 17.

60. Comparacencia del Sr. Ing. Jorgé Díaz Serrano, pp. 3-18. President José López Portillo stated that the actual costs of the gasduct were less than estimated in the State of the Union message; no figures were given in Excelsior, September 2, 1978.

61. Comparacencia . . . .

62. Excelsior, September 2, 1978, p. 32A.

63. July 1977 Netherlands oil price; U. S. oil price from U. S. Department of Energy, Energy Information Administration, International Petroleum Annual, March 1978, pp. 29, 38. August 1977 U. S. gas price from U. S. Department of Energy, Monthly Energy Review; January-November 1977 Netherlands natural gas price, Parra Ramos and Parra, International Crude Oil and Product Prices (N. P. , N. D.).

64. Current speculation includes the possibility of shipping Mexican crude to Caribbean refineries currently handling Middle East petroleum on a swap basis, especially if foreign rates begin to make European refining of Mexican crude competitive with European refining of Middle East crude.

65. Heberto Castillo, Interview, July 1978.

66. See Tables 6. 1 and 6. 2.

67. Boletín Energéticos, March 1978.

68. Boletín IIE 1, no. 4 (August 1977): 3.

69. Ibid. , p. 4.

70. Juan Eibenschutz, Interview, July 1978.

71. Pablo Mulas, "Fuentes Alternativas," El Economista Mexicana, March-April 1978, p. 20.

72. Boletín Energéticos, March 1978; Energéticos: Demanda Sectorial Analisis y Perspectivas.

73. Boletín IIE, No. 2 (June 1977), p. 4.

74. Howard Green, "Geothermal Electricity: A Promising as Well as Problematic Basin Energy Alternative," Caribbean Basin Economic Survey 4, no. 5 (September-October 1978).

75. Boletín Energéticos, March 1978; Boletín IIE 1, no. 3 (July 1977): 1.

76. Ibid. , p. 5.

77. Ibid. , p. 11.

78. Ibid. , pp. 4, 11; Boletín Energéticos, October 1977, p. 14.

79. Wilson, Energy, p. 253.

80. Boletín IIE 1, no. 3 (July 1977): 13.

81. Boletín IIE 1, no. 7 (November 1977).

82. Spurgeon M. Keeney et al. , Nuclear Power Issues and Choices (Cambridge, Mass.: Ballinger, 1977), p. 87; Boletín IIE, 1, no. 7 (November 1977): 18; Arnulfo Morales Amado, Interview, July 1978.

83. Boletín IIE 1, no. 7 (November 1977): 2-4.

84. Keeney, Nuclear Power Issues, p. 393.

85. Ibid. , p. 395.

86. "Interview with Ing. Guillermo Robles Garibay, Coordinador Ejecutivo del Proyecto, El proyecto necleoélectrico de Laguna Verde, Ver. ," Boletín IIE 1, No. 7 (November 1977).

87. Arnulfo Morales Amado, La Energía Nuclear (México, D. F. , 1978), p. 44.

88. Ciencia y Desarrollo, No. 21 (July-August 1978), p. 93.

89. Science 201 (August 11, 1978): 51; Excelsior, October 20, 1978.

90. Interview, July 1978.

91. New York Times, July 30, 1978; Electrical World, October 1977.

92. Morales Amado, Interview; also see Morales Amado, La Energía Nuclear, op. cit. , pp. 44-45.

93. Excelsior, July 30, 1978; November 22, 1978; December 12, 1978.

94. Audiencia Pública, March 30, 1978, shift 34, p. 1; El Universal, March 17, 1978.

95. Excelsior, November 17, 1978.

96. Ciencia y Desarrollo, November-December 1978.

97. Audiencia Pública, March 14, 1978, shift 15, p. 2.

98. Ibid.

99. Compare Morales Amado, La Energía Nuclear, pp. 78-80.

100. Gutierrez Santos, Algunas Hipótesis; animal power is included in traditional sources.

101. Interview with Dr. Mulás, July 1978.

102. Boletín IIE 2, no. 6 (June 1978): 3.

103. Ibid. , pp. 4-5.

104. Norman L. Brown and James W. Howe, "Solar Energy for Village Development," Science, February 10, 1978, p. 653.

105. Interview, July 1978.

106. Boletín IIE, No. 2 (June 1977), p. 12; and, Dr. Mulás, Interview.

107. Dr. Mulás, Interview.

108. Dr. Usmani, Interview, March 1978.

109. Excelsior, August 16, 1978.

110. Dr. Mulás, Interview.

111. Excelsior, December 9, 1978, p. 19B.

112. Luis Unikel, in collaboration with Crescencio Ruiz Chiapetto, Gustavo Garza Villarreal, El Desarrollo Urbano de México: Diagnostico e Implicaciones Futuras (México, D F. : El Colegio de México, 1976), p. 324.

113. Mario Alberto Chavez Gonzalez, head of Departmento de Planeación Social de la Subsecretaría del Mejoramiento del Ambiente México, "Legal Protection of the Environment in Mexico," in Ignacio Carrillo Prieto and Raúl Nocedal, Legal Protection of the Environment in Developing Countries (México, D. F. : Universidad Nacional Autonoma de México, Instituto de Investigaciones Juridicas, 1976), p. 297.

114. Lucio Cabrera Acevedo, "Demographic and Legal Aspects of Pollution in Mexico," in ibid. , p. 306.

115. Ibid. , p. 307.

116. Excelsior, November 26, 1978, p. 19A.

117. Chavez Gonzalez, "Legal Protection," p. 298.

118. Julian Jurgensmeyer and Earle Blizzard, "Legal Aspects of Environmental Control in Mexico: An Analysis of Mexico's New Environmental Law," in Albert E. Utton, ed., Pollution and International Boundaries (Albuquerque: University of New Mexico Press, 1973), and Mario Alberto Chavez Gonzalez, "Legislación Mexicana Sobre el Medio Ambiente: Su Aplicación en el Golfo de California," Natural Resources Journal 16, no. 3 (July 1976): 475-82.

119. Excelsior, November 26, 1978, p. 19A; December 11, 1978, p. 5A. The State of Mexico planted 18 million trees in 1978 as a conservation measure.

120. Lucio Cabrera Acevedo, "Legal Protection of the Environment in Mexico," California Western International Law Journal 8, no. 1 (Winter 1978): 34.

121. Jorgé Díaz Serrano, Memoria (Mexico City, 1978).

122. David L. Scott, Pollution in the Electric Power Industry: Its Control and Costs (Lexington, Mass. : Lexington Books, 1973), pp. 74-77.

123. Excelsior, August 28, 1978; see also Michel and Allub, "Petróleo y Cambio Social."

124. Scott, Pollution in the Electric Power Industry.

125. Cabrera Acevedo, "Demographic and Legal Aspects," p. 311.

126. Keeney, Nuclear Power Issues, pp. 195, 211, 213, 230, 239-40.

127. Electrical World, February 1978.

128. Scott, Pollution in the Electric Power Industry.

129. Arnulfo Morales Amado, "Uranio," El Economista Méxicano, March-April 1978.

130. Scott, Pollution in the Electric Power Industry.

7

NIGERIA
Ernest J. Wilson III

The national energy profile of Nigeria presents an opportunity to examine critical policy issues of energy utilization and supply in a less developed, industrializing country. There are five key reasons to study the energy sector of Nigeria. First, Nigeria is the seventh largest exporter of petroleum in the world, producing around 2.2 million barrels per day. With a population of roughly 80 million people and reserves estimated to last only another 20 to 30 years at current production rates, Nigeria has emerged as one of the leaders of the price "hawks" within OPEC. In its international behavior, and in the domestic politics that conditions that behavior, Nigeria provides us with an example of energy policy making by a key OPEC member. Because of the confluence of international market conditions and internal political dynamics the country has experienced erratic production and earnings swings, showing among the greatest stop-and-go policies in OPEC. Second, Nigeria stands on the threshold of industrialization. Manufacturing grew 9.1 percent between 1960 and 1970, and 13.4 percent between 1970 and 1977, with value added in manufacturing standing at $941 million in 1975, up from $529 million five years earlier. [1] These are precisely the years when domestic energy demand is likely to spiral, even more rapidly than annual increases in GNP. Electricity, for example, has increased by about 22 percent annually, and per capita consumption of energy (in kilograms of coal equivalent) nearly tripled between 1960 and 1976. [2] At these early stages of economic development energy inputs are critical for the success of commerce and industry, whether the country is an exporter or an importer, and throughout the 1970s Nigeria has been both, importing as much as 40 percent of its furnished petroleum products from abroad. Third, of all the countries surveyed in this volume, Nigeria has the largest per-

315

centage of people in the traditional sector where energy supplies are necessarily restricted to wood and animal wastes due to low per capita income ($420 in 1977).[3] This huge noncommercial energy sector presents special problems to the policy maker. Fourth, the politics of energy in Nigeria are highly visible, very complex, and analytically challenging. Energy policies there are made in a national context of a federal, decentralized political system with 19 states and an active, socially mobilized, ethnically diverse, and politicized population. This population, which frequently acts through its regionally based ethnic groups, sees itself, despite significant internal class and regional cleavages, as the preeminent regional and continental leader. More and more Nigerians also see their state, the most populous and the richest in Black Africa, as an important actor to be reckoned with in the world arena as a whole. Oil has played a major part in the emergence of this new national self-image. Finally, at the start of a new decade Nigeria is poised on the edge of a great national experiment: after 13 years of military rule the country has just returned to civilian rule.

THE PATTERN OF ENERGY USE

The pattern of energy use in Nigeria is very much a product of Nigeria's three-quarters of a century of colonial history as an agricultural colony (and not an energy-intensive mining one), of its large rural population, its growing cities, and its newly exploited oil reserves.

Although energy analysts have given much attention to the Nigerian oil industry and its critical export role as a major supplier to the world market, there is no single sustained work examining the important issue of domestic energy allocation as well as international supply.[4] In fact, of course, issues of production and export and issues of domestic allocations of energy are but two sides of the same many-sided dilemma of national energy policy. Each is an important element in the Nigerian national energy profile considered as a whole. Policies in one directly affect policies in the other. There are also linkages through the state personnel making the key decisions. In addition the macroeconomic goals set by the government inform, in important ways, the separate strategies of the petroleum export industry and the domestic energy industries. Perhaps most importantly, the decisive factors of demographic concentration, geography, capital absorption capacity, the level and structure of gross national product, and the complex mosaic of national politics all act to shape both international and domestic energy policy. In Nigeria the role of politics and the state is critical. As with most developing countries, the bulk of energy production and distribution in all major sectors is in the hands of the government. This means that the state is intimately involved in energy policy, which means that in turn politics plays a large part in the conception and execution of national energy policies. This holds true for both domestic and international energy policies. Yet as we shall see, the structural factors of Nigeria shape policy outcomes in different energy subsectors in different ways, acting through different channels with different specific results.

Still, if we can find a single thread running throughout both halves of recent Nigerian energy policy, it is a shift toward the distribution and allocation of energy goods and services, away from a preoccupation with production issues. This shift toward distribution is a child of the 1970s; and in this shift the allocation of energy and energy-generated resources lies at the very heart of national energy policy in Nigeria, and in the center of the shift lies politics. One of the purposes of this chapter is to argue that politics, as well as technology, resources, and the market, is an important independent variable in the setting of Nigerian energy policy, in both the domestic and international policy arenas.

This chapter also argues that if Nigeria (and other monoexport developing countries) is to achieve self-sustained economic development, then a shift from production concerns to distribution concerns is an important shift but an insufficient one. The country must move toward a much more self-conscious third phase in which national energy policy is founded on the bedrock of transforming oil as a wasting resource into a national infrastructure promoting more balanced economic growth. Lanuching such a policy will be the task for the 1980s.

The historical perspective on commercial energy use in Nigeria reveals the shift away from coal and toward petroleum and the increases in commercial fuel use per capita, which parallels energy patterns in most other countries. The most dramatic expression of change in the energy sector between 1945 and today was, first, the decline of coal and the rise of oil and hydropower as the principal sources of supply and, second, the rapid increase in consumption in existing primary demand centers as well as in secondary and tertiary centers. In the power sector this new pattern of supply and consumption in turn encouraged the construction of long-distance transmission lines that in the late 1960s and early 1970s were linked into a regional and, ultimately, a national electricity grid. In the period of a single generation the dominant position of coal as the national fuel was eroded by its principal competitors, oil and gas (see Table 7.1). Coal plummeted from its top position of 64 percent of the national energy market in 1955 to the bottom position in 1972 of a mere 6 percent of the total market. [5] The cause of this decline was that oil was seen as a more desirable fuel by Nigerian policy makers, and it was imported in grow-

TABLE 7.1

Respective Shares of Energy Resources in Nigeria's Energy Market, 1955–72

Source of Energy	Percentage Share of Energy Market											
	1955	1960	1964	1965	1966	1967	1968	1969	1970	1971	1972
Coal	64	31	25	22	17	3	3	3	2	5	6
Hydropower	4	3	3	2	3	3	3	19	22	19	20
Natural gas	—	—	2	6	8	10	9	4	5	7	8
Petroleum products	32	66	70	70	72	84	88	77	71	69	66

Source: Central Bank of Nigeria, Annual Report and Statement of Accounts, various years, cited in Second National Development Plan (Lagos: Ministry of Information, 1970), p. 63.

ing quantities to replace domestically available coal. Oil and gas jumped from 32 percent and 0 percent, respectively, in 1955, to 66 percent and 8 percent of the market in 1970.[6] Hydropower (generated by falling water at dam sites) actually fell between 1955 and 1968 from 4 percent to 3 percent, then it shot up to 19 percent in 1969 with the commissioning of the Kainji Dam in that year.[7]

Beyond the source of energy supply, its geographic distribution is an important fact to weigh in assessing the development performance of countries in today's technological age, just as it was in assessing the development prospects of nineteenth-century Europe. The distribution of naturally occurring fuels will also affect the pattern and rhythm of development between regions within the same country, just as it will affect development patterns between countries. In Nigeria the geography of energy supply is uneven, and this affects, and is in turn affected by, the structure of national and regional politics. In Nigeria, coal is located in the Eastern Region, while hydroelectric supplies are found in the Northern and North-Central Regions. Petroleum and associated gas supplies are in the Eastern and Mid-Western regions. Table 7.2 indicates the primary fuel reserves estimated to be within Nigeria.

The Nigerian national energy distribution profile, like that of other countries, has its distributional asymmetries. Yet because Nigeria is still underdeveloped, and still with the distortions of a colonial political economy, the asymmetries in energy are especially stark.

TABLE 7.2

Primary Fuel Reserves in Nigeria

Resource	Unit	Known Reserve	Probable Reserve	Possible Reserve
Crude oil	Million tons	200	600	1,200
Natural gas	Million cubic meters	280	400	800
Coal	Million tons	360	500	800
Lignite	Million tons	75	150	300
Thorium	Thousand tons	15	20	25
Hydroelectric potential	Billion kwh per annum	17	20	20

Source: Ignatius Ukpong, "Impact of Foreign Aid on Electricity Development in Nigeria," Journal of African Studies 1 (1974): 276.

They are stark between the entire commercial sector and the entire noncommercial sector, but there are also striking and politically salient imbalances within the commercial sector. These distributional asymmetries can be seen along spatial dimensions, sectoral dimensions, and social dimensions. Spatial dimensions include urban/rural disparities, as well as disparities among different national regions or states. Since regional differences frequently coincide with ethnic ones they are especially politically significant in the Nigerian context. Sectoral dimensions refer to energy differences between types of national economic activity—industry, agriculture, transport, and so on. Social asymmetries are far more difficult to measure quantitatively because of poor or nonexistent data, but they are no less real. This dimension refers to differences in access to or consumption of energy among income groups.

Petroleum and electricity are the major two commercial fuels and provide the best indicators of fuel use. The distribution of petroleum by product and by state reveals the heavily skewed character of energy consumption in the country. Of course, these energy differences in turn reflect differences in the level of urbanization and economic development. The spatial and sectoral distribution of commercial energy is expressed in Table 7.3.

The sectoral figures on petroleum product use demonstrate how dependent the country is on gasoline for transportation (which may include private trucks used as taxis and buses), and on diesel, used for the railroad and by the National Electric Power Authority (see Table 7.4). These categories constitute about 19.8 and 13.7 million barrels respectively.

The same forces of uneven spatial and sectoral development that emerge in the petroleum figures cited above are also apparent in the figures available for electricity distribution. [8]

Thus far we have directed our attention to the commercial fuel sector—oil, gas, coal, hydro, and electricity—but we need to say something about the traditional, noncommercial sector as well. The vast majority of the population relies on this sector; yet, paradoxically, there are almost no data on this sector for Nigeria, nor for any other African country. [9] There are at least two reasons for this, one technical, one political. There are real technical difficulties in collecting data on both the demand and the supply side. There are tens of millions of separate, unconnected suppliers and consumers of the traditional fuels: wood and animal wastes, charcoal and sunlight. There are few direct policy instruments easily available to the government since these energy forms are not very susceptible to price, tax, or subsidy programs, and it is hard to renew the supply (trees). In fact, one of the greatest "energy crises" in parts of Nigeria is the growing deforestation problem. Many areas are inaccessible and there

TABLE 7.3

Nigerian Domestic Consumption of Petroleum Products, January–December 1977

Products	Refinery Offtake	Total Imports	Lagos State	Sokoto State	Niger State	Kaduna State	Kano State	Borno State	Bauchi State	Gongola State	Benue State
Liquefied petroleum gas (m. tons)	10,046	13,341	12,529	107	3	1,263	529	104	351	27	52
Aviation spirit	—	9,646	6,765	—	—	419	208	40	—	26	51
Motor spirit											
Premium	772,162	1,514,831	474,497	33,302	27,072	131,494	131,873	46,595	18,204	12,822	35,742
Regular	205,931	149,147	56,560	18,589	3,733	28,044	21,727	20,621	5,774	3,907	5,812
Dual-purpose kerosene											
Household	381,792	522,370	120,532	14,223	5,767	40,265	68,250	13,012	4,662	2,901	13,310
Aviation turbine	3,711	—	267,926	1,699	—	3,212	63,894	3,001	—	—	—
Automotive gas oil											
Gas oil	633,960	875,526	314,869	16,845	19,890	46,191	80,977	30,189	18,855	18,855	12,348
Diesel oil	142,289	33,168	161,636	34,283	3,921	28,517	27,776	35,204	12,592	13,541	19,119
Fuel oil											
High pour	322,517	35,856	91,876	—	—	—	—	—	—	1	—
Low pour	266,113	45,619	153,985	3,571	27	60,405	16,055	821	1,482	696	248
Lubricating oils	51,125	88,018	42,130	2,831	601	6,111	10,091	1,976	15,863	410	2,132
Greases (000 kgs)	384	2,742	1,280	46	—	154	150	47	406	2	32
Petroleum jelly, waxes, etc. (000 kgs)	135	72,094	12,998	33	1	99	2,031	86	7	1	—
Bitumen & asphalt, etc. (m. tons)	—	225,539	261,781	710	2,754	2,849	8,164	645	718	275	739
Others	28	94,031	67,545	379	374	2,172	1,979	110	34	78	204

(continued)

Table 7.3, continued

Product	Plateau State	Kwara State	Ogun State	Ondo State	Oyo State	Bendel State	Anambra State	Imo State	Cross River State	Rivers State	Total Consumption
Liquefied petroleum gas (m. tons)	133	191	181	83	1,060	1,377	1,003	348	1,210	1,241	21,792
Aviation spirit	145	–	–	1,248	66	52	26	–	14	86	9,176
Motor spirit											
Premium	83,800	69,751	96,891	91,298	278,831	209,350	182,092	116,814	87,741	99,331	2,227,500
Regular	25,902	11,525	21,466	14,775	38,756	31,432	20,343	11,455	7,514	10,145	358,080
Dual Purpose Kerosene											
Household	32,649	17,512	30,008	19,005	70,077	52,163	48,385	37,979	35,334	50,348	676,382
Aviation turbine	2	–	–	–	1,319	1,778	111	131	1,228	1,468	345,769
Automotive gas oil											
Gas oil	25,027	42,917	31,827	29,698	45,343	124,365	46,127	43,983	44,056	102,093	1,097,137
Diesel oil	36,027	20,654	13,351	8,730	32,993	68,074	26,560	14,849	33,456	17,684	608,967
Fuel oil											
High pour	–	–	102,394	82	–	–	–	–	–	54,756	194,353
Low pour	2,017	10,234	7,406	1,395	6,418	29,207	22,435	2,288	9,494	56,636	384,820
Lubricating oils	3,927	1,733	3,708	1,384	9,540	13,541	87,928	3,230	4,407	8,539	220,082
Greases (000 kgs)	80	17	43	17	172	262	240	164	261	234	3,607
Petroleum jelly, waxes, etc. (000 kgs)	268	2	15	2	216	280	80	331	91	162	16,703
Bitumen & asphalt, etc. (m. tons)	523	1,143	505	39	7,569	23,689	15,242	7,631	25,889	36,676	397,541
Others	681	202	139	983	7,091	1,633	1,481	848	199	501	86,333

Notes: All figures are in thousand liters unless otherwise stated. The figures under "Refinery Offtake" indicate the total withdrawal from local refinery for both domestic consumption and export. Import figures of lubricating oil, greases, bitumen and asphalt, and others include those for private importers.

Source: Oil industry sources in Nigeria, 1979.

TABLE 7.4

Petroleum Product Use in Nigeria by Sector
(in 1,000 barrels)

Class	Gaso-line	Kero-sene	Diesel	Fuel Oil	Lubri-cants
Manufacturing			274	3,820	110
Marine			1,370	425	50
Aviation		2,302			
Private transportation	19,769				520
Public transportation			12,052		320
Others		5,370			
Total	19,769	7,672	13,696	4,245	1,000

Source: Oil industry sources in Nigeria, 1978.

is considerable variety in fuel efficiency, from village to village. The second reason reflects institutional and political reality. Traditional agricultural and rural populations are the hardest to reach with any kind of public service or marketed commodity. Also, such communities lack sufficient political clout to put alternative energy sources and rural supply on the national agenda.

There are clear asymmetries that exist within the Nigerian energy picture. The southern part of the country consumes far more than the north, industry more than agriculture, and urban areas more than rural areas, and only a few large urban areas consume a huge proportion of all energy consumed. Beyond these raw data lie social and political interests. The differential perception of these asymmetries by different interest groups in the country can express itself in the political arena through demands for greater distributional equity.

In summary, we have seen that the Nigerian energy sector runs the gamut from such energy choosers as modern industry to rural cattle herders, and to many energy supply choices, from tree branches and animal wastes to the power of falling water, coal, and oil. Government energy policy is almost entirely directed toward the modern energy economy, which is the most capital intensive and efficient but which can provide energy services to a relatively small if growing portion of the population.

The distribution of energy consumption in Nigeria, as every-

where, directly reflects the distribution of economic activities. The
energy structure of demand very nearly reflects the economic structure itself, and it is to the impact of the economic structure on the
energy sector that we now turn. Economically, Nigeria exhibits the
uneven development—the dual economy—that characterizes all underdeveloped countries, and of course this shapes the energy sector. Its
per capita GNP was only $420 in 1977, with that figure greatly distorted
through the oil revenues. The average growth rate per capita was 3.6
percent between independence in 1960 and 1977.[10] Although agriculture
accounted for only 34 percent of GDP in 1977, it accounted for about
60 percent of the work force.[11] The overall sectoral composition of
the economy is shown in Table 7.5.

TABLE 7.5

Percentage Distribution of Nigerian GDP, 1971/72-1974/75

Sector	Year			
	1971/72	1972/73	1973/74	1974/75
Agriculture	32.0	27.9	24.7	23.4
Mining	39.3	43.4	45.1	45.5
Manufacturing	4.1	4.8	4.8	4.7
Electricity and water	0.3	0.4	0.4	0.4
Building and construction	4.1	4.7	5.4	5.7
Distribution	8.1	7.4	6.9	6.7
Transport and communication	1.9	2.1	2.1	2.3
Services*	10.1	9.3	10.4	11.3

Note: The valuation is constant 1974/75 prices.
*Services comprise general government, education, health, and
other services.
Source: Third National Development Plan (Lagos: Ministry of
Economic Development, 1975.

Typically is is the growth of urbanization and industry that generates much of the increased demand for energy. According to the
World Bank, GDP grew 6.2 percent between 1970 and 1977, with industry growing 10.3 percent, manufacturing 13.4 percent, and agriculture -1.5 percent.[12] By 1977 the bank estimated that industry accounted for 43 percent of GDP, but manufacturing broken out sepa-
rately accounted for only 9 percent of total GDP.[13] Clearly much of

the growth then was in the petroleum export sector. Thus, while the oil revenues do skew the picture, there has been nearly a doubling of manufacturing since 1960, and these new industries have placed great demands on the production, supply, and distribution of commercial energy. We can recall that energy consumption per capita, in kilograms of coal equivalent, went from 34 in 1960 to 94 in 1976. [14] The national figures for urbanization are also revealing in this regard (see Table 7.6). It should be remembered that while the growth rates are not as high as in some other countries, the high original base with a population of 80 million means that pressures on social services and energy services will be great.

TABLE 7.6

Urbanization in Nigeria, 1960 and 1975

Urban population	
As percent of total population	
1960	13
1975	18
Average annual growth (percent)	
1960-70	4.7
1970-75	4.6
Percent of urban population	
In largest city	
1960	13
1975	17
In cities of over 500,000 people	
1960	22
1975	33
Number of cities of over 500,000 people	
1960	2
1975	5

Source: World Development Report, 1979 (Oxford: Oxford University Press for the World Bank, 1979), p. 164.

Since industry is the largest consumer of electrical energy and one of the largest consumers of petroleum products, and because there was a wide disparity in regional consumption rates, we should not be surprised that there are considerable asymmetries in the spatial distribution of industries (see Table 7.7). Lagos state, with 2.6 percent

TABLE 7.7

Nigerian Regional Distribution of Industries, 1971

State	Population (Percent of National)	Industries		Employment		Location Quotient*
		Number	National Percent	Number	National Percent	
South Eastern	8.3	28	3.2	5,047	3.5	0.42
North Eastern	14.0	14	1.6	1,067	0.7	0.05
Mid-West	4.5	64	7.4	8,627	5.9	1.31
Benue Plateau	7.2	48	5.5	3,297	2.3	0.32
North Western	10.3	16	2.8	2,205	1.5	0.15
East Central	11.2	149	17.1	8,445	5.8	0.52
Kano	10.4	41	4.7	9,099	6.3	0.61
North Central	7.3	37	4.3	18,996	13.1	1.79
Rivers	2.8	11	1.3	1,442	1.0	0.36
Kwara	4.3	21	2.4	5,441	3.7	0.86
Lagos	2.6	273	32.0	67,884	46.7	17.96
West	17.0	163	13.7	13,895	9.5	0.56
Total Nigeria		865	100.0	145,445	100.0	1.00

*National percent of employment divided by population.

Note: Percents do not always equal 100.

Source: Nigeria: Abstract of Statistics (Lagos: Federal Office of Statistics, 1972).

of the population, had 32 percent of the country's industries located there and 46 percent of its employment. North Eastern state, with 14 percent of the country's population, had only 1.6 percent of its industries. These differentials are of course reflected also in energy demand.

Per capita income is an important influence on energy consumption, with rises in per capita income associated with increases in per capita energy consumption, especially in the developing countries. Wage and income levels grew much faster for commerce and manufacturing than for other sectors, and we would anticipate that workers in these areas, all other things being equal, would consume greater amounts of energy than their counterparts in agriculture, for example.

Compared to many countries in the world, Nigeria's energy profile is remarkably complex on both the demand and the supply side. For in Nigeria there are many energy choices available from which to choose a source of supply, and on the demand side there are many energy choosers. On the supply side there is a wide array of primary energy resources on which any number of alternative or complementary national energy strategies may be based. There are large oil and gas reserves that make Nigeria the fifth largest exporter in the world. There are more than ample supplies of hydropower for electricity, and there are coal reserves as well. Solar energy is plentiful in this tropical country, and wind power and other more exotic energy forms look promising for commercial exploitation. On the demand side, among the choosers and consumers, where groups organize themselves within the country to secure reliable and low-cost supplies, the Nigerian energy profile is even more complex. Consumers and potential consumers, whether private firms, individuals, villages, or state agencies, all compete for scarce energy supplies, and they are all conditioned by the peculiar features of the Nigerian political economy.

One important conditioning feature is that the state plays a dominant role in the energy sector domestically, especially in electricity and coal where it has an effective monopoly, and to a slightly lesser extent in petroleum distribution. Another conditioning feature is Nigeria's federal system. There are 19 relatively autonomous states, a weak executive, and expanding state and federal bureaucracies. There are separate energy-related agencies for each of the states and at least half a dozen energy actors at the federal level. These numerous public and private actors must now work under a new (1978) Constitution that tries to combine elements of the U.S. presidential system with the cabinet system of Britain.

Underlying these changing institutional and juridical features of the Nigerian political economy are the more permanent social and political features of Nigeria, which express themselves along several prominent axes: regional, ethnic, class, and national. [15] These together form the basis of an active pluralist political system based within regional strongholds, where groups are highly mobilized and where the political cultures encourage competition. Competition and the distribution of scarce resources have been viewed typically by most participants as proceeding on a regional basis. Since the major ethnic groups—the Hausa-Fulani in the north, the Yoruba in the west, and the Ibo in the east—are also regionally based, the distribution of these resources regionally carries strong ethnic implications and the potential for ethnic conflict. Politics in Nigeria, then, usually leads to divisions along the following lines: region versus region, ethnic group versus ethnic group, urban versus rural, public versus private, employer versus employee, and foreign versus domestic.

British colonialism has left a political economy legacy in which foreign enterprise, whether in banking, trade, or manufacturing, still effectively controls the "commanding heights" of the economy. [16] In many respects, postcolonial Nigeria is very much a typically dependent, peripheral country in which multinational corporations (MNCs), through their alliance with local elites, can effectively dominate public economic policy. Neither the state officials nor the local businessmen have been able to establish effective control over MNCs as has happened to a greater extent in other export-oriented developing countries like Brazil, Mexico, Korea, or Algeria. This lack of effective state control leads to incredible political stalemates, in which state-directed large-scale projects languish on the drawing boards because the effective executive power to reach and impose a national, legitimate decision is missing. There is no effective, nationally dominant political coalition. The energy sector is a good example of such nondecisions.

What is critical here, and absolutely essential to our understanding of Nigerian energy policy, is that this condition is a function of political stalemate. The political power of the central government grew in some respects, but effectiveness did not grow apace. In the energy sector, where the state has such a preponderant role, the apparent inefficiencies and delays cannot simply be reduced to a lack of skilled manpower, a lack of capital, or to inexperience. Much of the explanation lies in the character of national politics as they play themselves out in the energy sector, and it is the play of politics that can in turn be used to explain the desultory pace at which Nigeria moved to indigenize its work force and control more closely the behavior of the oil companies. This situation can be resolved only through political means, and in the early 1970s steps were taken to begin building such a political capacity. [17]

DOMESTIC ENERGY POLICY

To talk about a national growth policy in the Nigerian context is to say both very little and a great deal. A great deal because the thrust of regional, national, and individual expressions regarding national policy has been an unrestrained call for quick and widespread economic growth. On the other hand, to say that there has been a growth policy is to err on the side of rhetoric. Under the laissez-faire doctrine of the first two plans, the government played a caretaker role. Only in the last five years has there slowly emerged a notion of a coherent national growth policy. [18] More typical is to see growth policy as a series of discrete projects with little explicit coordination between them. Still, there has been a general orientation toward building up infrastructure and the service and industrial sectors, with very few resources committed to agriculture. Despite occasional pronouncements to the contrary, until very recently growth-producing investments and allocations were directed toward already established cities. As one would expect, the energy sector followed these general growth orientations of the political economy as a whole.

From the 1940s through the late 1960s, the key domestic energy issues in all subsectors (coal, hydro, oil and gas, electricity) was to build up energy productive capacity in order to provide sufficient supplies for the major urban and industrial consumers. Here we can think of the Ughelli and Afam gas turbines (then the largest in Africa), and especially the commissioning in 1969 of the huge Kainji Dam, called the pivotal centerpiece for the entire Five-Year Development Plan, because of its importance to the economy and because it was the single largest project in the plan. This orientation was essentially a supply orientation in the way it defined the problems to be solved and in its commitment of scarce capital, labor, and administrative resources.

By the opening of the new decade of the 1970s, however, the government began to address the allocative aspects of energy, and to ensure that existing supplies were more widely and evenly spread. These were in large measure a response in the Second Development Plan (1970-75) to the growing popular demands on all levels of government to expand and improve service delivery beyond the confines of the large cities and industrial centers. The sources of this new popular demand lay largely in the social mobilization caused by the Nigerian Civil War (1966-70) and in the floodgates of pent-up economic and political demands released at the war's end. The rural-urban migration rate accelerated and cities were swamped with spiralling demands for social services, including water, transport, electricity, and fuel. Not only did the numbers of people demanding services and distributive equity increase, but their means became more assertive. The number of strikes rose dramatically at the end of the war, from 32 in 1968-69 and 53 in 1969-70 to 143 in 1970-71. [19]

The Second Plan set forth five rather broad goals for the nation, and its constituent sectors, to move toward the plan period. They were to establish Nigeria as a united, strong, and self-reliant nation; a great and dynamic economy; a just and egalitarian society; a land of bright and full opportunities for all citizens; and a free and democratic society. [20]

During the first chaotic civilian Republic (1960-66), these goals were honored as much in the breach as in policy during the free-wheeling, chaotic, and frequently vindictive and vituperative years that followed independence. By the formulation of the Third Development Plan under the military government, the planners apparently realized that even if the goodwill and political means were available it would be almost impossible to implement and measure goals so broadly stated. While not repudiating them, the Third Plan spelled them out in greater detail. They were: increase in per capita income; more even distribution of income; reduction in the level of unemployment; increase in the supply of high-level manpower; diversification of the economy; balanced development; and indigenization of economic activity. [21]

Each of these domestic objectives contains a kind of double implication for the energy sector. First, to be implemented, several of these objectives would require improved infrastructural inputs, not the least of which would be energy. Both balanced development (whether along sectoral or geographic lines) and diversification of the economy will require, at a minimum, more energy production and, in a fuller sense, a different kind of energy strategy. In other words, if the economy is to change course to follow a particular national growth strategy, then, as an important part of the national economy, the energy sector also must be redirected. Second, the energy sector too must be provided with a new strategy that follows national priorities. A more vertically and horizontally diverse economy requires a more vertically and horizontally diverse energy sector. The structure of effective demand will have changed.

In terms of the energy sector the Second Development Plan marks the beginning of the shift from production to distributional concerns, although it was expressed in inchoate and preliminary terms. Written in 1969-70 during the waning months of the war, the plan spelled out the government's attitude toward the mining and petroleum sector. The state called for:

"Active state participation in mining operations" (Prior to this time government oil policy was essentially passive; hereafter they took more cues from OPEC.)

"Diversification of mineral products" (For the petroleum sector this would mean expanding gas production and sale.)

"Policy development in the area of conservation of the country's mineral resources"

"Research into efficient extraction methods. "[22]

Once again these concerns as expressed in the Second Plan reveal a limited notion of the sector. The focus here is on production and conservation. In a later chapter the government does announce its intentions to move more into control of refinery and product distribution but admits that

> the activities of the oil prospecting and producing companies are . . . so shrouded in secrecy that discussion of this important sector has always been in terms of generalities. The industry is entirely private, except for Government's partial interest in the refining branch. Any meaningful Government policy regarding the petroleum industry has, therefore, not been possible beyond broad guidelines with respect to (i) production, distribution and pricing of crude and refined petroleum products; (ii) government revenues from the industry and (iii) recruitment of Nigerians into the industry. [23]

The government also announced that it would take up the "vexed question of countrywide uniform prices for petroleum, bearing in mind that the differential in transport costs is an important element in the price structure and a vital factor in the availability of petroleum products in all parts of the country. "[24] The plan then raised some of the looming distributional questions in petroleum, but the resolution of how equitably to allocate energy resources to competing regions, sectors, industries, and ethnic and income groups would remain a hot topic throughout the decade of the 1970s.

For the electricity sector the primary focus on production peaked a year earlier with the coming on-stream of the 320 MW Kainji Dam on the Niger River. The electricity sector had held much of its expansion plans in check pending the commissioning of the dam, which, at a cost of around 85 million Nigerian pounds, was by far the largest single item in the Second Plan, and was popularly called the keystone of that plan. [25]

The pressures for greater energy distribution, which were just making themselves felt in 1970, intensified over the next four or five years. Demands were expressed for regular access to electricity and petroleum by national and local interest groups such as Chambers of Commerce and other businessmen's associations, by individual firms, villages, and rural communities, and by state officials. Their demands were expressed through press articles, editorials, and letters to the editor, in annual company reports, in local manifestos, and in village improvement association minutes. Interviews conducted by the author during the mid-1970s showed these concerns for reliable sources of energy at reasonable prices. The emphasis of most was on the security of supply; most were willing to pay for energy services (whether kerosene in rural areas or electricity in towns and cities), but the problem was having a supply of the product that was dependable. One long-time analyst of Nigeria estimated that fuel and light accounted for only 1. 5 percent of household expenditures nationally and 3. 5 percent in cities. [26] These arguments and demands, carried through public and private channels and backed up by political pressure, merged with the perception of national government officials that more needed to be done in the energy sector to satisfy public demand. These considerations were included in the preparation of the Third Development Plan for 1975-80.

The exploratory movements toward distributional policies of the Second Plan were firmed up in the Third Plan, and there was a greater sense of the important backward and forward linkages between energy and other sectors of the economy. The plan called for the following in the petroleum sector: (1) manpower development and accelerated transfer of technology; (2) achievement of internal self-sufficiency in the supply of petroleum products; (3) effective distribution network for petroleum products; (4) export of petroindustry products; (5) commercial utilization of gas; and (6) involvement in the oil services subsector. [27]

These new priorities, which did allocate state resources in new directions, typify the shift toward distributive issues. We will simply point out a few of their implications. At a minimum they indicate a major commitment to shift control from expatriate MNC control toward Nigerian control. This holds in points 1, 3, 4, and 6 quite explicitly, since the government held that these should be areas of operation for either the Nigerian state or individual Nigerian entrepreneurs. There are also regional implications: Where should training schools be located (and hence from what ethnic groups are they most likely to draw their students)? Will the public sector or the private sector control the new petrochemical industry? Will new facilities be based around cities or in rural areas? Each of the potential answers to these questions creates new distribution patterns that in turn create new patterns of job creation and income distribution. Since the stakes are high, politics will come into play.

The oil embargo had little impact directly on Nigeria. In most countries of the world 1973-74 provided a double shock—a cyclical downturn in the world economy and the simultaneous fourfold increase in oil prices. This disrupted economic growth in most countries more than the very short-term "embargo." Nigeria did benefit from higher prices, and it experienced these years differently than most. The economy, released from its wartime constraints, and with reconstruction, grew rapidly if unevenly. Reserves rose from 409.1 million naira* in 1973 to 3.5 billion naira at the end of 1974.[28]

Just as the price increases affected Nigeria positively, contrary to the experience of most states, so there was little incentive to conserve energy in an oil-booming economy. Energy conservation simply isn't on any serious agenda, with rather good reason. There are multiple fuels available, and in the Nigerian context conservation has a rather different meaning. The term calls to mind efforts to set the most effective rates to exploit (and conserve) the national reserves of oil and gas. Much of the same holds true for policies of recycling and environmental protection from energy excesses.

Throughout its history and until the late 1970s Nigeria obtained its petroleum supplies from two sources. Like most countries of the world Nigeria imported oil. Roughly 40 percent of its refined product supply came from abroad in 1974. These products were imported from a variety of sources.

The remaining supply to meet Nigerian demand came from Nigeria's own oil wells. In this respect Nigeria was not dissimilar in result to countries like Brazil, India, or the United States, which have some domestic supply, although insufficient amounts to meet their entire national demand. However, Nigeria is the fifth largest oil exporter in the world. The reasons for this anomaly were severe infrastructural bottlenecks compounded by political stalemate. The major policy issues facing the government's energy officials during this period were how to eliminate their oil supply bottlenecks while diversifying their domestic supplies through expanded gas production. Such bottlenecks hurt financially. It was far more expensive to rely on imported refined products than to refine domestically produced crude in national refineries.

The Nigerian domestic oil and gas industry in all its stages from production to refining and distribution is a perfect example of Nigeria's shifting balance between public and private, local and foreign participation in the economy as different actors jockey for market shares and political influence. The result is that the contours of the Nigerian petroleum industry have been remarkably fluid over the five years between 1975 and 1980.

There are several sources of this fluidity, including the demon-

---

*The current exchange value of the naira is 1.63 naira = $1.00.

stration effect of other OPEC governments' successes in controlling the domestic oil market, the higher prices, tighter market and greater potential of rewards, and new self-confidence and muscle by both local entrepreneurs and local state officials. Another factor, however, lies also in a lack of elite consensus over national oil policies, leading to both stalemate and bitter feuding.

Historically owned and controlled by foreign companies that distributed and sold petroleum products throughout the country, the domestic supply network was targeted by the Nigerian government, which since 1975 had pushed for greater local control. This move by the government was part and parcel of the larger nationalist policy of 1972 to reserve certain categories of business activity for Nigerians. One schedule of businesses (the less capital-intensive) were 100 percent reserved, while other more complex areas like banking and manufacturing were to be given over 40 percent to Nigerians. The political impetus came from Nigerian businessmen anxious to break into lucrative merchant import and export areas, but the policy was eventually extended to the petroleum sector as well.

Nigerian participation in the domestic oil service subsector that actually does the road hauling of products for the large marketing companies has grown and well over three-fourths of the oil tank truck business is now in Nigerian hands. [29] Whether the government or private businessmen will purchase the shares of oil-related companies forced to comply with the Indigenization Decree is a source of considerable friction in Nigeria. It is another expression of the large number of actual and potential actors in the Nigerian energy scene that complicates its politics and its policies, making coherent policy difficult. Elements of the private Nigerian entrepreneurial stratum are anxious to force the government to require that more contracts and oil business go to private Nigerian businessmen. For example, at a national meeting of the Nigerian Chamber of Commerce, Industry, Mines, and Agriculture called to discuss "Investing Nigeria's Oil Wealth: A Unique Challenge for Balanced Development," a local Nigerian oil consultant called for greater private Nigerian participation in the business. He criticized government decisions to keep control of the industry in its own hands as "a one-sided monopolistic situation which is alien and detrimental to the community."[30] There are counterarguments on the side of some (but not all) government officials. The permanent secretary in the Ministry of Mines and Power (then responsible for the petroleum industry) argued that "the important thing is not to create Nigerian millionaires as it were by mere designation through the government handing over oil bearing concessions to what must necessarily be only a handful of people, with no defensible criteria of previous oil industry experience."[31] While directed more at offshore production than domestic distribution and servicing, the general concern about distributive equity behind the comment is part of the reason the state officials dispute private-sector control of the industry. Another reason, one suspects, is that it would mean their ceding power over the country's critical resource to other actors.

Prior to 1975 there were six petroleum marketing companies in the country: Shell (the largest), Total, British Petroleum (BP), Mobil, Agip, and Esso. Agip and Total also distributed gas products (primarily for cooking), as did Nidogas, Cotsgas, Sungas, Utilgas, and Grenigas.

In that year the Nigerian Enterprises Promotion Decree of 1972 was finally extended to parts of the domestic oil and gas distribution industry. Interestingly, the first to sell 60 percent of its shares was Shell, which besides being the largest domestic marketer is also by far the largest oil producer in the country. Sources within the Nigerian petroleum industry surmised that Shell's early willingness to cooperate hinged on its feeling that cooperation on the domestic front would help them on the international production side where the real money was being made. [32] Accordingly, the federal government bought out a controlling 60 percent of Shell and renamed it the National Oil and Chemicals Company. The second company to sell out was the smallest in Nigeria, Esso, which had not been doing well and would not be severely affected by government takeover. It was renamed Unipetrol. In 1979, following on the heels of the nationalization of its production shares, BP's domestic operations were taken over by the government and renamed African Petroleum. These three firms represent about 32 percent of the domestic oil market. [33]

Four other companies, by way of contrast, did not sell their shares to the federal government, but rather through the local stock exchange largely to private individuals. There the 60 percent shares could be purchased by anyone interested; apparently, most of the buyers were private citizens but there appear to have been some state government purchases as well. Thus the domestic arm of Mobil, Total, Texaco, and Agip have conformed to the indigenization order by selling to private citizens.

There are other sources of domestic oil industry politicization over the distributive impact of national energy policies. These include pricing (which affects different classes of consumers differently depending on their income) and the distribution of energy infrastructure. And as in so many other disputes, regional differences provide much of the basis for these differences. And because so many issues align along a north-south split, each substantive issue takes on a symbolic character, often with detrimental effects on the oil industry.

Until the mid-1970s consumers in the north had to pay more for their oil than their southern counterparts. This was due to the higher transport costs in the more demographically dispersed north, and because it is farther from the national sources of supply located in the southern coastal regions. The higher northern prices led groups to protest. They exerted political pressure to compensate northern consumers for market differences. Northern leaders argued that prices should be equalized, since they were being penalized for a fluke of geography, and that the oil price differentials would further retard the already less economically developed north. It is far from clear empirically that fuel price differentials gave any competitive edge to the south, but that wasn't really the issue. The critical issue is that the disputes introduced into the fuel issue had their origin in long-standing regional inequalities and political differences. For these reasons a government decree rendered the price more equal between the regions. * A Price Equalization Board was authorized to administer the program. The domestic distributors had complained that they were being forced to bear the burden unfairly of government policy and demanded explicit subsidies to compensate them.

The early 1970s saw Nigeria in the anomalous position of being the world's ninth largest producer of petroleum and yet importing over 40 percent of its petroleum products for domestic use. The technical origins of this dilemma lay in the small size of the original refinery built in 1965 by a consortium led by Shell-BP. Built at Alesa-Eleme near Port Harcourt in the oil fields, and operating under the name of the Nigerian Petroleum Refining Company (NPRC), Shell-BP held 50 percent and the Nigerian government the rest. Built with a capacity of 35,000 barrels per stream day (bpsd), it was felt that this would adequately cover Nigerian demand in the coming year. The onset of the Civil War shut down the refinery almost immediately, and it

> resumed production in May 1970 after extensive rehabilitation in which its operating capacity was increased to 50,000 bpsd from 42,000 bpsd to match internal demand which had begun to show signs of rapid growth. Beset by some technical problems, the refinery, in spite of its increased capacity, was unable to meet the internal demands of all the products in manufacturing, and resort was made to importation of certain grades of petroleum products, albeit on a very modest scale, to augment the inadequate supplies. [34]

The "modest" incremental trickle soon turned into a flood of imported

---

*An exact parallel of this pricing dispute also came up in the electricity sector, with a similar political resolution.

oil as demand for refined products (as with electricity) shot way beyond the plans of the NPRC. Five thousand bpsd were added to capacity in 1972, yet demand continued to spiral. According to interviews
conducted in Nigeria it is clear that after 1972 and 1973 serious thought
was given to building a second refinery, the most likely site being
Warri, again near the oilfields.

However, a dispute developed as to whether the second refinery
would be built in the south (that is, at Warri) or in the north. There
was a long and expensive delay while this political (and not, it should
be made clear, a technical or financial) stalemate was fought over.
The political impasse was breached only when it was decided that
while the second refinery would indeed be built at Warri, the third
would be constructed in the north. Yet here again political rationality
prevailed in the short run, since the northern Kaduna facility will be
fed by crude shipped through an expensive pipeline, where it will be
refined and the product sold; then, because of the demand structure
in the less developed north the remaining products will have to be re-
shipped back to the south for consumption or export. Economic efficiency may catch up with political expediency, but in this case politics
preceded.

In the mid-1970s, however, the press ran frequent reports of
petroleum shortages in various parts of the nation, and the Nigerians
experienced their own version of the "oil crisis." The origins of this
crisis clearly had more to do with domestic bottlenecks than with
OPEC strategies, but they did generate considerable antagonism on
the part of the affected populace. According to one industry report
the causes of the 1974-75 distributional shortages were: inadequate
berthing space at the Apapa Petroleum Wharf; subnormal performance
by the Nigerian Railway Corporation; inadequate refining capacity; and
abnormally high growth consumption in the northern states after the
introduction of the uniform pricing policy. This policy is expected to
stimulate high growth demands in the hinterland areas. [35] Steps were
taken, as reported in the Third Development Plan, to improve each
of these bottlenecks.

An important fact to keep in mind here and in the discussions
that follow is the systematic nature of the problems faced by energy
industries in developing countries. Although Nigeria is a large international supplier, it shares problems of product distribution with other
developing countries that import all of their petroleum supplies. It is
interesting to note that only one item on the list above could be solved
through improved performance of the Nigerian petroleum industry; the
others are problems faced by most developing countries. In the industrialized states one is accustomed to the difficulty in making energy
policy because, on the demand side, there are so many market uncertainties. Developing countries share those problems, but in addition

they founder on the shoals of insufficient basic infrastructure such as roads and ports. By consequence distributional asymmetries are worsened.

To reduce distributional asymmetries in the electricity subsector, a subsector that in 1970 had only 230,000 consumers and most of those concentrated in urban areas, the national government started on a rural electrification program. [36] With modest beginnings in the Second Plan, by the Third Plan the anticipated oil revenues gave the government the wherewithal to launch a national program. By the publication of the Third Plan's Second Progress Report the government was able to report that 40 percent of the countrywide electrification scheme of 105 towns had been completed. [37] The low completion rate of many other projects, however, had led the central government to permit the 12 (later 19) states to set up their own projects. They have the option of tieing into the national grid, or if too distant from the grid, they may be supplied by isolated generators. The 12 state governments planned to spend a total of $235.6 million by the end of the plan period. The pressure for rural and improved town electrification has been immense, and the governments at both levels have had to take serious cognizance of it. I have described these pressures elsewhere. [38]

The production and distribution of coal have declined since the war, but efforts are under way to restore the industry. Coal was already on the decline when the difficulties of the war worsened its position within the energy sector. Production in the last year before disruption—1965—was 740,000 tons and fell to 346,000 tons in 1975. [39] Coal has become somewhat more attractive for domestic use and export with the higher prices of oil. At the moment, however, there does not appear to be a large political constituency or major market for coal.

Energy politics in Nigeria in the 1970s were based simultaneously on the social mobilization and the release of repressed demand that followed the end of the Civil War, and the rapid rise in oil-generated income. The one provided the impetus and the other the means to expand energy allocations. Energy expansion paralleled the expansion of other goods and services and occurred within the context of an expansion of industry and commerce, which were the major customers for the new energy expansion.

The national energy sector, however, is not the automatic and reflexive outcome of "economic development," with each stage of energy sector expansion lock-stepped into a specific stage of economic production. [40] Final demand for energy supply by firms and individuals in the Nigerian market does shape the broad contours of what can sustain the energy market in an economic sense. However, political demands and responses, through their ordering of competing priorities, play an important role in filling in the distributive details of the energy system. In each plan period and in each year the sectoral, technical, and resource possibilities are shaped—as the rural electrification program, the Petroleum Price Equalization Law, and the refinery sitings were shaped—by political demands expressed through the state and federal agencies.

Just as self-interested political demands are found on the domestic side of the Nigerian energy sector, they are also found on its international side. However, the number of actors, their institutional bases, their constituencies, and the arenas in which actors compete are different.

INTERNATIONAL ENERGY POLICY

When we speak of international energy policies we mean governmental actions over taxation and pricing, production levels, and marketing of either oil or gas in refined or crude form destined for export. [41] More broadly, in developing exporting countries, the issues are often reduced quite simply to those of ownership and control. The distinction between domestic and international policies is somewhat artificial since elites and constituencies, institutional arenas and concrete decision rules overlap considerably. With these caveats in mind the distinction is nonetheless a useful one. We will first treat the ownership and control issues before returning to our theme of distribution. By way of introduction, Table 7.8 shows the variety of Nigerian export markets and the importance of Nigerian oil for some of these markets. It should be noted that while Nigerian crude is only 5.3 percent of total U.S. demand, Nigeria is the second largest foreign supplier and sells the United States over 15 percent of its total imported oil supply.

In Nigeria, as in all other oil-exporting countries, the oil industry has certain characteristics that are useful to keep in mind. The industry has tended to be foreign dominated, an enclave sector, and based upon a wasting resource. [42]

The oil industry in Nigeria and all other developing countries has until very recently been foreign dominated. Through their monopolistic control over production technology and training, buttressed by the state power of their home governments, the seven large, vertically integrated companies dominated the world market for petroleum from well to pump. The juridical mechanism through which this economic and political power was organized between the company and the oil country was the concession. The concession granted to an oil company the right to ownership of the oil and in effect nearly unlimited right to control every aspect of the petroleum operations, including exploration, production, pricing, and marketing. This principle of the concession was not challenged until the early 1970s.

TABLE 7.8

Nigeria's Crude Oil Exports, 1972–77
(thousand barrels per day)

Destination	1977	1976	1975	1974	1973	1972
North America	812.5	746.5	538.9	703.3	567.9	457.3
of which:						
Canada	—	8.5	11.1	2.9	32.3	35.7
United States	812.5	738.0	527.8	700.4	535.6	421.6
Latin America	386.6	222.8	234.2	241.1	278.3	115.4
of which:						
Brazil	34.8	18.1	—	—	5.3	20.0
Trinidad & Tobago	—	—	—	—	13.7	—
Uruguay	—	8.5	3.7	5.6	13.1	8.3
British Territories	129.6	196.2	171.8	175.0	227.1	87.1
Western Europe	785.1	997.2	860.4	1,110.8	1,002.9	1,080.0
of which:						
Belgium & Luxembourg	3.4	8.9	13.0	9.0	17.2	12.1
Denmark	1.3	5.1	19.3	10.1	21.8	29.3
France	170.1	184.9	196.0	212.5	262.4	285.1
Germany (West)	115.4	121.3	108.5	138.4	61.9	72.9
Italy	26.7	14.8	19.8	67.6	46.4	77.6
Netherlands	297.7	439.7	239.4	315.7	266.5	208.4
Norway	14.8	11.3	14.3	11.1	5.7	11.7
Sweden	16.5	14.0	22.0	13.9	22.3	31.1
United Kingdom	135.0	192.2	213.7	320.5	297.7	329.0
Africa	46.0	38.1	32.2	34.1	30.4	25.8
of which:						
Ghana	19.4	17.7	12.3	—	12.1	5.2
Ivory Coast	13.6	8.0	10.6	—	7.9	10.7
Senegal	4.1	6.2	5.1	—	4.6	4.7
Sierra Leone	3.6	3.9	4.2	—	4.6	4.5
Asia and Far East	—	8.6	47.6	90.1	98.6	77.6
of which:						
Japan	—	8.6	47.6	90.1	98.6	76.5
Total	2,030.2	2,013.2	1,713.3	2,179.4	1,978.1	1,756.1

Note: No effort has been made to reconcile balances.
Source: OPEC Annual Statistical Bulletin 1977, September 1978, p. 71.

The development and control of the oil industry in Nigeria began with a monopoly held by Shell-d'Arcy Petroleum Development Company, which acquired exploration and production leases in 1938. [43] The name was switched to Shell-BP when British Petroleum joined as partner, and the company maintained effective monopoly control over the territory until the early 1960s when Mobil, Tenneco, and others acquired concessions. Shell-BP still is the largest producer in Nigeria by far. [44] Today, every major oil company is or has been active in Nigerian fields. Commercial oil was discovered in 1956, and production began in 1959. In that same year a Petroleum Profits Tax Ordinance fixed a tax rate of 50 percent of the company's net profits. [45] Despite an adjustment in 1967 basing government revenues on posted prices, the essential basis for government-firm relations remained the oil concession, in which the company set production, price, and distribution. With the new tax and pricing agreements in Teheran and Tripoli in 1970-71 the historic power relationship between country and company began to shift in favor of producer countries like Nigeria. [46] In Nigeria the "oil revolution" was in one sense launched with the creation of a new national instrument, the Nigerian National Oil Corporation (NNOC). This moved the state beyond concessions and directly toward participation. The decree gave NNOC formal authority to engage in all aspects of the industry, including exploration, production, and marketing. [47]

This formal NNOC authority was not matched by a pool of trained Nigerian personnel to implement the decree fully, and NNOC has been slow getting off the mark. Just as the Nigerian leadership was dovish on pricing, it was also dovish and apparently reluctant to push the oil companies to accelerate the share of Nigerians in top positions. [48]

The personnel shortage of NNOC, its own internal and interministerial squabbles effectively described in the work of Terisa Turner, [49] and the apparent lack of general policy orientation have left NNOC as a manager of state participations in the already producing companies, and not as a truly operational national oil company involved in all segments of the industry. By 1975 it had increased its participation from 35 percent to 55 percent, and in 1979 it forced the companies to agree to a 60 percent participation agreement, thereby bringing it in line with other large indigenized industries in other sectors like banking. In some newer, smaller entrants into the field such as Ashland Oil, the government enjoys a 65 percent-35 percent participation.

The story of Nigeria's search for greater control is centered not only on the push from concessions to participation, and greater leverage over pricing and production levels, but also more and more state control over marketing practices as well. Nigeria, like other OPEC countries, has expanded government-to-government deals with consuming countries, bypassing the traditional transport and marketing role of the majors. As recently as 1973 the seven majors lifted and marketed the vast majority of the world's international oil. By early 1980 their share organized through long-term contracts had fallen to only 42 percent. [50] According to the International Energy Agency, fully one-sixth of OPEC production is now sold in such government-to-government deals, and another sixth sold on the high-priced spot market. [51] Nigerian policy is part of an international trend in the oil market, and in this area it is likely to continue as such.

Nigeria in 1979 established direct deals of 100,000 barrels per day (bpd) with the national oil companies of Hungary, India, Portugal, Spain, Uruguay, and West Germany. Petrobrás from Brazil has also negotiated increases in its existing Nigerian import levels. [52] The reasons for the shift to direct government-to-government deals lie in producer countries' desires to prevent profit-taking by the majors, to diversify their distribution sources to prevent overreliance on a few firms, and to gain better leverage with their large, powerful customers. This may help them translate oil wealth into more generalizable power. Said President-elect Shagari: "We should like to see [the countries like the United States] generally contributing towards the development of Nigeria and of Africa in general . . . then, of course, it is our objective to reciprocate by providing them with the necessary oil that they require." [53] Not only will these government-to-government deals increase, but the chief of staff of the Supreme Military Headquarters in Lagos said just before the civilians took over that he would like to see the number of oil companies operating in Nigeria tripled. [54] This too would make the country less dependent on any single company.

Thus we see while moving in fits and starts, the NNPC* has gained greater control than it had in the past over the key production decisions of exploration, production, price, and marketing. The true test, however, will be whether NNPC can transform itself, like Algeria's Sonatrach, from an essentially supervisory company overseeing work contracted to others to a genuine operating company. Not only must the benefits of the oil industry be captured from the control of the MNCs, but these benefits must be distributed beyond the narrow confines of the sector itself. [55] This raises fundamental issues of social and economic equity. In this sector distribution raises special problems because of the enclave character of the industry. By enclave sector we mean a highly self-contained industry that has very few direct linkages with the rest of the economy. There are few products that the local economy can contribute to the sector (backward linkages), and only limited contributions that the industry itself can make to the

---

*The NNOC was renamed the NNPC in April 1977.

rest of the economy, whether in terms of jobs, the creation of ancillary products, or as inputs to other sectors.

Employment in 1976 in the oil sector was limited to only about 1.3 percent of the modern employment sector. [56] This represented about 4,500 people directly engaged in the industry and about 15,000 in oil service industries. [57] Cumulative expenditures on local purchases and contracting work, local rents, wages and salaries paid locally, and so on, amounted to about 950 million naira. [58] This is a drop in the budget for a country of over 80 million people. Thus the best way to think of the industry is as a highly modern, technologically complex, and capital-intensive one grafted onto an otherwise traditional, agricultural, and poor country.

One notable effort to build auxiliary industries is the government's decision to go ahead with a $500 million petrochemical complex to be built by the U.S. firm Pullman-Kellogg. This complex will be a major user of natural gas that is now flared, to yield 1,000 tons each of ammonia and compound fertilizer and 1,500 tons of urea. [59]

The greatest impact of the sector is in its contribution to government coffers, and hence to spending for economic and political purposes. By the late 1970s oil was contributing about 90 percent to government revenues and about 84 percent to foreign exchange earnings. This last figure should be contrasted with 1969 figures of 15.4 percent, and the latter figure was up from about 7 percent in 1963.

Oil has come to account for almost half the country's gross domestic product. Very few would deny that oil has made significant contributions to aggregate growth in Nigeria. Nor would many suggest that these contributions have been equitably distributed. Fabulous, instant fortunes have been visible in Lagos, Kano, and Port Harcourt, while being hard to document. Commentators from all sides of the political spectrum now call, however, for greater equity in the distribution of oil revenues, not the least of whom is the new President Shagari. [60]

Returning to our concern with the spatial, sectoral, and social distributional aspects of energy policy, Table 7.9 shows the priorities assigned to the four major national sectors in the Third Plan and the actual spending for them.

A point of bitter contention during the oil years has been the design of a politically acceptable formula for dividing oil revenues that could pass the self-interested scrutiny of the major national and regional actors. The original formula, relying on a principle of derivation, allocated considerable monies to the two producing states along the Bight of Benin's Delta Area. This soon proved unacceptable to the majority of other states (and to many top civil servants in Lagos, who felt that the oil wealth should be nationally controlled from Lagos, and should be distributed more equitably to other parts of the country).

TABLE 7.9

Percentage Distribution of Public Expenditure in Nigeria, 1975-76 and 1976-77

Sector	Plan Proportions, 1975-76	Actual Proportions, 1975-76	Plan Proportions, 1976-77	Actual Proportions, 1976-77
Economic	70.03	53.57	61.66	52.53
Social	11.31	20.21	11.50	16.94
Regional development	8.18	8.68	13.84	11.91
General administration	10.48	17.54	13.00	18.62
Total	100.00	100.00	100.00	100.00

Source: Second Progress Report on the Third National Development Plan (Lagos: Ministry of Economic Development, 1977).

Slowly, oil came to be seen as a "national" commodity, and the derivations principle slowly fell under an alternative formula. In the process, the political center in Lagos was vastly strengthened. (The successful completion of the war also strengthened Lagos' hand.) There were, of course, losers as well as winners, for the formula finally agreed upon was to divide revenues for the states into two pools, 50 percent to be allocated by population, 50 percent by state.[61]

The final characteristic of the industry relevant here is that petroleum is a wasting resource. This raises unique technical and political economy issues. There is first a need to find an optimal technical rate of extraction, which varies from well to well. Second, governments must determine the optimal social rate of extraction that can maximize benefits to the economy and the society as a whole. Too rapid a rate may generate more revenues than the national infrastructure is capable of channelling and investing effectively, it will generate inflation with new money chasing the same quantity of goods and services, and it may create regional imbalances. In addition, as oil is sold for dollars, and dollars or other hard currencies lose their value through inflation, then the value of the national patrimony has declined through large dollar holdings. It may be more rational from the OPEC country's perspective to keep oil in the ground, since its value is likely to appreciate, rather than sell it for depreciating dollars. The Third Plan recognizes this when it says: "In the relatively short time that the economy will enjoy a surplus of investible resources it is intended that maximum effort will be made to create the economic and social infrastructure necessary for self-sustaining growth in the longer run when resource scarcity may recur."[62] However, there has been very little consistency in moving from recognition of the problem of a wasting asset to concrete public policy. Government investments have been made largely in basic infrastructure roads, power lines, communications, and only marginally on the productive economic base necessary to carry the country beyond oil.

On the whole, how effective were the Nigerian efforts to increase state control over the oil industry and at the same time respond to popular and elite demands for energy services and revenue distribution? The results are mixed. State performance was of course shaped by the same features that shape energy policy in all developing nations' oil export sector: as a foreign-dominated enclave industry based on a valuable but wasting resource. Yet there were national peculiarities. Nigerian performance was in many ways at odds with the performance during the same 1973-79 period of all other OPEC countries, with wide swings in production levels and earnings.

The 1973 price hike sent Nigeria's balance of payments soaring, from 808.1 million naira prior to the boom in 1972 to 5.2 billion naira in 1974.[63] In an effort to capture even more revenue, production was pushed up in 1974 to almost 2.3 million bpd.[64] The policy decision to increase prices and production seemed a good one at the time, especially since Nigerian light crude was very much in demand because its low sulfur qualities enables it to be used in the United States and other developed countries with stringent antipollution controls. However, the rapid and unexpected jump in oil prices combined with an already impending worldwide cyclical recession led to considerable softening of demand in the industrialized world, the major consumers of OPEC oil. Nigeria, stuck with high prices and high production, got caught rather badly and earnings fell from over 1 billion naira to 4.2 million naira in 1975 as production was finally cut back from 2.2 million bpd to 1.8 million bpd.[65]

During a period of rising demand, as in the early 1970s, exporters' hands are strengthened, but with slack demand and the real price of oil flat, advantage shifts more to the companies. Still, during this period the NNOC increased its equity share in the majors to 55 percent and began its policy of taking over private local distribution companies as discussed above. However, they were not able to accomplish all they wanted as market conditions worsened.

Real earnings did continue to decline, however, and what followed between 1975 and 1979 was a series of somewhat contradictory policy moves. First crude prices were reduced in early 1975 for both buy-back oil and direct sales in an effort to make Nigerian oil more competitive; yet at the same time they increased the income tax and royalty rates from 65.7 to 85 percent and from 16.75 to 20 percent respectively.[66]

> At the end of the year, however, the government reversed its price policy in an effort to increase revenues. Both selling and posted prices were increased by about 38 cents a barrel and credit terms were tightened. Company profit margins were reduced from 50 to about 30 cents a barrel, closer to Middle East levels. However, as investment per barrel of production is higher in Nigeria, the companies deemed this lower margin to be too low to justify new investments and exploration activities declined substantially.[67]

The number of rigs drilling in Nigeria declined between 1974 and February 1977 from 29 to 13. Seismic activity decreased from 75 party months in 1973 to 45 by 1977.[68]

Contributing to Nigeria's problem was new competition from British North Sea and U.S. Alaskan North slope oil, both of which also have low sulfur contents, and which enjoy transport cost advantages, being nearer major markets.

Throughout this same period the mania of oil money and petro-naira was running rampant throughout Nigeria. The economy was badly overheated, inflation was skyrocketing, and imports were soaring for luxury items and food as well as capital goods. This was also the period of the infamous cement glut, when profiteers made millions on government cement contracts, resulting in, among other things, a Lagos harbor clogged with cement-carrying vessels. Much of the money during this period was not spent wisely. Another contributory factor was a highly inflationary wage settlement—the Udoji Report—that was politically expedient but economically disastrous. Nigeria's current account balance fell from a $4.5 billion surplus in 1974 to $700 million in 1975, a particularly precipitous drop (see Table 7.10). This drop is larger than many other OPEC nations' and reflects the population's size and demands, as well as undisciplined governmental policy.

By 1978 there was a balance of trade deficit of $4 billion, and Nigeria was obliged to declare an "austerity budget" and to move vigorously into the world money market, where for a time they found it difficult to put together a loan. In 1978 the head of state, General Obasanjo, had to announce a curtailed budget, slashing recurrent expenditures 10 percent and demanding "sacrifices for all."[69]

With recognition that the balance-of-payments scissor was getting serious and that exploration rates were plummeting, the state oil

TABLE 7. 10

Distribution of OPEC Current Account Surplus, 1974-75
($ billions)

	1974	1975
Total	61. 5	29. 4
Saudi Arabia	22. 9	18. 0
Iran	10. 9	4. 7
Kuwait	7. 3	4. 4
United Arab Emirates	3. 1	3. 1
Qatar	1. 3	0. 8
Algeria	0. 4	-2. 5
Ecuador	0. 0	-0. 2
Gabon	0. 2	0. 0
Indonesia	0. 4	-2. 2
Iraq	2. 4	1. 0
Libya	1. 9	-1. 4
Nigeria	4. 5	0. 7
Venezuela	6. 2	3. 0

Source: Morgan Guaranty Trust Company, World Financial
Markets, January 21, 1976, p. 7. Reprinted with permission.

officials again took action in the oil industry. The government again
reversed itself by giving more lenient tax credits, amortization benefits, and more favorable expensing of drilling and exploration costs. [70]
In addition, to align itself more effectively with the soft international
market it moved prices downward. The short-term impact of these
moves, combined with the falling dollar, was a serious drop in revenues in the neighborhood of several hundred thousand dollars per day. [71]
The government was having a difficult time finding the right balance-
of-price levels, company incentives, and production levels that would
match the market. The military government of the time had to try to
satisfy the growing expectations of itself and its 80 million people for
more goods and services by squeezing the companies, while not squeezing so hard that they seriously reduced exploration and production.
This was an uncertain process at best with only a small cadre of
trained and experienced Nigerian managers and engineers.

By February 1978 production had fallen to a level as low as any
in the postwar recovery period, to 1. 57 million bpd. [72] Again, prices
were cut by 21 cents per barrel. [73] With the United States as Nigeria's

largest customer since British oil came on-stream, Nigeria was also especially vulnerable to capital losses through dollar inflation; 1978 was a bad year.

Fortunately for the Nigerian oil industry, the turbulence in Iran and its reduction of production and export cutbacks diminished world supply and made Nigerian crude a much-sought-after commodity. So sought after, in fact, that Nigeria could add on surcharges for its premium oil above the OPEC ceiling. Earnings for 1979 soared over 50 percent.

The fits and starts in Nigerian oil policy between 1973 and 1979 can only partially be explained by the ebb and flow of the international oil market. It would appear that the shifts in levels of production were greater for Nigeria than for other OPEC countries. In Table 7.11 we see that there was a 26.2 percent decline in crude oil production between 1977 and 1978. This is by far the greatest shift of any OPEC member. Then between June 1978 and June 1979 again we find the greatest percentage shift of any country outside of the troubled Iran.

We can only surmise the reasons for this greater see saw Nigerian rate of change. A conventional economic explanation can provide part of the answer. New "sweet" oil supplies came on-stream on the supply side, and imports were sluggish on the demand side. A technical learning curve explanation is also valid. Much of the blame must be placed at the feet of faulty government planning. The government simply misread market signals, overpriced and overproduced its oil, and the country's earnings and development projects paid the cost. It is likely that the coincidence of moving to take over a complex, powerful, and recalcitrant industry, and the unprecedented turbulence in a market known since the 1950s for its tranquility, overloaded the capacity of the new technicians, managers, and the military policy makers to respond adequately. * Sources in industry and government suggest that at critical junctures the advice of the technicians in the NNPC regarding production and price levels was overridden by the military, which was intent on quickly expanding revenues to pay for even more quickly expanding imports. [74] Other sources suggest too that mistakes were made in the then overconfident NNPC. There is perhaps a broader political explanation to be offered as well, which compliments the learning curve-market condition arguments. It is quite likely that the absence of a unified political elite in Nigeria, one that could transcend regions, public and private sectors, and religion,

---

*To find parallels one could trace the turbulence in production and earnings when other oil-exporting governments took control of the oil industry.

TABLE 7.11

Changes in Crude Oil Production, 1978 versus 1977 (thousand barrels per day)

Country	First Half, Percent Change
OPEC	-8.9
Saudi Arabia	-17.5
Kuwait	+0.8
Iran	-0.6
Iraq	+8.3
Libya	-12.5
Abu Dhabi	-14.5
Venezuela	-0.5
Nigeria	-26.2
Indonesia	+0.3
Algeria	-8.0
Non-OPEC	+7.7
United States	+7.9
Western Hemisphere	+4.2
Mexico	+11.5
United Kingdom	+38.3
Canada	-1.8
Non-communist world	-3.2

Country	June 1979	Average for the First Half		
		1979	1978	Percent Change
Saudi Arabia[a]	8,500	8,997	7,573	18.8
Iran[a]	3,900	2,564	5,583	-54.1
Iraq[a]	3,500	3,299	2,398	37.6
Kuwait[a]	2,300	2,294	1,697	35.2
UAE	1,859	1,824	1,826	-0.1
Algeria[a]	1,225	1,225	1,225	
Libya	2,017	2,068	1,876	10.2
Nigeria	2,400	2,419	1,672	44.7
Indonesia	1,615	1,607	1,680	-4.3
Venezuela	2,251	2,343	2,052	14.2
Other OPEC[b]	1,474	1,509	1,241	21.6
OPEC	31,041	30,151	28,825	4.6
United States	8,655	8,678	8,646	.4
Canada	1,534	1,486	1,275	16.5
Mexico	1,420	1,399	1,131	23.7
United Kingdom	1,740	1,522	984	54.7
Norway	356	368	351	4.8
Other non-OPEC	4,254	4,213	3,967	6.2
Non-OPEC	17,959	17,667	16,354	8.0
Free world total	49,000	47,818	45,179	5.8

[a] Includes estimates.
[b] Includes all Neutral Zone production.
Note: There are slight discrepancies in the totals.
Sources: West Africa, September 25, 1978, p. 1908; Platt's Oilgram News 57, no. 165 (August 1979).

exacerbated the effect of these constraints. This Nigerian immobil-isme has been described by businessmen, bureaucrats, academics, and journalists alike. [75] It signifies the failure of the existing region-ally based elites—whether in khaki or pinstripe or traditional robes—to reconcile political differences over the distribution of scarce re-sources in Nigeria. Because the country structurally has the second largest population of any OPEC country it faces incredible stresses not felt in less populous states with larger petroleum reserves. Rec-onciliation and compromise have proven difficult to achieve. Until such time, region versus region, public versus private, urban ver-sus rural, rich versus poor cleavages will continue to exert their hold over international as well as domestic energy policies.

ENERGY AND THE ENVIRONMENT

The saliency of pollution as a discrete policy problem is asso-ciated with an advanced threshold of industrialization and the negative externalities that are associated with energy-intensive industrial pro-duction. Nigeria is perhaps the most developed country of Black Afri-ca, it has a number of import-substitution and oil-related industries, and it is grouped by the World Bank as a middle income country. Yet the "environment" simply has not emerged in Nigeria as an issue. It is not consistently raised by interest groups, nor by the press in gov-ernment statements. The level of industrialization, and the density of production that elsewhere places energy pollution on the public a-genda are absent. [76]

In fact, the solicitous concern of the already industrialized and already polluted nations for developing-country environmental prob-lems is a double-edged sword:

> Africa is a "late-comer" in terms of contemporary eco-
> nomic growth and technological change: it is suspicious
> that concern for the environment is the latest strategy
> devised by the rich states to prevent its industrialization.
> Moreover, Africans have long perceived concern among
> the rich states over populations problems to be a way of
> limiting their own growth and power potential; likewise
> they see the recent rise of the ecological issue to be
> merely another way of perpetuating their subordination. [77]

Given these attitudes, and the absence of even moderate energy-related pollution, energy and the environment is of very little political moment in Nigeria.

FUTURE PROSPECTS

It may be that the technical and managerial lessons of the last decade have been learned, and that the 1980s will see a surer and more effective Nigerian energy sector. This is especially so in the oil industry. It may become, in other words, the Nigerian oil industry rather than an oil industry operating in Nigeria. There are some signs that this is taking place. The NNPC has been reorganized and given larger responsibility and a new structure. Under its new chairman, A. K. Hart, the new structure will have five major divisions: petrochemicals, tankers and transportation, marketing, exploration, and exploitation. The new president, Alhaji Shehu Shagari, has appointed Y. A. Dikko, former general manager of the electricity corporation, as his special adviser on energy matters and it appears that they are taking direct responsibility in energy.

Nonetheless, the success or failure of Nigerian energy policy, in such a politicized and plural society, will not hinge on organizational and personnel appointments alone. There are at least two limits that must be considered. One is the physical limit to Nigerian oil. The U. S. Department of Energy estimates that there is a 95 percent probability that Nigeria will discover at least 1.8 billion new barrels of oil and a 5 percent probability that 22.8 billion still await discovery. [78] Hydroelectric resources are ample, with the following five hydro sites probably constituting the most promising generation points: Shiroro, with 100 MW, Ikom with 400, Jebba with 500, Makurdi with 600, and Lokoja with 1,950 MW. However, with electricity maximum demand expected to reach at least 2,100 MW in 1984/85 and 4,900 by 1994/95, thermal plants at Port Harcourt, Warri, Onitsha, Kaduna, and possibly a 500-plus-MW nuclear plant at Lagos will have to be considered as well.

The other perhaps more important limitation is the human (and infrastructural) limit to a broadly successful program of energy for national development. There are severe limits on the number of competent electricians, mechanics, and engineers that can be trained, apprenticed, and certified each year. The pool of prospects is large, but the demands for services are even larger and ever growing. Even if all training objectives are met, there remain problems channelling individuals into specifically targeted energy subsectors. The problem, unfortunately, is not restricted to insufficient people to build and maintain infrastructure in the energy sector. National road networks, railways, telecommunications, and ports all contribute to or subtract from the successful working of the energy sector. The government will have to push harder to attract workers into the energy sector. Here too the 1980s will find problems of compensation, with the public sector perhaps finding it difficult to offer salaries high enough to attract and retain qualified personnel who have private-sector options. Related to this is the political problem of government pressure on foreign energy firms to replace expatriates with Nigerians. Evidence in other countries suggests that companies will indigenize their management when forcefully and consistently pressured to do so.

Within these natural resource and human resource constraints we can identify several key energy issues that will be important through the 1980s. On the export side gas is the key issue. Other issues will include expanding refining capacity to export finished oil products, and strengthening the NNPC. Finally, the linkages between the enclave export sector and the rest of the political economy must be more effectively rationalized. On the domestic side, electrification will continue to pose a nearly insoluble set of problems.

For electricity the dilemma is how to expand constantly distribution of the product when supply of trained personnel is not expanding as rapidly. There is a Hobbesian choice: to stop rural electrification expansion and instead consolidate present service levels, which are generally acknowledged to be poor; or to expand services further, playing catch-up with staff and materials and guaranteeing the delivery of poor service. A related problem in the domestic market is to smoothly bring on-stream new generating capacity. Nigeria may see a shift toward production headaches again in the mid- to late 1980s. Planners must determine the optimal sequence of electricity-generating sources based on hydro, oil, gas, coal, or nuclear power.

Another related issue will be for the government to construct by the end of the decade an institutional and policy framework for debating the question of nuclear power. There have already been some considerations given to nuclear power for the next 20 years, including discussion of a 600-MW plant for Lagos. Pressure to go nuclear will likely come as much from political and strategic as from energy considerations. There is the international status that derives from operating a nuclear plant, and there is the fact that Nigeria's archenemy and principal rival for influence on the continent—South Africa—already has a nuclear capability.

At the other end of the energy spectrum, the wealth of domestic commercial energy resources has tended to deflect concern away from renewable energy resources, although there is likely to be some expanded interest over the next decade. Today there is some interest in solar energy at the new National Science and Technology Development Agency, and research is being conducted at the universities of Lagos, Ahmadu Bello, Ife, and Ibadan. As solar photovoltaic cell prices decline, attention to solar energy will grow in Nigeria, but through the 1980s the bulk of the resources for energy will continue to go toward conventional commercial energy sources.

The negotiations currently under way for the development of a gas export industry in Nigeria have been tenaciously conducted, with some European, but especially U. S. , firms taking part. The prices offered by the Nigerians are "significantly higher than the price for Mexican gas. " The U. S. companies seem somewhat willing to proceed, but they and the Nigerian government are uncertain whether the U. S. Department of Energy will give its approval for the higher negotiated price level. This will be an important issue over the next few years as more contracts are negotiated. The U. S. decision will of course partially be determined by economic considerations, and partially by political ones. According to Platt's Oilgram News, the NNPC in Lagos "is anxious to win a political decision from the U. S. government" about buying Nigerian gas before proceeding with the projected plants. [79] Some American officials apparently are concerned precisely about the reliability of Nigeria as a supplier in view of potential U. S. - Nigerian political differences over South Africa. It is likely, however, that some deal will be signed and approved for both the United States and Europe. The NNPC has already put together a consortium to build a 2 million cubic feet per day facility, consisting of NNPC (60 percent), Shell/BP (20 percent), Agip (7. 5 percent), Phillips (7. 5 percent), and Elf (5 percent).

At some point over the next decade Nigeria will have to give serious thought to its need to develop an export strategy for nonpetroleum manufactured products. It will have to decide how to transform this wasting resource oil into an export platform that can provide long-term growth for the economy. Brazil, India, and other middle-income countries have had to make this leap from production for the local market and the export of raw materials to the export of labor-intensive finished products. A special Nigerian problem is that its currency is so overvalued by petroleum that its exports will be expensive for international buyers. This too will require the political will to shift resources from the easy short-term profitability of commerce to the more complex but in the long term more sustainable manufacturing sector. In Nigeria as in Brazil and Mexico, this will probably require considerable intervention on the part of the state.

To recapitulate and conclude on a comparative note, we have seen that Nigeria has as much in common with all other developing-country energy policy dilemmas and choices as it has with those uniquely associated with OPEC. As in other developing countries there are serious problems of infrastructural bottlenecks within the energy sector, including inadequate refining capacity, poor distribution systems, and recurrent electricity blackouts due to old equipment and insufficient maintenance. There are also the genuine capital shortage problems all countries face. Despite its vast earnings, Nigeria cannot, for example, economically satisfy all its citizens' desire for electricity. There are intersectoral coordination problems of guaranteeing that a newly completed industrial plant will not sit idle waiting for its power supplies. There are the difficulties too that poorer countries face in making choices between which energy sources to develop. One nearly intractable problem is the extent of the noncommercial energy sector and its resistance to easy government solutions. Here Nigeria is like India, Kenya, or other non-OPEC countries (NOPEC).

However, Nigeria is still a large exporter of oil and shares common features with other oil-exporting nations. On the credit side there are considerable reserves, international political clout, and a still unrealized potential for development. On the debit side there are other less desirable features. "Ports and railroads are clogged with imports, the trade deficit has widened precariously, raw materials are in short supply, government spending is at record levels, capital is scarce, as is experienced personnel, and inflation is conservatively placed at 19 percent for 1979. "[80]

This description of another oil-exporting developing country, Mexico, is an apt picture of Nigeria at many points during the postwar oil boom, with its own booming inflation. Unlike Mexico, Nigeria didn't have a semiindustrialized economic infrastructure in place before the oil revenues started pouring in.

These are the kinds of problems and opportunities confronting Nigeria in the future, with its huge population and its wasting oil reserves. These domestic and international decisions will require settlements in the 1980s. The instability of the past has hurt the country generally and the energy sector specifically through revenues and investment lost to civil unrest, military spending, or withheld because of uncertainty over returns in a turbulent environment. Gavin Williams, an incisive commentator on Nigerian affairs, has written that the Nigerian state has

> not been able to institutionalize procedures for regulating class conflict and determining income distribution. This in turn requires the establishment of constitutional arrangements supported by public sentiments which legitimated and effected bourgeois domination, guaranteed national unity and regulated competition among the bourgeoisie. The military state has been no more able to resolve this issue than the politicians who preceded it. [81]

Now there is a civilian government again and new constitutional arrangements. Most observers and participants, including the new president, recognize the need for rationalizing politics in Nigeria, and they realize that competing groups want to determine the ideological and power basis for that rationalization. In a recent interview

President Shagari said that Nigeria "wants to create an atmosphere of confidence for investors and industrialists . . . and that means peace and stability."[82] In the absence of energy supply constraints in the forseeable future, and with ample capital resources if managed wisely, the politics of Nigerian energy will be the politics of distribution along spatial, sectoral, social, and international lines.

NOTES

1. World Development Report, 1979 (Oxford: Oxford University Press for the World Bank, 1979), pp. 128, 136.

2. Ibid., p. 138.

3. Ibid., p. 126.

4. The overall energy sector remains unexamined in its entirety. Ignatius Ukpong writes about domestic electricity policy; Terisa Turner, Eno Usoro, and others about the oil export sector; while S. A. Madujibeya, Scott Pearson, and Gregory Emembolu treat the international aspects of oil as well as some of its domestic linkages. The only full sector survey of which the author is aware is the comprehensive if dated effort of Mourtada Diallo, written in 1967 before the Kainji Dam's completion and the skyrocketing of oil production. See Ignatius Ukpong's "Economic Consequences of Electric Power Failure in the Greater Lagos Area," Nigerian Journal of Economic and Social Studies 15 (March 1973); Terisa Turner, "The Transfer of Oil Technology and the Nigerian State," Development and Change 7 (1976); Eno J. Usoro, "Foreign Oil Companies and Recent Nigerian Petroleum Policies," Nigerian Journal of Economic and Social Studies 14, no. 3 (November 1972): 301-14; S. A. Madujibeya, "Oil and Nigeria's Economic Development," African Affairs 75, no. 300 (July 1976): 284-315; Scott R. Pearson, Petroleum and the Nigerian Economy (Stanford: Stanford University Press, 1970); Gregory Emembolu, "Future Prospects and the Role of Oil in Nigeria's Development," Energy and Development 1, no. 1 (Autumn 1975); Mourtada Diallo, "Energy Resources and Utilization," in A. A. Ayida and H. M. A. Onitiri, eds., Reconstruction and Development in Nigeria (London: Oxford University Press, 1971), pp. 556-99. See also the author's "Political Economy of Public Corporations in the Energy Sectors of Nigeria and Zaire," unpublished Ph.D. dissertation, University of California, Berkeley, 1978.

5. Second National Development Plan (Lagos: Ministry of Information, 1970), p. 63.

6. Ibid.

7. Ibid.

8. Plan for Electrical Power System Development, Motor Co-lumbus, Consulting Engineers, Vol. I (Baden, 1975), p. I-5.

9. For useful discussions of these issues in Africa, see James W. Howe et al., Energy for the Villages of Africa (Washington, D. C.: Overseas Development Council, 1977); and Kofi Bota, Jay Weinstein, and Jesse Walton, eds., Proceedings of the African Solar Energy Workshop (Atlanta: Resource Center for Science and Engineering, Atlanta University, 1979).

10. World Development Report, 1979, p. 126.

11. Ibid., pp. 130, 162.

12. Ibid., p. 128.

13. Ibid., p. 130.

14. Ibid., p. 138.

15. Victor A. Olorunsola focuses on ethnic axes in Victor Olo-runsola, ed., "Nigeria," The Politics of Cultural Sub-Nationalism (Garden City, N. Y.: Anchor Books, 1972). Richard Sklar targets the class axis in his Nigerian Political Parties of Power in an Emergent African Nation (Princeton, N. J.: Princeton University Press, 1963).

16. Barbara Calloway, "The Political Economy of Nigeria," in Richard Harris, ed., The Political Economy of Africa (Cambridge: Schenkman, 1975). See also E. O. Akeredolu-Ale, Underdevelopment of Indigenous Entrepreneurship in Nigeria (Ibadan: Ibadan University Press, 1978).

17. A useful and many-sided critique is the collection Nigeria's Indigenization Policy (Ibadan: Nigerian Economic Society, 1974).

18. Sayre Schatz has a useful characterization of these changing emphases in his Nigerian Capitalism (Berkeley: University of California Press, 1977), p. 7.

19. Robin Cohen, Labour in Nigeria (London: Longmans, 1974), p. 194.

20. Second National Development Plan, p. 32.

21. Third National Development Plan (Lagos: Ministry of Economic Development, 1975), p. 29.

22. Second National Development Plan, p. 135.

23. Ibid., p. 162.

24. Ibid., p. 163.

25. Edwin Dean, Plan Implementation in Nigeria 1962-1966 (Ibadan: Oxford University Press, 1972), p. 166.

26. Peter Kilby, Industrialization in an Open Economy (Cambridge: Cambridge University Press, 1969), p. 25.

27. Third National Development Plan, p. 138.

28. Annual Report and Statement of Accounts (Lagos: Central Bank of Nigeria, 1974), p. 90.

29. J. J. Akpieye, "Refining and Distribution of Petroleum Products in Nigeria " (Lagos: Annual Nigerian Oil Seminar, 1975), p. 3. Mimeo.

30. M. O. Kagha, "Nigerian Petroleum Industry," Proceedings of the Sixth Annual Conference (Lagos, 1974), p. 29.

31. Ibid.

32. Interview with an oil company official active in Nigeria, 1980.

33. Ibid.

34. Akpieye, "Refining and Distribution," p. 1.

35. Ibid. , p. 4.

36. E. O. Ilumoka, "Electricity Supply in Nigeria" (Lagos: National Electric Power Authority, 1973), Appendix III.

37. Second Progress Report on the Third National Development Plan, 1975-80 (Lagos: Ministry of Economic Development and Reconstruction, 1977), p. 73.

38. Ernest J. Wilson III, "Public Corporation Expansion in Nigeria: The Interplay of Political Interests, State Structure, and State Policy," in Pearl Robinson and Elliott P. Skinner, eds. , Transformation and Change in Africa (Washington, D. C. : Howard University Press, forthcoming).

39. The Role of Foreign Governments in the Energy Industries (Washington, D. C. : Department of Energy, 1977), p. 297.

40. The most careful comparative analysis of this subject is Joel Darmstadter, Joy Dunkerly, and Jack Alterman, How Industrial Societies Use Energy: A Comparative Analysis (Baltimore: Johns Hopkins University Press for Resources for the Future, 1977).

41. John Blair, The Control of Oil (New York: Vintage Books, 1978).

42. Michael Tanzer has an excellent discussion of these and related features in The Political Economy of International Oil and the Underdeveloped Countries (Boston: Beacon Press, 1969).

43. Pearson, Petroleum and the Nigerian Economy, p. 15.

44. Shell/BP had 56 percent of Nigerian production in 1978. West Africa, February 20, 1978, p. 364.

45. Douglas Rimmer, "Elements of the Political Economy," in Keith Panter-Brick, ed., Soldiers and Oil (London: Frank Cass, 1978), p. 151.

46. Edith Penrose, "The Development of Crisis," in Raymond Vernon, ed. , The Oil Crisis (New York: W. W. Norton, 1976).

47. Decree No. 18, 1971.

48. Kagha, "Nigerian Petroleum Industry," p. 30.

49. Turner, "The Transfer of Oil Technology."

50. "National Oil Companies Crowd the 'Seven Sisters,'" New York Times, December 30, 1979.

51. Ibid.

52. African Business, December 1979, p. 37.

53. West Africa, September 17, 1979, p. 1707.

54. West Africa, July 23, 1979, p. 1320.

55. I am indebted to Robinson Hollister of Swarthmore College for pointing out some of the international trade and economic policy implications of this notion.

56. Madujibeya, "Oil and Nigeria's Economic Development," p. 286.

57. Ibid. , p. 285.

58. Ibid. , p. 287.

59. African Business, December 1979, p. 77.

60. West Africa, January 14, 1979, p. 53.

61. Ali D. Yahaya, "The Creation of States," in Panter-Brick, Soldiers and Oil, p. 216.

62. Third National Development Plan, p. 30.

63. Africa, No. 86 (October 1978).

64. Ibid.

65. Ibid.

66. Role of Foreign Governments in the Energy Industries, p. 296.

67. Ibid.

68. West Africa, February 20, 1978, p. 364.

69. West Africa, May 1, 1978, p. 849.

70. Africa, No. 86 (October 1978), pp. 106-07.

71. Ibid. , p. 107.

72. West Africa, May 1, 1978, p. 849.

73. West Africa, May 15, 1978, p. 938.

74. Interviews in Nigeria and the United States with oil industry personnel and government officials.

75. One can read all four groups' comments on this endemic problem in the annual publications of the conference proceedings of the Chamber of Commerce, the Manufacturers Association of Nigeria, the Nigerian Institute of Management and the Nigerian Economic Society.

76. Cynthia Enloe, Politics of Pollution in Comparative Perspective (New York: David McKay, 1975), p. 27.

77. Timothy Shaw and Malcolm Grieve, "Africa and the Environment: The Political Economy of Resources" (Sherbrooke, Quebec, 1977), p. 3.

78. Report on the Petroleum Resources of the Federal Republic of Nigeria (Washington, D. C. : U. S. Department of Energy, October 1979).

79. Platt's Oilgram News 57, no. 227 (November 1979).

80. New York Times, December 9, 1979.

81. Gavin Williams, "Nigeria: A Political Economy," in G. Williams, ed. , Nigeria: Economy and Society (London: Rex Collins, 1976), p. 51.

82. Washington Post, January 24, 1980.

8

TAIWAN
John Franklin Copper

THE PATTERN OF ENERGY USE

In view of Taiwan's heavy dependence on energy imports, it is ironic that the island was once viewed as attractive for local, usable sources of energy. However, with one of the world's fastest growing economies in recent times, Taiwan has outstripped the capacity of domestic energy resources to meet its large and growing needs. This section will examine traditional sources of energy and energy use juxtaposed with post-World War II changes in the quantity of energy used and new patterns of energy use.

Since coal lies near the surface in many areas in Taiwan, it was used in early times as a fuel for heating and cooking. [1] Petroleum and natural gas also were known centuries ago, but were not used to any meaningful extent until the nineteenth century. Around 1865 two Americans took up the coal and camphor business in the northern port city of Keelung and made coal the island's first major export. [2] Oil wells were drilled just over ten years later by another American businessman who succeeded in finding sizable deposits of crude. By the 1870s coal mines were operating in Taiwan with the aid of machinery. The biggest of these mines, just a short distance from Keelung, provided coal for foreign ships. [3] Some coal was also shipped to Fukien Province in China and became a valued energy source for the Chinese

---

I acknowledge gratefully a research grant from the Pacific Cultural Foundation that made possible the research and interviews upon which this chapter is based.

359

navy. By 1876 the coal mines around Keelung were producing 200 tons a day. At nearly the same time small amounts of oil were marketed locally for use in lamps and medicine, until local authorities, fearing calamitous imbalances of nature caused by the oil wells, shut the business down. Subsequent attempts to develop an oil business in the 1870s and 1880s by Americans did not succeed for the same reason. [4]

In the early twentieth century, coal was mined at various sites in Taiwan and was used in steam locomotives, ships and launches, opium factories, arsenals, coke making, and brick and tile factories. [5] Because of unfavorable geologic conditions, however, and little improvement in the use of machinery, production did not expand very much and coal mining did not survive as an export-oriented business. Meanwhile, domestic demand increased. During the Japanese occupation from 1895 to 1945, coal was Taiwan's chief source of energy. The Japanese developed petroleum for commercial use, but production was never sufficient even for local needs. Furthermore, imports were minimized to conserve foreign exchange and to facilitate more petroleum imports to the main Japanese islands. [6]

Taiwan's impressive economic growth under the Japanese, which produced the highest standard of living and level of economic development in Asia exclusive of Japan, came to an abrupt halt at the end of World War II. In the early period of Nationalist Chinese rule, Taiwan was used as a base of operations for continuing the war against the Communists; rather than expand industry to produce war materials as the Japanese had done during World War II, the Nationalists dismantled factories and shipped them to the Mainland, or sold them to buy weapons abroad. Thus energy demand in Taiwan did not register an increase until about 1953, when Taiwan's economic boom started.

Because of rapid economic growth after 1953, especially in the past 15 years, coupled with an inability to develop local energy resources, Taiwan has become extremely energy dependent. Through the decade after 1953 the economy grew at an annual figure close to 7 percent, and during the next decade at a rate of nearly 10 percent. [7] Recent and present economic growth rates have been similar—9.9 percent in 1977, 14 percent in 1978, and 8 percent in 1979. In addition a number of other variables have caused a steep rise in energy consumption beyond that fostered simply by rapid economic expansion.

First, Taiwan's post-World War II economic growth has been based largely on the expansion of industry, which has been accompanied by transfers of investment and labor from the agricultural to the industrial sector. From 1953 to the present, the size of the labor force in agriculture increased only slightly, while during the same period the labor force in nonagricultural pursuits nearly tripled. [8] During this time the agricultural sector's contribution to the economy, as a proportion of GNP, went up seven times while that of the industrial sector

increased almost 20 times. Recent estimates put the contribution of
the agricultural sector at 13.4 percent, manufacturing at 29.6 percent, and services at 50.2 percent. [9]

A second factor underlying increasing rates of energy consumption is the type of industry that has contributed most to economic
growth. Taiwan's leading industries are textiles, electrical products,
metal products and machinery, plywood and other wood products, and
plastics, all of which are large consumers of energy. The textile industry is, by a large margin, the largest user of electricity in Taiwan. [10] All the others also use sizable quantities of electricity, in addition to other forms of energy, while the plastics industry also uses
petroleum as a basic raw material. With regard to total energy consumption, the chemical industry leads, followed by producers of non-
metallic mineral products, transportation, and textiles. [11] Textiles
and transportation show the largest increases in energy use: 20.1 and
15.3 percent, respectively, each year through the late 1960s and early
1970s. [12] Now, however, chemicals and transportation are moving a-
head because high labor costs in the textile industry and stiff competition from nations that have managed to maintain lower costs have
caused Taiwan's textile factories to experience hard times.

A third factor is the rapid rise in per capita income, together
with a general leveling of incomes or reduced income disparity, which
have produced a mass consumer-oriented society. Average per capita
income has risen steadily from $132 a year in 1952 to $1,720 in 1979.[13]
Equalization of incomes has resulted from a more efficient tax system,
high levels of employment, and a shortage of labor in certain sectors
of the economy. [14] According to the gini coefficient, a commonly used
means of measuring income inequality, Taiwan has a more equitable
distribution of income than almost all other developing countries, as
well as the United States or Japan.[15] The result has been more buying power in the hands of consumers, and thus consumer energy use.
For example, from 1952 to 1972 the percentage of homes with electricity grew from 35.4 to 99.7 percent, while the number of homes with
television sets increased from zero to 94.5 percent. [16] Other appliances, such as refrigerators, ovens, and air conditioners, also came
into widespread use.

Finally, Taiwan's population growth has exhibited one of the highest rates in the world since the end of World War II. The population of
the island in 1945 was about 6 million. In 1949 more than a million people moved from the Mainland when Nationalists armies lost to the Communists. This, coupled with large natural rates of increase because
of improved public health standards and a lower death rate, brought
the total population of Taiwan to over 18 million in 1979. [17] While the
birth rate has dropped markedly in recent years, the population will

continue to grow for some time, which, together with Taiwan's income equality, will no doubt contribute significantly to the continued growth of energy demand by households and consumer-oriented businesses.

All of the above-cited factors produced a 10 percent annual growth in energy use during the first decade after 1953 and over 10. 4 percent growth in the past decade and a half. [18] In 1977, with a land area one-thirteenth that of California, Taiwan became one of the world's major consumers of energy, using the equivalent of 22. 9 million kiloliters of oil each year, or 3. 2 times the consumption in 1961.[19] The growth of energy use is ahead of GNP by nearly 2 percent, an unfavorable ratio likely to continue for some years.

Thus in a very short span of time Taiwan has been transformed from an energy self-sufficient nation to one heavily dependent on foreign sources. The transition was especially marked in the 1960s and 1970s, creating for Taiwan very serious new problems. In 1954 Taiwan imported 18.5 percent of its energy from abroad. With the development of local sources—coal and water power particularly—imports did not rise very much for a decade, constituting only 26 percent in 1961. In the years immediately following this import situation changed, and by 1977 Taiwan imported 78 percent of its total energy. [20] Rising dependence resulting from growing energy use coincided with the inability to find new local energy sources or to increase production of the old ones. For example, coal production increased from 2. 1 million tons in 1954 to 5. 1 million tons in 1967;[21] but it peaked in 1967 and declined in subsequent years, to 3. 3 million tons in 1973. [22] Coal production is still declining and is likely to remain that way. There were 188 producing coal mines in 1977, but only ten of these mined more than 5, 000 tons a month. [23]

Hydroelectric power was also a major source of energy in the 1950s and production doubled in the two decades following 1953. There continues to be growth in the production of water—generated electricity, but not enough to cope with increased demand. Thus hydroelectric power has declined as a percentage of total energy production. In 1960 it was the main source of electricity; by 1977 it provided only 14 percent of total need. [24] Natural gas is produced locally but accounts for only a small proportion of all energy use. [25] In 1977 it amounted to only 8 percent of total energy consumption. [26]

The rise in energy demand and the paucity of domestic resources mean that imported petroleum and nuclear power have had to make up the deficiency. It seems improbable that Taiwan will be able to reverse the trend of importing larger quantities of energy in the near future, thus assuring foreign exchange and dependency problems, not to mention a possible cutoff of supplies.

Growing reliance on foreign energy sources has also been influenced by changing consumption patterns with respect to the kinds of energy used. Until 1961 coal was the major source of energy in Taiwan, supplying 57 percent of energy needs, with oil providing 26 percent, hydropower 16 percent, and natural gas 1 percent. [27] By 1977 petroleum was supplying 78 percent, with coal following at 10 percent, natural gas at 8 percent, and hydropower at 4 percent. [28] Although nuclear power stations are reportedly going to supply over 40 percent of Taiwan's electricity in the 1980s, and perhaps more by the end of the century, it is likely that petroleum will continue to be the most important source of energy and that reliance on it will increase. [29] Hydroelectric power will decrease in relative importance; coal and natural gas will probably decline as a percentage of total energy use (see Table 8.1). Other sources are being used, such as geothermal, solar, and marsh gas, but their future is uncertain. The advantages and the potentials relating to these forms of energy will be discussed below. Suffice it to say here that they will not play a major role in the immediate future.

The current pattern of energy use, as well as trends in conservation, industrialization, and pollution, suggests that the proportion of energy use relative to economic growth should decline, or at least not exceed the rate of economic growth as has been true in the past. Compared to 1961, the percentage of energy used by industry in 1977 dropped from 63 to 57 percent, agriculture from 5 to 4 percent, residential and other uses from 23 to 18 percent. [30] During the same period, however, transportation increased its share from 6 to 10 percent, and industries using petroleum as a raw material, especially the petrochemical industry, pushed their share up from 3 to 11 percent. [31]

DOMESTIC ENERGY POLICY

Taiwan was hit hard by the steep rise of oil prices in 1973. A healthy trade surplus for several years before 1973 gave way in 1974 to a deficit of $1.3 billion, more than half of which was due to imports of petroleum from Saudi Arabia, Kuwait, and Iran. [32] The total amount of trade in 1975 declined for the first time since 1969; in fact, it was a drop far more significant than any since the end of World War II. [33] The oil crisis also provoked serious inflation: 35 percent in wholesale goods and 47 percent in consumer goods in 1974 alone. [34] In a little more than two years after OPEC increased the price of petroleum, retail prices in Taiwan rose well over 50 percent, while in some specific cases the increase was 100 to 200 percent. [35] Also, an unusual situation developed in which unemployment and prices rose simultane-

TABLE 8.1

Long-Term Energy Supply and Demand in the
Republic of China, 1961-87
(figures in percents unless otherwise noted)

	1961	1976	1977	1987
Energy supply (million kiloliters of oil equivalent)	5.3	23.0	23.6	52.6[a]–65[b]
Percent of which is imported	26	76	76	73[c]–83[d]
Coal	57	11	10	5–15[e]
Hydroelectric power	16	5	4	2–3[f]
Natural gas	1	8	8	5–15[g]
Petroleum	26	76	78	40–80[h]
Nuclear	—	—	—	10–20[i]
Energy use (million kiloliters of oil equivalent)	4.7	21.2	22.9	56–60[j]
Per capita (liters)	425	1,286	1,384	

Notes and Sources:

[a]This is an official estimate based on an annual GDP growth rate of 7.8 percent and a yearly increase in energy use of 8.5 percent.

[b]Author's estimate based on higher economic growth rates and energy use rates. (Note: Taiwan's GNP growth rate in 1978 was 13.7 percent.)

[c]Official estimate apparently assuming future natural gas finds and offshore oil discoveries.

[d]Author's estimate, which does not assume the above.

[e]Author's estimate. Coal production will probably drop in the next decade, thus accounting for only about 5 percent of total energy supply. The 15 percent figure assumes that coal will be imported. (It has been government policy not to consider coal importation, but recent evidence—including plans to build coal-fired power plants—suggests a change in policy.)

[f]Author's estimate.

[g]Author's estimate. Lower figure assumes few or no new discoveries of natural gas and no imports of LNG. Higher figures assume both.

[h]Estimate based upon possibilities in other areas, new energy sources, uncertain future prices of imported petroleum, and unknown offshore oil potential.

[i]Author's estimate.

[j]It is assumed that supply of energy will have to continue to exceed the use due to the fact that Taiwan exports and has a large and growing petrochemical industry.

ously. With respect to overall economic growth as measured in GNP, the rate of growth dropped from 11.6 percent in 1973 to 0.6 percent in 1974, recovering only to 2.8 percent in 1975.

However, Taiwan adjusted to the oil price revolution and growth returned to double digits after 1975—11.8 percent in 1976. This resilience was due partly to the fact that Taiwan's Asian competitors, notably Japan and South Korea, were experiencing similar or worse problems. It was also a result of astute government policies aimed at keeping economic growth high despite rising energy costs, while limiting unnecessary consumption. The problem of rising wages, and thus of higher prices for Taiwan's exports, was offset in part by qualitative improvements in exported products combined with government intervention to keep investments at a high level and to maintain the undervalued status of Taiwan's currency. Additional incentives were given to export industries, and at the same time a greater push was given to the "ten large projects"—big industrial concerns or infrastructure projects that had been started in 1972 under government supervision to guarantee continued economic development and modernization. [36]

As it turned out, the oil crisis had very little impact on the amount of energy Taiwan imported. In fact, after a short period of adjustment, imports rose rapidly once again. Nor did the crisis have much effect on conservation, the quest for domestic energy sources, or the types of energy used, except for the higher priority given to nuclear power facilities, the construction of which was already under way. The reasons for this lack of response to the global energy crisis lie in economic growth policies as they relate to political problems and to Taiwan's specific energy situation.

Economic development, which reached "take-off" in 1953, placed Taiwan among the top nations of the world in terms of economic growth by the 1960s. This is remarkable in view of Taiwan's scarcity of natural resources, and the fact that it is severely overpopulated, more so even than Holland, Europe's most crowded country, and far more so than either Japan or China. Economic success was facilitated in the initial phase by U.S. aid; in fact, Taiwan was one of the rare countries where U.S. aid actually worked to promote rapid economic growth. Also instrumental was American advice on economic planning. Because of U.S. influence, Taiwan linked its economy to the world marketplace through international trade, by specializing its industry for the export market. By the 1970s it had become the world's number-one nation with respect to the percentage of GNP in foreign trade. [37]

Impetus for economic growth has come also from a number of other sources, including an intense desire on the part of the people for the material rewards that accompanied economic development. Most important, however, the government discovered that economic success to a large extent offset Taiwan's diplomatic failures vis-a-vis the People's Republic of China, including expulsion from the United Nations in 1971, with the subsequent loss of diplomatic ties with most countries. [38] Those breaking with the Republic of China after 1971 maintained trade and other ties; in fact, nearly all increased their volume of trade with Taiwan. [39] Thus Taiwan's expanding trade made it possible to substitute economic for diplomatic ties. The growing economy also attracted foreign investment, a significant amount of which went into energy conversion facilities and other energy-related projects. Here lies the explanation for Taiwan's lack of concern for conservation. Continued rapid growth was seen by the government and the people of Taiwan as intimately related to survival as a nation. Because of its economic success, foreign investment was available to expand even more—especially through the pursuit of infrastructure projects related to energy, which please foreign investors.

Economic development served also to bridge the gap between Mainlander Chinese (those who came from China in 1949 and their descendants), who monopolize positions in government and education, and the locally born Taiwanese, who control business. Cooperation between the two groups has been a quid pro quo to economic growth, and both have benefited from the success. In fact, feelings of difference or dislike between the two groups have been blunted by the emergence of common attitudes associated with Taiwan's new consumer-oriented culture. [40] This in turn has made it impossible for Peking to try to divide Taiwan's population in an effort to pressure the government to negotiate Taiwan's "reversion."

Judging from remarks made in early 1979, President Chiang Ching-kuo considers economic development more than ever the key to Taiwan's survival in the wake of U.S. recognition of the People's Republic of China and the announced termination of the defense alliance between the United States and the Republic of China, together with President Carter's statement that "acknowledges" Taiwan to be a part of China. [41] Business and trade ties no doubt will continue to support the view that Taiwan is separate from China. Because of Taiwan's more rapid development, its capitalist and trade-oriented economy, and its Western-style materialist culture, a convincing argument can be made that it should not become part of the People's Republic of China. What is more, trade and other economic ties will continue to substitute for diplomatic ties, reflecting that Taiwan is an independent, autonomous state, allaying fears of the population, and promoting confidence in the government.

Concerning continuing rapid economic growth as it relates to Taiwan's energy needs, it is necessary to make special comment on the "ten projects" and the new "twelve projects," which are getting under way. [42] The "ten projects" include a shipbuilding enterprise, Taiwan's first nuclear power plant, an integrated steel mill, a petrochemical complex, a freeway running the length of the island from north to south, a new international airport, a new rail line, electrification of the present main north-south rail line, and two harbors. Seven of these were completed by the end of 1978. The new "twelve projects" include two more nuclear power stations, additional capacity for the steel mill, a rail line around the island, three new highways, and a number of other industries or improvement projects. Nearly all the new projects are energy intensive and are certain to expand Taiwan's energy needs in the future. The nuclear power plants will help to relieve dependence on imported petroleum, but they will probably not offset the increased need for power as a result of continued growth, or will succeed only in the short run. In any case nuclear fuel has to be imported. Taiwan's plans clearly are in the direction of increasing energy use, with more imports of energy, mostly in the form of petroleum and nuclear fuel.

In light of these trends, it is instructive to examine Taiwan's domestic energy resources. Clearly the island is not well-endowed and future prospects are not good. Economically recoverable coal reserves amount to about 226 million tons. [43] Most of this is located in the northern part of the island, but the seams are thin and deep in the ground, and there is little prospect of further mechanization of the extraction process. The price of coal therefore is high in spite of the government's heavy taxation on petroleum-derived fuels to help keep coal competitive. Production has declined steadily, as noted earlier, with almost no hope of further discoveries. [44] The absence of facilities for importing coal virtually assures that it will continue to be replaced by other forms of energy. [45]

As we have seen, the Japanese drilled for oil in Taiwan but never found enough to meet domestic needs. [46] Deeper drilling in 1959 aroused some hope that further reserves might be found, but so far there has been little but disappointment. [47] Taiwan's estimated reserves as of January 1978 were 12 million barrels, the lowest of any country in the Asia-Pacific region except Thailand and South Korea, where exploration has just begun, and three countries with no proven reserves—Bangladesh, Singapore, and Sri Lanka. [48] As of July 1978, Taiwan had 66 producing wells, but daily production was only 4,300 barrels, most of which was condensate, and production had declined 15 percent since the previous year. [49]

Offshore exploration and drilling offer more promise, though both technical and political problems have impeded progress. In 1969 Taipei entered into negotiations with several U.S. oil companies to explore for offshore oil in the Taiwan Strait and to the island's north and south. The government of the Republic of China claimed the continental shelf off the coast of China on the basis of its claim to be the

legal government of China, and it offered various of these areas to
U. S. companies for exploration. However, an agreement signed with
Amoco and Gulf in September 1970 revealed a reluctance on the part
of the companies to go along with Taipei's claim; instead, Amoco and
Gulf assumed a median line drawn between Taiwan and the Mainland.
In 1975, when drilling by another U. S. company was about to start in
a zone to the far north of Taiwan, claimed by Taipei on the strength
of its claim to China rather than its legal jurisdiction over Taiwan,
the U. S. State Department intervened to stop the operation. U. S. companies were pressured subsequently not to drill in offshore areas that
were not close to Taiwan. [50] In these areas, despite favorable geologic and seismic conditions, results so far have been discouraging,
and only small amounts of condensate oil have been produced.

Up to 1959 natural gas was thought to offer little hope of providing more than minute quantities of energy, even though it had been
a commercial source of energy prior to World War II and remained in
use through the 1950s. In 1959 a new and deeper well brought increasing local supplies to market. Reserves of natural gas are presently
estimated at 30 billion cubic meters. In recent years the location and
recovery of new sources has about equalled rising demand while falling
behind the overall increase in energy use. [51] Thus it appears that in
the near future natural gas will play about the same role, or possibly
a less important role, as it is playing now—that is, about 6 percent of
total energy supply. At the moment, demand for natural gas exceeds
the supply, and at the present rate of use, assuming no new discoveries, Taiwan has an estimated 11-year supply. [52] Natural gas has
been found offshore and present evidence suggests that gas finds are
more promising than petroleum discoveries. Meaningful offshore
sources remain in the realm of speculation, however. [53]

A similar situation prevails with respect to hydroelectric power.
There are 30 hydroelectric stations producing an aggregate of nearly
1. 5 million kilowatts of electricity in Taiwan. [54] A project under study
in central Taiwan would double the production of hydroelectricity, but
it is close to a tourist area and is handicapped by environmental problems. Total potential for hydroelectric energy in Taiwan is said to be
around 5 million watts, an amount not likely to be realized in the near
future. [55] Dams are costly, they can be built only where electricity is
not needed, and they entail the sacrifice of farm land or damage to the
environment. Thus the trend away from hydroelectric energy seems
likely to continue; the decline has been from 16 percent of total energy in 1961 to 4 percent in 1978. [56]

Of the novel forms of energy used in Taiwan, geothermal, solar,
and marsh gas (biogas) are the most promising, though all are in the
experimental stage in terms of use on a large commercial scale. Judging from results so far, they can be only partial solutions to increase

energy demand, if that. Energy sources such as wind and tidal power have been discussed, but they are in a very early stage of experimentation. The same is true of biomass. The last three seem to offer greater commercial possibilities, but their practical use is farther away in time.

Work on geothermal energy started in 1966 in a volcanic area at the northern tip of the island. [57] Subterranean waters of sufficiently high temperature at depths around 1,500 meters indicated that commercial possibilities existed. In 1978 the potential for production of electricity in the area was estimated at 100 to 500 million watts. [58] On the other hand, no well was found that could produce sufficient quantities of hot water to generate electricity for commercial use. Moreover, all the wells drilled had the problem of corrosion due to the presence of sulfuric acid in the water. As a consequence, most of the wells in the area now utilize the available steam instead of hot water—for lumber drying kilns, greenhouses, and laboratories, but not for the generation of electricity. In recent years other uses have been found for geothermal steam in the area, such as in the paper and salt industries, fish drying, and poultry raising. [59] Electricity production, however, has been abandoned as unworkable.

An islandwide survey in 1972 found subterranean hot water in large quantities in other areas in north central Taiwan and corrosion was not so marked in these areas. This find led to the construction of a pilot power generating plant in 1977 with a capacity of 1,500 kilowatts. [60] Though this project was successful, the potential for generating commercial quantities of electricity was found to be meager. There is thermal power potential in a few other areas but it is not possible to be optimistic about large-scale electricity production. [61]

Solar energy is already being used and appears to offer considerably more potential in Taiwan, for the southern part of the island is located in what is known as the "maximum solar energy band," and there are sizable tracts of land not being used for agriculture because of the heavy salt content in the soil, in addition to areas that are too mountainous to cultivate. [62] However, the use of solar energy for electricity generation is still in the experimental stage. Solar water heaters are being used in large buildings and private homes in many areas of Taiwan, but the number of cloudy days in Taipei and other cities in the northern part of the island makes solar energy somewhat less promising than the intensity of the sun and the length of the summer might suggest. While solar power in heating and air conditioning has possibilities, there has been no effort as yet to mass produce solar equipment for commercial use. The obstacle standing in the way of solar heating where appropriate is the accurate prediction of future energy costs and the sizable investment required. [63]

Marsh gas (biogas) produced from hog manure is another source of energy with some potential. Marsh gas contains 70 percent methane and is relatively nonpolluting; it is also easy to produce, store, and use. More than 8,000 households in rural Taiwan are using it currently as a fuel for cooking and heating. [64] Marsh gas also has potential for generating electricity locally and even for running farm machinery and cars. In fact, there are some real commercial possibilities for the use of marsh gas. There are over 6 million pigs in Taiwan because of the high demand for pork. This means that the widespread conversion of hog manure to marsh gas could equal one-third of the present natural gas consumption. [65] However, the use of marsh gas is still in a twilight zone between experimentation and development. There are uncertainties about its economic viability, even in rural areas, because of costs related to production and storage equipment. Furthermore, marsh gas does not have much promise for commercially generated electricity or for the production of energy usable in large industries, cities, or homes and apartments.

Both wind and tidal energy have been discussed seriously in Taiwan. Both appear to be feasible at some future time, depending on the magnitude of price increases for other energy sources and the development of the appropriate technology. Difficulties related to storage, conversion, and transmission of electrical energy from winds or tides, however, must be overcome. Physical conditions are generally good. The western coast of Taiwan receives fairly strong, sometimes quite strong, winds. However, it is feared by some that typhoons might destroy wind-harnessing equipment. Others believe the typhoons themselves might be used. Ocean currents around Taiwan are strong and a number of areas show promise for harnessing tidal energy. Biomass has potential for development in Taiwan's offshore ocean space, but a good deal more research will be needed on methods and application.

INTERNATIONAL ENERGY POLICY

As of 1979, approximately 83 percent of Taiwan's energy was imported. The portion of energy produced locally was still decreasing, and there was little hope of this situation changing. Hence energy imports and dependence are serious problems for Taiwan on at least two levels: economic, because of energy costs; and with respect to foreign policy, because Taiwan needs some assurance of future imports. These problems are especially vexing in view of the intense competition faced by Taiwan in the export market, and its declining diplomatic influence abroad. In this connection one must consider China's desire to absorb Taiwan and the potential availability of Chinese energy resources.

Petroleum and nuclear fuel are Taiwan's two chief types of imported energy, and there are two main sources: the Middle East and the United States. More than 90 percent of Taiwan's imported petroleum comes from the Middle East, with half of that coming from Saudi Arabia. [66] Virtually all processed nuclear fuel is purchased from the United States, although raw uranium has been obtained from other sources and Taiwan is looking around for other sellers of processed fuel. Each of these energy sources, so crucial to Taiwan's economic stability, involves separate difficulties requiring different solutions. Hence both must be examined in greater depth.

High energy costs did not become a serious matter until 1974. In that year petroleum imports increased by 10 percent while costs went up by 133 percent. [67] Taiwan's import bill for petroleum in 1969 came to $17.9 million; in 1975 it reached $623.7 million; and in 1976 it passed the $1 billion mark. [68] The jump in oil import costs between 1975 and 1976 alone was 68.5 percent, due to the return of Taiwan's economic growth rate to double digits. By 1977 petroleum had become Taiwan's largest import in dollar terms. [69] A period of serious inflation resulted after 1974 that made adjustment and recovery seem difficult, but effective government controls succeeded in neutralizing the effects of high oil costs in most realms of the economy. In a few energy-intensive industries, such as textiles, which was hurt also by rising labor costs, the impact seems to have been permanent. In other cases, the government was able to help businessmen move into other types of production such as electronics, or to merge with more viable companies. In some cases the government encouraged larger and more efficient manufacturing in spite of the danger of monopoly practices.

In the meantime, Taipei was able to open up markets for Taiwan's products in the Middle East. This offset to some extent the balance-of-payments deficit with countries in that region. Prior to 1973, Taiwan's exports to the Middle East were in the $10 to $20 million range. By 1976 they had increased to $389.7 million. [70] Saudi Arabia became Taiwan's tenth-ranking importer in 1978. [71] However, this trade compensated only in part for the large fuel import bill from the area, which totaled $712.5 million in 1975 and $1.19 billion in 1976. [72] Nor did it reverse the trade deficit, which grew from $504 million in 1975 to $800 million in 1976. [73]

The oil crisis also fostered a serious overall trade deficit in 1974, to the amount of $1.3 billion, caused by the increased costs of imported oil. [74] Taiwan's imports maintained a high level of dependence on the United States and Japan. About 70 percent of imports originated from those two countries, which passed on rising costs of their exports because of inflation induced there by the energy crisis. [75] Taiwan found a market in the two countries for well over 50 percent of its exports, but at this time the United States and Japan experienced recession and a decline in buying power. [76] Moreover, Taiwan's purchases from the United States and Japan fell into the category of needed machinery, equipment, and technology, while U. S. and Japanese purchases were more consumer goods. Soon Taiwan was able to diversify and boost its exports to the United States while restricting the flow of imports from Japan. Also new markets were found, most notably in Western Europe. Thus, by 1976, a favorable balance of trade had been reestablished and foreign trade was again growing rapidly, which permitted the purchase of increasing quantities of imported oil at higher prices.

At the present time, Taiwan's export products are improving in quality and Taipei is sitting on a comfortable $6.5 billion of foreign reserves, which ought to help the country manage further energy price hikes. [77] It is uncertain whether this situation will obtain in the future. There are clearly some reasons for doubt. In 1978 some 40 percent of foreign sales went to the United States, with which Taiwan has maintained a favorable balance of trade for several years. [78] For some time, American businessmen have pressured the U. S. government for a limitation on a number of Taiwan's imports. If the U. S. balance of trade remains in deficit, Taiwan's exports may well be hit with regulations. Also, with the prospects of trade with China, many American businessmen are drawn to what looks like a market of formidable size. This latter situation also obtains in Taiwan's economic relationship with Japan. On the other hand, Taiwan's export products continue to build a reputation for quality and price increases have been kept to a minimum.

To insure future supplies of petroleum, Taipei has been maintaining close relations with most Middle Eastern countries, especially Saudi Arabia. In 1970, King Faisal of Saudi Arabia visited Taiwan and reportedly established a close friendship with Chiang Kai-shek and Madam Chiang. [79] When the oil embargo struck in 1973, Taiwan was excluded from Saudi Arabia's list of countries denied oil. Subsequently Taiwan was given a guarantee of future oil supplies in addition to low-interest loans. Also Saudi Arabia invested in Taiwan's economy to the tune of $150 million by mid-1977, mostly in construction and communication projects. [80] Ties between the two countries have been strengthened by their common hostility toward communism, a bond of loyalty between their leaders, and a mutual sense of morality in the conduct of foreign relations. Taipei has pleased Saudi Arabia by sending aid missions to Middle Eastern countries, apparently at Saudi Arabia's request. [81] Chinese firms from Taiwan are also active in Saudi Arabia with 2,000 engineers and technicians working on construction projects valued at $650 million as of late 1977. [82]

Taiwan's relationship with Israel has been affected by this close liaison with Saudi Arabia. For instance, Taiwan's purchase of missiles from Israel in 1977 has been kept as quiet as possible so as not to offend the Saudis. [83] Taipei's relations with Saudi Arabia may also explain why Taiwan rejected an offer to buy Israeli fighter planes in 1978, although the official reason was that they were not significantly better than aircraft already in service, and that Taiwan hoped to buy American planes. Some other Arab countries have pressured Saudi Arabia to establish ties with the People's Republic of China, but so far it has ignored such requests.

Taipei also has close ties with Kuwait, once its first but now its second most important supplier of petroleum. Through the 1960s and up to 1977, Kuwait provided about 50 percent of Taiwan's oil compared to Saudi Arabia's 40 percent—figures that have now been reversed. [84] Taipei has offered Kuwait technical and other advice and has had some success exporting to Kuwait, but not nearly enough to pay for the oil bought there. Taipei's decision to buy more oil from Saudi Arabia relates to the breaking of diplomatic relations when Kuwait decided to recognize the People's Republic of China in 1970. Despite the lack of official diplomatic ties, Taiwan's relations with Kuwait remain good.

Taiwan's third largest supplier of petroleum (11.3 percent in 1977) is Indonesia. [85] Indonesia is Taiwan's only important supplier outside the Middle East and has the potential to become an even bigger supplier in the future due to its proximity and good political relations. After an attempted coup in 1965, in which Peking was implicated, Indonesia broke diplomatic ties with China. Indonesia has made no move to restore those ties and its attitude toward Peking is one of caution. Although the Indonesian government chose not to recognize the Republic of China, Taiwan has made heavy investments in Indonesia and purchases increasing amounts of Indonesian oil. [86] While Indonesia is capable of shoring up to some extent a break in Taiwan's oil lifeline to Saudi Arabia and Kuwait, this could be only a partial solution.

In the event that petroleum imports should be cut, Taiwan has in storage sufficient reserves to last about three months. [87] Reserves are augmented somewhat by the fact that Taiwan's own tankers carry more than 70 percent of its imported oil. [88] Thus there is always a considerable amount in transit. Also Taiwan purchases more petroleum than it uses in order to market gasoline and other products in South Korea, Japan, Okinawa, Hong Kong, Guam, Indonesia, and the United States. This could be diverted to domestic use in the event of short supplies. Because of building excess refining capabilities, Taiwan soon expects to get 15 percent of its domestic oil consumption from foreign companies in return for refining crude for these same companies. [89] The development of its own petrochemical industry has similarly helped to reduce Taiwan's dependence on foreign countries for a variety of petroleum products, an advantage whose price is greater dependence on imports of the basic raw material.

Taiwan's second most prominent source of imported energy is nuclear. Studies on nuclear power started in 1955 and the decision to build nuclear power plants was made in 1968. [90] The decision was made for several reasons. First, it was argued that nuclear energy would be comparatively cheap. Second, Taiwan needed alternative sources of energy so as to avoid an excessive dependence on Middle Eastern petroleum, which seemed inevitable with the growth of energy use and Taiwan's insufficient indigenous energy resources. Third, it was argued that nuclear power would mitigate air pollution, which by 1968 was becoming a serious problem in Taipei and other large cities in Taiwan. Fourth, some people thought that nuclear power plants would give Taiwan an edge should it become necessary to build nuclear weapons, a option considered important in light of the nuclear capabilities of the People's Republic of China. Other perceived advantages were the ties with U.S. companies, thus strengthening the American commitment to Taiwan's future, and the relative ease of transporting nuclear fuel by airlift should Taiwan be blockaded by China.

In 1969 Taiwan Power Company awarded General Electric contracts for a nuclear steam supply and nuclear fuel fabrication systems, and gave Westinghouse a contract for a turbine and generator set. Bechtel Corporation did the studies on site location feasibility, and so on. Construction was begun the following year on a site on the north coast of the island. Two more sites were decided on later, one in almost the same area as the first, and the other at the southern tip of the island. All three sites have two nuclear power systems. One of the nuclear generators at the first site, Chingshan, started producing electricity in 1977, and the second in 1979. The two generators at the Kuosheng site are expected to begin working in 1981 and 1982. The Maanshan site will start operations in 1984 and 1985. The three sites together are expected to produce more than 5 million kilowatts of electricity, or more than 40 percent of Taiwan's needs. [91] Another 14 reactors are planned for three sites by the end of the 1990s.

For the moment it appears that Taiwan's decision to build nuclear power plants was a wise one. Electricity produced by the plant now in operation is cheaper than that generated by thermal units fueled by petroleum. One source declares the savings to be at least $55 million a year, a figure that is expected to grow as other nuclear installations are put on-line. [92] On the other hand, Taiwan's nuclear power program is not without its difficulties. First, Taiwan depends solely on the United States for nuclear fuel and may not be able to rely on continued supplies, or may not want to. South Africa, which sells uranium and is developing fuel enrichment services, might be a future alternative source. [93] Australia and Canada have also been mentioned frequently as possible suppliers. * Second, while Taiwan has built re-

---

*In April 1980 Taiwan signed a contract with South Africa to purchase 4,000 tons of uranium.

processing facilities so that its nuclear fuel can be converted, it has also pledged not to inaugurate reprocessing because of the implications for nuclear weaponry. Thus its reprocessing facilities are officially "closed."[94] Any move to develop reprocessing capabilities would have serious political repercussions. Third, security, safety, and pollution problems associated with nuclear energy have increased with new concerns and doubts in the United States and elsewhere about nuclear energy.

At the present time Taiwan has invested very little abroad to insure a flow of energy, though it seems to be moving in that direction rather quickly. Should a developing country want to drill for oil without dealing with large oil firms or with Western nations, Taiwan can provide drilling equipment. Its tanker fleet is also capable of helping oil-producing countries market their petroleum.

Taiwan has some indirect investment in several Middle Eastern countries through aid missions and through underbidding other nations on construction contracts. Taiwan has also purchased for experimental purposes some raw uranium that could be enriched and used in nuclear plants, with the help, of course, of foreign enrichment facilities. It has invested in Indonesian oil operations through individual business connections. The reason for Taiwan's meager investments abroad to support an energy supply is that energy use has escalated far ahead of efforts at investment abroad. It is also true, unfortunately, that more of the countries with energy supplies do not need Taiwan's investment. Conversely, Taiwan has been reluctant, in part because of competition, costs, and risks, to commit resources to exploration for energy in other countries.

ENERGY AND THE ENVIRONMENT

Pollution of various kinds has become a serious matter in Taiwan despite the fact that little is said or written about it. It is not so much that the government or business wishes to conceal the problem, but rather because it is a problem that has appeared so suddenly in its present form. Dust, soot, and sewage problems are old and most people have learned to ignore them. Pollution caused by energy conversion and use, however, is new and the scope of its effects on human life is not fully realized.

Even a brief visit to Taiwan makes one immediately aware that air and water pollution are serious problems. Taipei and Kaohsiung are among the world's worst cities with regard to visible air pollution.[95] Fortunately both, as well as most other cities in Taiwan, enjoy a constant breeze that reduces the volume of dirty air. The Tamsui River, running through Taipei to the Taiwan Strait on the western coast, is laden with many kinds of pollutants; and it is not unlike other rivers in Taiwan. Once again, Taiwan is lucky to have steady rainfall, which dilutes waterborne pollution and washes most of it away. Despite the benefits of wind and rain, however, people living in Taiwan's cities now experience more frequent colds and serious respiratory diseases; lung cancer, pneumonia, and bronchitis are increasing markedly in frequency. [96] In addition residents everywhere on the island have discovered that fish from rivers and coastal areas near river deltas are frequently inedible or unsafe and that the environment has been damaged in a variety of other ways due to industrialization and increased energy use.

There has been some government action: outlawing the use of coal for heat and cooking in Taiwan's large cities and some efforts to reduce the count of microorganisms in water. Such regulations have made air and water in some ways cleaner, but new problems are becoming serious faster than the old ones are being resolved. New pollutants, especially chemicals in water, are more dangerous than those of the past and not so easy to control. City air in Taiwan is much more polluted than it has been in the past despite the lower levels of coal dust in the air, which makes the air appear cleaner.

An important reason for both government and citizen inaction toward the problem is the belief that pollution control is expensive and would make Taiwan's products less competitive on the world market, thus interfering with economic growth. Since nearly everyone in the country is proud of the nation's economic achievements and benefits directly from them, there is a reluctance to do anything that might endanger continued growth. It is an unhappy truth that countries with which Taiwan is competitive in textiles and electronic products, to mention only two lines of exports, do not concern themselves with pollution standards, an indifference that could give a country temporary price advantages over more scrupulous competitors. [97]

Also, Taiwan has urbanized so quickly that the long-range effects of air and water pollution have not been realized yet; nor is there much awareness of the long-range environmental impact of industrialization and energy conversion. The problems of pollution are very visible in Taiwan, but a clear public perception of them has not yet emerged. Insensitivity to direct environmental impact is accompanied by an understandable unconcern about the indirect effects or global environmental problems. Outside of academic circles almost nothing is said about problems such as pollution of the sea around Taiwan, genetic damage caused by pollution, or modification of climatic patterns. Young people are absorbed in practical and personal matters, more so than in most countries, and citizens' groups are not as active as they are in more highly industrialized states. The media also show little interest in pollution, being more concerned with political and economic issues.

All of this is not to say, however, that nothing is being done in the realm of pollution control. In addition to the restriction on coal burning in large cities, the government has raised the cost of auto and other vehicle license fees in an attempt to regulate the growing flood of motor vehicles in the cities. The effectiveness of the measure has been minimized by economic growth and higher standards of living and the resulting demand for vehicles. Also, while there are pollution control standards for all vehicles, laws cannot be enforced adequately because of the rapid increase in the number of vehicles, the absence of bureaucratic machinery to oversee the standards, and a shortage of mechanics able to repair pollution control equipment. The government has also tried to limit dust pollution by restricting building licenses and the use of certain kinds of construction materials and equipment. Also, more streets have been paved. These measures remain insufficient, however, considering the magnitude of the problems.

With regard to the pollution caused by industrial energy conversion, there are regulations, which are fairly well enforced, on the production of sulfur dioxide and nitrous oxide, probably the two worst pollutants. [98] With regard to the former, power plants and large factories are required either to install emission control devices or to use low-sulfur fuels. Taller smokestacks are also mandatory, even though it is realized that stacks merely spread particulates over a larger area to affect more people. In many cases, depending on winds and regions, particle pollution is swept out to sea, and many new industries as well as power plants are now being located near the west. Nitrous oxide is controlled by effecting more complete combustion in power plants and factories, but since it is a less troublesome problem, regulations on combustion modification apply only to new equipment. Like most other countries, there are no laws governing the chemical composition of particles or the emission of trace metals. [99] City governments, most notably in Taipei, have discouraged migration into urban areas, confined factories to areas away from the center of the city, instituted zoning laws, and built parks for greenery and insulation against noise.

Government regulations concerning water pollution resulting from thermal and effluent waste from power plants and factories are preventive rather than just remedial. In other words, factory and power plant construction is prohibited in areas likely to pollute streams and rivers. Little, however, is being done to stop the pollution going on from previously built power stations and factories. Also, the "preventive" policy consists primarily in the location of new facilities in areas where they will pollute the ocean rather inland water systems. The general feeling is that little can be done to improve the rivers and that the effort to do so would undermine economic progress. [100] In short, there is some impetus to prevent Taiwan's rivers from becoming more polluted, but little pressure to clean them up. Some concern has been registered by Taiwan's fishermen about water pollution that has killed or contaminated fish in various coastal areas, but their complaints have had little or no impact upon governmental action.

One reason for the decision to build nuclear power plants was the belief that pollution caused by them would be less of a problem than that produced by fossil fuel plants. So far this seems to have proven true. The two earliest sites are on the northern coast away from population centers, but still close enough for the inexpensive transmission of electricity. The choice of sites also minimized the threat of earthquakes and thermal water pollution in Taiwan's rivers. Much the same arguments are applied to the third nuclear station to be built at the southern tip of Taiwan. [101]

The prospects of thermal pollution of oceanic waters around the nuclear plants has not been taken seriously, but it is certain that ecosystem disruption will occur for some distance. The general attitude is that a sacrifice had to be made, and that this is less serious than the kind of environmental damage that would be caused by fossil fuel plants.

The disposal of spent nuclear fuel is a virtually nonexistent problem. Under the agreement with the United States, through which processed fuel is obtained, all used fuel will be returned. Not only does this arrangement prevent the use of spent fuel for weapons, it also rids Taiwan of disposal headaches. With regard to safety, Taiwan has followed the safety guidelines issued by the U. S. companies that built the equipment and oversee its use. In addition, the U. S. government and the International Atomic Energy Agency oversee Taiwan's nuclear plants, set safety standards, and make recommendations for the prevention of accidents and contamination. Other watchdog agencies are Taiwan's own Atomic Energy Commission, which sets safety and other standards, and the Taiwan Power Company. In the case of the former, guidelines are based on the U. S. Atomic Energy Commission's safety and environmental standards and are enforced in about the same way.

All of this means that there is constant monitoring of air, land, and water around the plant sites to detect radioactivity, and there is a regular check of safety and emergency procedures. [102] While there is no way to be sure that safety and pollution standards are being honored, several engineers and government officials responsible for the safety and pollution standards at Taiwan's first nuclear power station told this writer that better safety standards are used than in the older U. S. nuclear power stations and that even more safeguards are to be built into the second and third stations. They assert that Taiwan's nuclear power plants are safer than most of the world's nuclear plants because they are newer and are regulated by their private and government agencies, by U.S. agencies, and by an international organization. At the same time, officials interviewed on this subject boasted that Taiwan was able to build nuclear plants faster than most countries because there are no public fears or protests about the dangers of atomic power. Such confidence might suggest that safety standards could have been lowered or ignored as a shortcut to nuclearization, though there is no evidence to suggest that this is the case. In the opinion of this author Taiwan's nuclear power plants are probably safer because they are newer and they are inspected by local, U.S., and international agencies. In addition, the likelihood of sabotage seems less because of the general high level of concern about security. This is offset to some extent by the fact that all of Taiwan is subject to earthquakes. Because of high population density, an accident would probably cause more loss of life and injury than in most countries.

In summary, pressure on the environment is of less concern in Taiwan than it should be when one considers the rate of industrial expansion, the huge increases in energy conversion, and the high density of population. None of Taiwan's neighbors complain seriously about Taiwan's pollution load, except for some rather vague criticism from Peking about the misuse of resources and "ravaging of the land" because of Taiwan's capitalist system. It is certain that this will not remain true for long. International environmental standards may also soon play a role, and when this happens Taiwan's efforts to direct its pollution into the ocean or the air space above the ocean may prove counterproductive. In the meantime, in Taiwan not much is safe from "development" as economic growth presses on. Obviously some of the innovativeness of Taiwan's business leaders needs to be directed into energy saving and environmental protection.

FUTURE PROSPECTS

Assuming double digit economic growth in Taiwan for the next decade, energy use will continue to increase very rapidly. It has been estimated that demand for energy will increase at the rate of 11.91 percent a year through 1981, dropping slightly to 11.82 percent during the next six-year period to 1987.[103] If this projection is accurate (it should not seem startling in light of past growth rates), one can reasonably expect much higher energy costs for the nation and increasing environmental damage. Indeed, these twin problems could reach crisis proportions in the next decade. Without future discoveries of local energy sources—petroleum or natural gas—the dependency problem will become more serious. In fact, one observer believes that Taiwan will be importing 200 million barrels of oil a year by 1985. [104] This means a doubling of petroleum imports from 1979. Assuming that prices rise to $50 a barrel at that time (probably a very conservative estimate given recent spot prices), Taiwan's imports of oil will cost $10 billion a year, or double its entire GNP in 1969. The costs of nuclear fuel are not included in this estimate, and are also high and increasing. Needless to say, these energy costs cannot be met without continued rapid economic growth and an increasing volume of exports. Clearly there is considerable uncertainty in these trends.

Even if exports can be marketed to sustain double digit growth in the next decade, the pollution problem is likely to become a major inhibiting factor. No studies have been done anticipating how serious environmental damage may be in coming years in Taiwan. Nor is there any way to predict how soon there will be a public reaction to air and water pollution. However, it is the feeling of this writer that if frantic steps are not taken immediately the problem will reach crisis proportions within five years. Taiwan needs better planning to coordinate economic growth and environmental problems. In this connection, three related goals should be pursued: clean, replenishable sources of energy must be developed; future industrial development must be organized around a less energy-intensive industrial technology; and stricter environmental laws must be codified and enforced. The government and the citizenry must be made to realize that the environment is not just a "tradeoff," but the indispensable material base for any productive system. To destroy the environment is ultimately to destroy the economy dependent upon it.

The growing nuclear power industry in Taiwan will alleviate somewhat air and water pollution in inland areas, though it will generate offshore thermal and other kinds of pollution. Little can be done to reduce the environmental damage caused by Taiwan's nuclear power plants other than the careful enforcement of safety standards to prevent leaks. The polluting effects of energy use, however, can be reduced in two other ways. One is for Taiwan to switch to imported petroleum with a lower sulfer content. This point has been discussed in high government circles in Taiwan. It does not seem possible immediately, however, because of the country's heavy dependence on high-sulfur Saudi Arabian oil. Most other Middle Eastern oil is also high in sulfur content. Thus, Taiwan will probably have to limit its efforts in this realm to sulfur-removing processes. Second, liquid natural gas might be imported. Many of Taiwan's factories are equipped to use natural gas, while local supplies are insufficient to keep pace with rising demand. Thus importing natural gas would not require converting factories to its use in the short run. This option has not yet been given serious consideration, however, because of the high price of LNG. Its use would raise the cost of Taiwan's exports and undermine Taiwan's competitive edge in a number of markets. However, as air pollution worsens, LNG may become more attractive despite the cost.

Energy from the sun, wind, tides, and biomass, all clean and renewable sources, have considerable promise. The solar option, already in use in Taiwan, has the greatest promise, especially for water heaters, space heating, and air conditioning, probably in that order of feasibility. Solar water heating devices can pay for themselves in as little as nine months of use in the factories and office buildings of Taiwan.[105] The reservations holding back most large enterprises are initial investment costs and the lingering expectation that energy costs will drop in the near future because of some unexpected breakthrough in energy production. Solar water heating equipment is not mass produced at the moment and there is no widespread availability of maintenance and repair. Space heating and air conditioning systems driven by solar power are even farther away with respect to mass production and general use but are still in the realm of the immediately possible in terms of production, installation, and upkeep.[106]

There has been and still is serious discussion of solar energy for the generation of electricity, but officials in Taiwan are not yet convinced that it is economically feasible. In fact, it has been argued that, given the present state of technology, electricity can be produced in Taiwan's "maximum solar energy band" at about the same cost as for oil-burning furnaces; moreover, the potential for such an area is analogous to a large oil field, with the difference being that it cannot be depleted and causes little pollution.[107] Taiwan, however, is reluctant to make a commitment when the costs for failure are so high. The main obstacle is Taiwan's expectation that the United States and other advanced countries will further develop the solar cell in the next few years. In fact, the Taiwan government has been maintaining contacts with Chinese scientists in the United States who are doing some of the most advanced work on solar cells. In short, Taiwan feels that it doesn't have the financial resources to do sufficient basic research in this area and must wait for the United States and other countries to make a breakthrough. Once this is accomplished Taiwan may more quickly put solar energy into widespread use.

Wind energy appears to be workable in Taiwan. Wind energy power plants approach the cost of oil for the generation of electricity and do not suffer the disadvantage of corrosion that has impeded the development of geothermal energy.[108] The major problem with using the wind for electricity generation is the intermittent character of the source. This can be offset to some extent by burning waste in the wind generators to drive them when the wind velocity is inadequate, but there remain other obstacles. Even though wind energy seems less feasible than solar energy at the present time, and might be confined to the eastern part of Taiwan where there are no large cities and little industry, it is a renewable source of energy that seems worth adapting to the not unfavorable physical conditions of Taiwan.

Much the same can be said for tidal energy. While it is still in the experimental stage, it would be foolish not to pursue the option in the context of favorable conditions in Taiwan. Biomass also merits a financial commitment to bring it out of the realm of speculation as an energy source with substantial possibilities. [109] Hydrogen-producing algae that will grow in ocean water have already been developed in Taiwan. It has been estimated that culturing this algae in 60 square kilometers of ocean space could result in an amount of hydrogen equal in energy output to 100 times Taiwan's current natural gas production. [110] Furthermore, the algae can yield carbon in forms that can be mixed with hydrogen to form hydrocarbons usable in making plastics, medicines, and other products currently dependent on petroleum hydrocarbons.

The trend toward cleaner, less energy-intensive industries in Taiwan needs to be reinforced by deliberate policies and long-range planning. Fortunately the government seems to be aware that certain energy-intensive industries, such as textiles, are likely to decline as less developed countries with cheaper labor and indigenous sources of energy move into the market; but it is doubtful that such a development in itself will slow Taiwan's use of energy through the next decade. Overall growth momentum is likely to obliterate the positive effects of such unplanned reduction in energy use intensiveness.

The style of governmental planning in the energy field, like most other nations, has been improvisational and incremental with very strong political overtones. In fact, political considerations play a greater role in Taiwan's energy policies than most other nations. For example, the decision to build nuclear power plants had economic motives to be sure; but it was also part of the government's drive to maintain its credibility in the face of diplomatic defeats. It also makes Taiwan a potential nuclear power. The need for rapid economic growth, which demands increased energy use, prevents unemployment and spreads prosperity; it also placates a potentially restive and even revolutionary population. Considering the constant threat to Taiwan's sovereignty, these considerations are highly desirable and cannot be sacrificed even to prevent serious environmental damage. A guarantee from the United States of Taiwan's independence would no doubt encourage more rationality and foresight in economic and energy planning.

In the course of adjusting to the unhappy implications of energy dependence—which has happened faster and represents a more marked change from the past than in most other nations—Taiwan might well end up in the vanguard of those nations ready to try new kinds of energy in the interests of both sustainable economic activity and environmental balance. Taiwan has already demonstrated its willingness to experiment with cleaner and replenishable energy sources in ways that to many are regarded as novel and innovative. Given Taiwan's quick application of technology it may in coming years give application to some of the many techniques already developed to utilize better sources of energy and to conserve energy. Clearly, if Taiwan does not find cheaper and cleaner sources of energy soon and does not use its energy more efficiently, its future economic development prospects as well as its very survival can be doubted.

NOTES

1. James W. Davidson, The Island of Formosa (New York: Macmillian, 1903), p. 476.

2. Ibid. , p. 178.

3. Ibid. , p. 210.

4. Ibid.

5. Ibid. , p. 488.

6. Ibid. , p. 490.

7. For a good synopsis of Taiwan's economic growth, see Kung-chia Yeh, "Economic Growth: An Overview," in Yuan-li Wu and Kung-chia Yeh, eds. , Growth, Distribution and Social Change: Essays on the Economy of the Republic of China (Baltimore: University of Maryland Law School Occasional Papers/Reprint Series in Contemporary Asian Studies, No. 3, 1978).

8. Ibid. , p. 28.

9. Ibid. , p. 24.

10. Taiwan Statistical Data Book, 1976 (Taipei: Economic Planning Council of the Executive Yuan of the Republic of China), p. 91.

11. Ibid. , p. 90.

12. Ibid. , p. 91.

13. The first figures come from Ralph N. Clough, Island China (Cambridge, Mass. : Harvard University Press, 1978), p. 70, who cites various sources in Taiwan. The latter figure comes from The Far Eastern Economic Review Asia 1979 Yearbook, p. 303.

14. See Yuan-li Wu, "Income Distribution in the Process of Economic Growth in Taiwan," in Wu and Yeh, Growth, Distribution and Social Change, for further details.

15. New York Times, April 12, 1977.

16. See National Conditions (Taipei: Statistical Bureau of the Executive Yuan, Winter 1977), and Encyclopedia Britannica 1977 Yearbook, p. 498, for further details on consumerism. For details on the rapid electrification of Taiwan, see "Rural Electrification in Taiwan," Energy Quarterly (Taipei), October 1971.

17. See National Conditions, p. 11.

18. The Energy Situation in Taiwan, Republic of China (Taipei: Energy Policy Committee, Ministry of Economic Affairs, March 1978), p. 4.

19. Ibid.

20. Ibid. , p. 6.

21. K. S. Chang, "The Energy Situation in Taiwan, Republic of China—Past, Present and Future," Energy Quarterly, October 1974, p. 2.

22. Ibid.

23. The Energy Situation in Taiwan, Republic of China, p. 12.

24. Ibid. , p. 10. For further details on the problems of developing hydropower in the future, see David S. L. Chu, "Basic Concepts in Hydroelectric Development in Taiwan in the Future," Energy Quarterly, July 1974.

25. See Suyen Chain, "The Development of the Natural Gas Industry in Taiwan, Republic of China," Energy Quarterly, April 1971, for further details on locally produced natural gas.

26. The Energy Situation in Taiwan, Republic of China, p. 7. This figure is for 1977 but probably has not changed since that time.

27. Ibid.

28. Ibid.

29. According to present projections, atomic energy will provide 46 percent of Taiwan's needs in the 1980s, but reaching this figure depends on finishing nuclear plants on schedule. At the present time some delay may seem likely. If new nuclear plants are not built this figure will drop immediately, assuming projected increased demand continues. If new plants are constructed it is possible that nuclear power will provide up to 60 percent of Taiwan's energy needs. See the Christian Science Monitor, December 21, 1978, for an elaboration. Based on interviews this author has had with energy officials in Taiwan it seems unlikely that Taiwan will build more nuclear power plants after the completion of the third site.

30. The Energy Situation in Taiwan, Republic of China, p. 5.

31. Ibid.

32. Yuan-li Wu and Kung-chia Yeh, "Taiwan's External Economic Relations," in Wu and Yeh, Growth, Distribution and Social Change, pp. 192-93.

33. Ibid. , p. 193.

34. Ibid.

35. The author was living in Taiwan at this time and observed that the prices of some consumer goods doubled in less than six months. Part of this was caused by anticipated inflation and part by the fact that many people felt that the government would have to limit the supply of consumer goods, thus creating permanent price rises or rationing.

36. See Wu and Yeh, "Taiwan's External Economic Relations," pp. 193-94, for further details on government policy as it helped bring Taiwan out of a recession caused by high oil costs.

37. Taiwan's exports as a percentage of its domestic product increased from 8 percent in 1952 to 12 percent in 1959, 18 percent in 1965, 49 percent in 1973, and 52 percent in 1976. See Ibid., p. 192.

38. The author has argued elsewhere that Taipei has stimulated economic development and has even allowed its citizens to invest a-broad while continuing to attract foreign capital in order to increase economic ties with other nations, perceiving that this will strengthen foreign commitments to Taiwan's independence. See John F. Copper, "Taiwan's Strategy and American's China Policy," Orbis, Summer 1977.

39. Clough, Island China, p. 161.

40. See Richard W. Wilson, "A Comparison of Political Attitudes of Taiwanese Children and Mainlander Children on Taiwan," Asian Survey, December 1968, and Sheldon Appleton, "Taiwanese and Mainlanders on Taiwan: A Survey of Student Attitudes," China Quarterly, October-December 1970.

41. See, for example, Chiang's statements to the Tokyo Bureau Chief of U.S. News and World Report in Free China Weekly, July 22, 1979.

42. For further details, see "The Ten Major Construction Projects: Taiwan Government Sets National Priorities in Key Industrial Sectors," Asian Wall Street Journal, July 13, 1978, and the Far Eastern Economic Review Asian 1979 Yearbook, p. 306.

43. Chang, "The Energy Situation in Taiwan," p. 3.

44. Ibid.

45. Ibid. This is clearly reflected in the decreased use of coal and its present less important role as a source of energy in Taiwan.

46. See Note 1.

47. See "Mineral Resource Development in Taiwan, 1945-1974," Mining Research and Service Organization Report No. 150, October 1975.

48. Larry Auldridge, "World Oil Flow Gains Slightly, Reserves Dip," Oil and Gas Journal, December 25, 1978, p. 102.

49. Ibid.

50. For details on the above, see statement of Selig S. Harrison in Normalization of Relations with the People's Republic of China: Practical Implications, Hearings before the Subcommittee on Asian and Pacific Affairs of the Committee on International Relations, 95th Cong., September and October 1977.

51. Chang, "The Energy Situation in Taiwan," p. 5.

52. Far Eastern Economic Review, June 23, 1978.

53. Ibid.

54. "The Ten Major Construction Projects."

55. See David L. Cheng, "Development of Hydroelectric Energy on the Tachia Chi," Energy Quarterly, January 1971, for details on Taiwan's hydroelectric potential and various impediments to increasing hydroelectric utilization.

56. The Energy Situation in Taiwan, p. 7.

57. Sources on this subject include Weng-tse Cheng and Yung-juh Wu, "Origin and Potential of some Geothermal Areas in Taiwan," Energy Quarterly, July 1976; "Geothermal Energy in Taiwan, Republic of China," Mining Research and Service Organization Report No. 192, April 1977; K. K. Hwang, "General Review and Recent Progress of Geothermal Resources Studies in Taiwan, ROC," Energy Quarterly, July 1978.

58. Hwang, "General Review and Recent Progress."

59. The Energy Situation in Taiwan, p. 13.

60. Ibid.

61. See James T. Yen, "An Overview of the Energy Resources of the Republic of China," Energy Quarterly, October 1978.

62. Ibid. , p. 3.

63. This conclusion is based upon interviews conducted with officials of Ta Tung Corporation in Taipei. Ta Tung is presently conducting extensive research into the uses of solar energy.

64. Eric K. Y. Kang, "The Development of Marsh Gas Production from Hog Waste in Taiwan, Republic of China," Paper presented at the 10th Intersociety Energy Conversion Engineering Conference, University of Delaware, August 1975.

65. Ibid.

66. Far Eastern Economic Review Asia 1979 Yearbook, p. 303. Taiwan has not imported such a large portion of its petroleum from Saudi Arabia until just recently.

67. China Post (Taipei), January 3, 1975, cited in Thomas J. Bellows, "Taiwan's Foreign Policy in the 1970's: A Case Study of Adaptation and Viability," Asian Survey, July 1976.

68. China Yearbook, 1977 (Taipei: China Publishing Co. , 1977), p. 191.

69. Far Eastern Economic Review Asia 1978 Yearbook, p. 318.

70. China Yearbook, 1977, p. 193.

71. Far Eastern Economic Review Asia 1979 Yearbook, p. 305.

72. China Yearbook, 1977, p. 143.

73. Ibid.

74. Yuan-li Wu and Kung-chia Yeh, "Taiwan's External Economic Relations," in Wu and Yeh, Growth, Distribution and Social Change, p. 192.

75. Ibid.

76. Ibid.

77. <u>Far Eastern Economic Review Asia 1979 Yearbook</u>, p. 305.

78. This is based on a figure of 40.5 percent for the first eight months of 1978. See Ibid. During this same period the United States provided only 21.9 percent of Taiwan's imports.

79. See the <u>Christian Science Monitor</u>, October 29, 1976.

80. New York <u>Times</u>, September 17, 1977.

81. New York <u>Times</u>, July 11, 1977.

82. New York <u>Times</u>, September 17, 1977.

83. Ibid.

84. In 1976 Kuwait provided 52.0 percent of Taiwan's imported crude oil compared to Saudi Arabia's 38.4 percent. These data were provided by the U.S. Embassy in Taipei and are based on the first half of the year.

85. See Ibid. for source of data. It is noteworthy that Indonesia supplied only 3.4 percent of Taiwan's needs in 1976 but 11.4 percent in 1977.

86. Taiwan's investments in Indonesia are almost exclusively private and no figures on total amounts have been published, but it is assumed sizable. See Bellows, "Taiwan's Foreign Policy in the 1970s" for further details.

87. This figure was provided by a director of the China Petroleum Corporation in Taipei in 1977.

88. <u>China Petroleum Corporation Annual Report, 1976</u>, p. 13.

89. <u>Free China Weekly</u>, May 27, 1979.

90. For a good background analysis of Taiwan's decision to develop nuclear energy and the problems involved, see C. Y. Chu, "Nuclear Power Development in Taiwan," <u>Energy Quarterly</u>, April, 1974.

91. Clough, <u>Island China</u>, p. 83. Also see L. K. Chen, "Nuclear Power Development in Taiwan, Republic of China," <u>Energy Quarterly</u>, January 1972.

92. An "Official of Taiwan Power Company" quoted in <u>China Post</u>, February 21, 1979.

93. See Melina Liu, "Taipei Treads Lightly," <u>Far Eastern Economic Review</u>, January 6, 1978. According to this author, Taiwan has already purchased a sizable quantity of uranium from South Africa. Taiwan has also signed a contract with the government of Paraguay to explore jointly for uranium. If sizable deposits are found, Taiwan plans to sign a joint venture agreement for $100 million. See <u>Asia Research Bulletin</u>, July 31, 1978. (According to the government, Taiwan now has a seven—year supply of nuclear fuel. See <u>China Post</u>, February 7, 1979.)

94. Melinda Liu, "Accounting for the N-factor," <u>Far Eastern Economic Review</u>, December 17, 1976.

95. For a vivid description of some of the aspects of pollution

in Taiwan, see William Armbruster, "Letter from Kauhsiung," Far Eastern Economic Review, August 27, 1976.

96. For details on the problem of air pollution in Taipei, see Chin-yuan Chuang, "Research on Air Pollution in Taipei City," Energy Quarterly, January 1973.

97. Almost all of the officials that the author interviewed stated that something is being done about air and other types of pollution in Taiwan, but that it was not enough. They also mentioned export competition as one of the reasons. For further details on the cost of pollution control, see Tony T. L. Liao, "Environmental Impact on Power Development," Energy Quarterly, October 1972.

98. For further details, see Chuang, "Research on Air Pollution in Taipei City."

99. For further details on the above, see Thomas T. Shen, "Discussion of Environmental Pollution Control Regulations and Standards for Taiwan's Power Development," Energy Quarterly, October 1978.

100. Ibid.

101. See J. H. Chen and E. Lin, "Safety Designs of Taipei's Nuclear Power Plants," Energy Quarterly, July 1975, for further details.

102. P. C. Su and P. C. Chyen, "Radiation Protection in Taipei's Nuclear Power Project," Energy Quarterly, July 1975.

103. David Lan Cheng, "Present Power Systems and Future Power Development in Taiwan, Republic of China," Energy Quarterly, July 1978.

104. Yen, "An Overview of the Energy Resources."

105. In the summer of 1977 an official of Ta Tung Corporation showed me an experimental solar water heater that they had built that had paid for itself in this period of time. This company used hot water for employees' showers after each of three shifts each day—thus a large use of hot water. On the other hand this was located in Taipei where there are more cloudy days than in most other cities in Taiwan.

106. This is according to engineers at Ta Tung Corporation.

107. Yen, "An Overview of the Energy Resources," p. 11.

108. Ibid., pp. 11-12.

109. See Yih-yun Hsu, "Clean Fuels from Biomass," Energy Quarterly, January 1976.

110. Yen, "An Overview of the Energy Resources," p. 12.

SELECTED CONVERSION FACTORS

WEIGHT
1 kilogram (kg) = 1,000 grams = 2.205 pounds (lb)
1 metric ton = 2,205 lb = 1000 kg = 0.985 ton = 1.1023 short ton
1 short ton = 2,000 lb = 907.2 kg = 1 British long ton (2,240 lb)
1 mmt = 1 million metric tons
1 bmt = 1 billion metric tons
1 mmtce = 1 million metric tons of coal equivalent
1 ton of oil equivalent = 10^7 kilocalories (kcal)
1 ton of coal equivalent = 0.7 million tons of oil equivalent = 4.48
  million barrels (bbl) of oil = 25.19 trillion cubic feet of natural gas

LENGTH
1 kilometer (km) = 0.6214 statute mile
1 mile = 1.609 km

AREA
1 hectare = 2.47 acres = 0.003861 square mile
6.6 mou (mu in Pinyin) = 1 acre
1 square mile = 640 acres = 259 hectares

VOLUME AND CUBIC MEASURE
1 cubic meter (m^3) = 35.31 cubic feet (ft^3) = 0.02832 m^3
1 bm^3 = 1 billion cubic meters

ENERGY AND WORK
1 British thermal unit (Btu) = 252 calories (cal) = 0.0002931 kilowatt
  hour (kWh) = 1,055 joules
1 joule = 1 watt sec = 0.74 foot pound
1 kWh (kilowatt hour) = 3412 Btu
1 kilocalorie (kcal) = 3.968 Btu

POWER
1 kilowatt (kW) = 1,000 watts
1 megawatt (MW) = 1,000,000 watts
1 gigawatt (GW) = 1,000,000 kW, or 1,000 MW
1 horsepower = 746 watts

These conversion units have been chosen to facilitate the use of this book. Some of the numbers have been rounded off, and a few abbreviations scattered through the text have been clarified.

The base unit for energy is the kilowatt hour, which is the en-

1 quad = 10^{15} Btu = 300 X 10^9 kWh (thermal) = 170 X 10^6 bbl crude oil = 42 million tons of bituminous coal = 293 billion kWh of electricity

1 42 gallon bbl oil = 0.58 electrical megawatt hours

---

ergy used when a device rated at 1,000 watts operates for an hour (or a 100-watt device operates for ten hours).

The base unit for power is the watt. The power rating of an electrical device is found by multiplying the voltage by the current (in amperes), that is, a 125-volt appliance drawing ten amperes has a power rating of 1,250 watts (number of watts = voltage x current).

LIST OF ABBREVIATIONS AND ACRONYMS

AEC	Atomic Energy Commission (United States)
CANDU	Canadian heavy water reactor
CFE	Consejo Federal de Electricidad (Federal Electricity Commission, Mexico)
CIA	Central Intelligence Agency (United States)
FY	fiscal year
GDP	gross domestic product
GNP	gross national product
GWP	gross world product
HWR	heavy water reactor
IIE	Instituto de Investigaciones Eléctricas (Institute of Electric Research, Mexico)
INEN	National Nuclear Energy Institute (Mexico)
LDP	Liberal Democratic Party (Japan)
LNG	liquefied natural gas
LWR	light water reactor
MITI	Ministry of International Trade and Industry (Japan)
MNC	multinational corporation
NIEO	New International Economic Order
NNOC	Nigerian National Oil Corporation
NNPC	Nigerian National Petroleum Corporation (resulted from a merger in 1977 of NNOC with the Ministry of Petroleum Resources)
OPEC	Organization of Petroleum Exporting Countries
PEMEX	Petróleos Mexicanos (government oil firm)
PRC	People's Republic of China
SEPAFIN	Secretaria de Patrimonio y Fomento de Industria (Secretariat for Resources and Industrial Development, Mexico)
s. f. e.	standard fuel equivalent
UNDP	United Nations Development Program

GLOSSARY OF SELECTED TECHNICAL TERMS

For general reference, see the following:

D. Crabbe and R. McBride, eds. , The World Energy Book: An A-Z Atlas and Statistical Source Book (Cambridge, Mass.: MIT Press, 1979).

V. Daniel Hunt, Energy Dictionary (New York: Van Nostrand Reinhold, 1979).

Robert L. Loftness, Energy Handbook (New York: Van Nostrand Reinhold, 1978). This volume has detailed energy conversion tables.

ANTHRACITE: A hard, black, lustrous coal that burns efficiently and is valued for the quality of its heat.

BAGASSE: Fibrous material remaining after the extraction of juice from sugarcane; it can be used as a fuel.

Bbl/d: Barrels per day.

BIOMASS FUELS: The production of synthetic fuels, such as methane hydrogen, from plants and animal wastes.

BITUMINOUS COAL: Soft coal high in carbonaceous and volatile matter. When the volatile matter is removed by heating in the absence of air, the coal becomes coke.

BREEDER REACTOR: A nuclear reactor that produces more fuel than it consumes. The new fissionable materials are created by neutron capture in the fertile materials U-238 or Th-232. There are three types of breeder reactor: the liquid metal fast breeder (LMFBR); the gas-cooled fast breeder (GCBR); and the molten salt breeder (MSBR).

BY-PRODUCTS (residuals): Secondary products that have commercial value and are obtained from the processing of raw material, such as residues of tar and ammonia from gas production.

COAL GASIFICATION: The conversion of coal to a gas suitable for use as a fuel.

COAL LIQUEFACTION: The conversion of coal into liquid hydrocarbon and related compounds by hydrogenation.

COAL SLURRY: Coal in pulverized form suspended in water.

COKING COAL: The most important of the bituminous coals. It burns with a long yellow flame and produces an intense heat.

CONDENSATE OIL: A liquid hydrocarbon obtained by the combustion of a vapor or gas produced from oil or gas wells.

CONVERSION TO URANIUM HEXAFLUORIDE: In the preparation for uranium enrichment, yellowcake is converted to uranium hexafluoride (UF_6).

CRACKING PLANT: An oil refinery.

CRUDE: Oil in its natural state, before refining or processing.

CRUDE NAPHTHA: A light distillate made from the fractionation of crude oil.

DECONTAMINATION: The removal of radioactive contaminants from surfaces or equipment by cleaning or washing with chemicals.

DEPLETION ALLOWANCE: A tax allowance extended to the owner of exhaustible resources based on an estimate of the permanent reduction in value caused by the removal of the resource.

DESULFURIZATION: The process by which sulfur and sulfur compounds are removed from gaseous or liquid hydrocarbon mixtures.

DIESEL FUEL: Fuel used for internal combustion in diesel engines; usually that fraction which distills after kerosene. It is similar to gas oil.

DIRECT ENERGY CONVERSION: The generation of electricity from an energy resource in a manner that does not include transference of energy to a working fluid, such as magnetohydrodynamic conversion.

DISTILLATE FUEL OIL: Any fuel oil, topped crude oil, or other petroleum oil, derived by refining or processing crude oil.

ECOLOGY: The science dealing with the relationship of all living things with their environment and with each other.

ECOSYSTEM: A complex of the community of living things and the environment, forming a functioning whole in nature.

EFFICIENCY: The efficiency of an energy conversion is the ratio of the useful work or energy output to the work or energy input.

ENERGY: The capability of doing work. It takes such forms as potential, kinetic, heat, chemical, electrical, nuclear, and radiant energy. Some forms of energy can be converted into other forms, and all forms are ultimately converted into heat. Energy is measured in ergs.

ENERGY-GNP RATIO: This ratio is obtained by division of energy-growth (percent rate) over a given time period by GNP growth rate. It is not considered healthy for energy growth to outstrip GNP growth.

ENRICHMENT: A process by which the proportion of fissionable uranium isotope (U-235) is increased above the 0.7 percent contained in natural uranium. The principal method is gaseous diffusion.

ETHANOL: Ethyl alcohol or grain alcohol, which can be produced from biomass by fermentation.

EUTROPHICATION: The process by which a body of water becomes so rich in nutritive compounds such as nitrogen and phosphorous that algae and other plant life cause it to dry up and disappear.

EXPLORATION: The act of searching for potential subsurface reservoirs of gas or oil.

EXPLORATORY WELLS: A well drilled in search of a new, and as yet undiscovered, field of oil or gas, or with the expectation of extending the limits of a field already partly developed.

EXPONENTIAL GROWTH: Growth that occurs at a rate that is a constant percentage of the whole, thus resulting in doublings in shorter spans of time.

FABRICATION: The conversion of recovered fissile materials into a suitable form.

FISSILE MATERIAL: An element that can be converted into fissionable material by neutron irradiation, such as U-238 and Th-232.

FLARE GAS: Natural gas burned in flares at an oil field; waste gas.

FLY ASH: Small particles of air-borne ash produced by the burning of fuels.

FUEL: Any substance that can be burned to produce heat.

FUEL CYCLE: The steps involved in supplying fuel for nuclear reactors, which include mining, uranium refinement, fabrication of fuel elements, their use in a nuclear reactor, chemical processing to recover remaining fissionable material, reinrichment of the fuel, fabrication into new fuel elements, and waste storage.

FUEL OILS: Fuel oils are the petroleum fractions with a higher boiling range than kerosene, such as lighter oils (for example, Distillates 1, 2, and 4) used for residential and commercial heating of buildings and for transportation.

FUEL REPROCESSING: The processing of reactor fuel to recover the unused, residual fissionable materials.

FUSION: A nuclear reaction in which two light atomic nuclei unite (or fuse) to form a single nucleus of a heavier atom. The process takes place in the sun and in other active stars.

GAS, LIQUEFIED: Liquefied natural gas (LNG) is natural gas cooled to -160°C so it forms a liquid at approximately atmospheric pressure. As it becomes liquid the volume is reduced by a factor of some 600, thus allowing economical storage and transportation.

GAS, MANUFACTURED: A gas obtained by destructive distillation of coal, thermal decomposition of oil, or the reaction of steam passing through a bed of heated coal or coke, such as coal gas or blast furnace gas.

GAS, NATURAL: A naturally occurring mixture of hydrocarbons, found in porous geologic formations beneath the earth's surface, often in association with petroleum.

GASOHOL: A blend of 10 percent agriculturally derived ethyl alcohol and 90 percent unleaded gasoline that can be used as an automotive fuel.

GEOTHERMAL POWER: Power plants using hot water or steam that is stored in the earth from volcanic activity. Also, hot underground rocks can be used by fracturing them and dousing them with cold water, which produces steam that can be returned to the surface for steam production.

GROSS NATIONAL PRODUCT (GNP): The total market value of the goods and services produced by a nation before the deduction of depreciation charges and other allowances for capital consumption.

HALF-LIFE: The time in which half the atoms in a radioactive substance disintegrate, varying from millionths of a second to billions of years.

HEAVY WATER: Water containing the heavy isotope of hydrogen (deuterium). Heavy water is used as a moderator in some reactors because it slows down neutrons effectively and permits the use of natural uranium as a fuel.

HYDROCARBON FUELS: Fuels that contain an organic chemical compound of hydrogen and carbon.

HYDROELECTRIC PLANT: A plant in which the kinetic energy of falling water is used to turn a turbine generator producing electricity.

INCOME ELASTICITY OF ENERGY CONSUMPTION: See Energy-GNP Ratio.

ISOTOPES: Atoms with the same atomic number but different atomic weights. Since nuclear stability is governed by nuclear mass, one or more isotopes of an element may be radioactive or fissionable, while other isotopes of the same element might be stable.

ISOTOPE SEPARATION: The process of separating isotopes from one another, or changing their relative concentrations.

KEROSENE: The petroleum fraction containing hydrocarbons that are slightly heavier than those found in gasoline and naphtha, used today as fuel for gas turbines and jet engines.

LIGHT WATER REACTOR (LWR): Nuclear reactor in which water is the primary coolant and moderator. There are two commercial types: the boiling water reactor (BWR) and the pressurized water reactor (PWR).

LIGNITE: A low-grade coal intermediate between peat and bituminous coal.

LOAD: The amount of power needed to be delivered to a given point on an electric system.

MAGNETOHYDRODYNAMICS (MHD): Deals with phenomena arising from the motion of electrically conducting fluids in the presence of electric and magnetic fields. An advanced method for improving the efficiency of electricity generation from fossil fuels.

MAXIMUM PERMISSIBLE CONCENTRATION: The concentration of radioactive material in air, water, and foodstuffs that competent authorities have established as the maximum that would not create undue risk to human health.

METHANE: The lightest in the paraffin series of hydrocarbons. It is colorless, odorless, and flammable, and forms the major portion of marsh gas and natural gas. Can be made by biomass conversion.

METHYL ALCOHOL: The lowest member of the alcohol series. Also known as methanol and wood alcohol, since its source is the destructive distillation of wood. Can be used as a fuel for motor vehicles.

NAPHTHA: A petroleum fraction used, among other things, as a raw material for the production of synthetic natural gas.

NUCLEAR FUEL CYCLE: See Fuel Cycle.

NUCLEAR POWER PLANT: Any device, machine, or assembly that converts nuclear energy into some form of useful power.

NUCLEAR REACTOR: A device in which a fission chain reaction can be initiated, maintained, and controlled. The essential component is a core of fissionable fuel, surrounded by moderator, reflector, shielding, coolant, and control mechanisms.

OIL SHALE: An expression used for a range of materials containing organic matter that can be converted into crude oil, gas, and carbonaceous residue by heating.

OUTAGE: The period in which a generating unit, transmission line, or other facility is out of service.

PARTICULATES: Small particles of solid matter produced by the burning of fuels.

PETROLEUM: An oily flammable bituminous liquid varying from almost colorless to black. A complex mixture of hydrocarbons prepared for use as gasoline, naphtha, and other products by refining processes.

PHOTOSYNTHESIS: The process by which plants convert sunlight, carbon dioxide, and water into energy, releasing oxygen as a waste product.

PHOTOVOLTAIC CELL: A type of semiconductor device which converts sunlight directly into an electric current.

PLUTONIUM: A heavy, radioactive, man-made metallic element with atomic number 94. It is used for reactor fuel and in nuclear weapons.

POWER: The rate at which work is done, or the rate at which energy is transferred. Power is measured in units of work per unit of time. Typical units are the watt and the horsepower.

PRIMARY ENERGY: Energy in its naturally occurring form (coal, oil, uranium) before conversion to end-use forms.

PROBABLE RESERVES: A realistic assessment of the reserves that will be recovered from known oil or gas fields based on estimated size and on the reservoir characteristics of such fields.

PROVED RESERVES: The estimated quantity of crude oil, natural gas liquids, or sulfur that analysis or geologic and engineering data demonstrate with reasonable certainty to be recoverable from known oil or gas fields under existing economic and operating conditions.

PSI: Pounds per square inch.

RADIOACTIVE CONTAMINATION: Deposition of radioactive material in any place where it may harm persons, spoil experiments, or render products or equipment unsafe for some specific use.

RECOVERABLE RESERVES: Minerals and fuels expected to be recovered by present-day techniques and under present economic conditions.

RECYCLING: The reuse of fissionable material, after it has been recovered by chemical processing from spent or depleted reactor fuel, reenriched, and fabricated into new fuel elements.

REFINE: To cleanse or purify by removing undesired material. To make a substance usable by special processing.

REFINERY: A process in which crude oil is distilled into usable substances. Typical crude fractions are ether, methane, ethane, propane, butane, kerosene, fuel oil, lubricants, asphalt, tar, and paraffin.

REPROCESSING: Chemical recovery of unburned uranium and plutonium and certain fission products from spent nuclear fuel elements.

RESIDUAL FUEL OIL: A heavy fraction from the distillation of crude oil, often used as a fuel for power plants.

SMOG: A mixture of smoke and fog. A fog made heavier and usually darker by smoke and chemical fumes.

SOLAR POWER: Useful power derived from solar energy. Both steam and hot-air engines have been operated from solar energy.

SPILL: The accidental release of radioactivity. A large discharge of oil, as from a well or a tanker, into a body of water.

STRIP MINING: The mining of coal by surface methods as distinguished from the surface mining of ores, which is commonly designated as open-pit mining.

SULFER OXIDES: Compounds composed of sulfur and oxygen, produced by the burning of sulfur in coal, oil, and gas. They are harmful to man, plants, and animals, and may damage materials at sufficiently high concentrations.

SUPERTANKER: A very large oil tanker, now in the range of 500,000 dwt (dwt = dead-weight tons, or the total lifting capacity of a ship expressed in long tons).

SURFACE MINING: The mining of coal from outcroppings or by the removal of overburden from a seam of coal, as opposed to underground mining.

SWEETENING: The process by which petroleum products are improved in odor and color by oxidizing and removing sulfur containing and unsaturated compounds.

SYNTHETIC NATURAL GAS (SNG): A gaseous fuel manufactured from naphtha or coal, which has an energy content about that of natural gas.

TAR SANDS: Hydrocarbon-bearing deposits distinguished by the high viscosity of the hydrocarbon, which is not recoverable by ordinary oil production methods.

THERMAL EFFICIENCY: In power plants, the ratio of heat converted to useful energy to the heat input.

THERMAL POWER PLANT: Any electric power plant that operates by generating heat and converting the heat to electricity.

THERMODYNAMICS: The science and study of the relationships between heat and mechanical work. First law: Energy can neither be created nor destroyed (or, you can't get something out of nothing); Second law: Heat cannot pass from a colder to a warmer body without the additional expenditure of mechanical energy (or, energy systems always operate at a deficit).

TOKAMAK: An experimental fusion device.

TOTAL ENERGY: The use of packaged energy systems of high efficiency, in which gas-fired turbines or engines produce electrical energy and in which the exhaust heat is used for heating and cooling.

ULTIMATE RECOVERABLE RESERVES: The total quantity of crude oil, natural gas, natural gas liquids, or sulfur estimated to be producible from an oil or gas field as determined from a study of current engineering data. This includes all quantities produced to the date of the estimate.

URANIUM, NATURAL: A radioactive element with the atomic number 92 and an average atomic weight of about 238. The two principal isotopes are U-235 (0.7 percent of natural uranium), which is fissionable (capable of being split and thereby releasing energy), and U-238 (99.3 percent of natural uranium), which is fertile (having the property of being convertible to fissionable material).

VOLT: A unit of electrical force equal to that amount of electromotive force that will cause a steady current of one ampere to flow through a resistance of one ohm.

VOLTAGE: The amount of electromotive force, measured in volts, that exists between two points.

WASTES, RADIOACTIVE: Equipment and materials from nuclear operations that are radioactive and for which there is no further use. Wastes are generally classified as high level, low level, and intermediate, depending on the number of curies (basic unit to describe radioactivity in a sample, one curie being equal to 37 billion disintegrations per second) per gallon or cubic foot.

WAVE POWER: Using the sea's energy in advancing waves to generate power to drive a turbine.

WIND ENERGY: A form of solar energy, since winds are caused by
variations in the amount of heat the sun sends to different parts
of the earth. Electricity is produced when a windmill catches the
wind and revolves, rotating a turbine that powers an electric
generator.

WORK: Work is done whenever force is exerted through a distance.
The amount of work done is the product of force and distance.
The metric unit of work is the joule. When a force of 1 newton
is exerted through a distance of 1 meter, 1 joule of work is done.

TOPICAL INDEX

NAME AND SUBJECT INDEX

ABOUT THE EDITOR AND CONTRIBUTORS

KENNETH R. STUNKEL is professor of history at Monmouth College, West Long Branch, New Jersey. He received his B. A. , M. A. , and Ph. D. degrees from the University of Maryland. His publications include The Economic Super Powers and the Environment: The United States, the Soviet Union, and Japan, coauthored with Donald Kelley; Relations of Indian, Greek, and Christian Thought in Antiquity; and numerous articles in journals such as Japan Interpreter, Asian Profile, Philosophy East and West, Environment, Bulletin of the Atomic Scientists, American Historical Review, Journal of Religious History, Ohio University Review, Triveni: Journal of the Indian Renaissance, and American Behavioral Scientist. While continuing with work on traditional problems of comparative history and thought, he has added problems of energy and the environment to his research interests. Dr. Stunkel has lived, traveled, and taught extensively in East and Southeast Asia.

JOHN FRANKLIN COPPER is associate professor of international studies at Southwestern University, Memphis, Tennessee. He received his Ph. D. from the University of North Carolina. He is the author of China's Foreign Aid: An Instrument of Peking's Foreign Policy and, with W. R. Kintner, A Matter of Two Chinas: The China-Taiwan Issue in U. S. Foreign Policy, in addition to numerous articles and reviews on Asian and international affairs in such journals as Asian Affairs. Asian Quarterly, Asian Survey, China Report, Current Scene, International Studies, Orbis, and Pacific Community. Dr. Copper has lived, traveled, and taught extensively in East and Southeast Asia.

KENNETH PAUL ERICKSON is associate professor of political science and director of Inter-American Affairs at Hunter College, City University of New York. His book, The Brazilian Cooperative State and Working Class Politics, won the 1978 Hubert Herring Memorial Award of the Pacific Coast Council on Latin American Studies, and he is also author of "The Political Economy of Energy Consumption in Industrial Societies," in International Energy Policy. Dr. Erickson is coordinator of the Energy Policy Seminar at Hunter College.

DONALD R. KELLEY is chairman of the Department of Political Science, University of Arkansas. He received his B.A. and M.A. from the University of Pittsburgh and his Ph.D. from Indiana University. He is editor and contributor to Soviet Politics in the Brezhnev Era and The Energy Crisis and the Environment: An International Perspective, coauthor of The Economic Super Powers and the Environment: The United States, the Soviet Union, and Japan, and author of The Solzhenitsyn-Sakharov Dialogue: Politics, Society, and the Future. He has also contributed numerous articles to anthologies and journals such as American Political Science Review, Journal of Politics Soviet Studies, Polity, American Behavioral Scientist, and Canadian Slavonic Papers.

SAMUEL S. KIM is visiting professor of international affairs, Princeton University, and visiting research political scientist, Center of International Studies, Princeton University. He received his Ph.D. from Columbia University. His special interests are Chinese foreign policy and world-order studies. He is a member of the editorial board of Contemporary China. Dr. Kim's publications include The Maoist Image of World Order: China, the United Nations, and World Order; The War System: An Interdisciplinary Approach (coeditor and contributor); and China in the Global Community (coeditor and contributor). He is currently coediting (with Richard Falk and Saul Mendlovitz) a volume tentatively entitled Armament and Disarmament in Chinese Global Policy under auspices of the Center of International Studies.

LAURA REGINA ROSENBAUM RANDALL is professor economics, Hunter College, City University of New York. Dr. Randall is author of An Economic History of Argentina in the Twentieth Century and A Comparative Economic History of Latin America, 1500-1914, as well as editor and author of various publications on Latin American economic policy and history.

DAVID J. ROSEN is assistant professor in the Department of Human Ecology and Social Sciences, Cook College, Rutgers University, and the Department of Political Science, the Graduate School, Rutgers University. His research and teaching focus on energy and environmental policy internationally and in the United States. He has lectured and published widely on various aspects of energy policy. He received his doctorate from Rutgers University in 1975.

ERNEST J. WILSON III is an analyst with Resources for the Future, Washington, D.C. He received his Ph.D. from the University of California, Berkeley. He has been a Ford Foundation Fellow, American Political Science Association Fellow, and a Michael Clarke Rockefeller Fellow, among other fellowships. His publications include articles in <u>Africa Today</u>, <u>The Black Scholar</u>, <u>Review of Black Political Economy</u>, <u>Black World</u>, and in <u>Transformation and Change in Africa</u> (forthcoming). Dr. Wilson has traveled and worked extensively in sub-Sahara Africa.